STUDENT'S SOLUTIONS MANUAL

MILTON LOYER
Penn State University

to accompany

ELEMENTARY STATISTICS
TENTH EDITION

Mario F. Triola
Dutchess Community College

PEARSON

Addison
Wesley

Boston San Francisco New York
London Toronto Sydney Tokyo Singapore Madrid
Mexico City Munich Paris Cape Town Hong Kong Montreal

Copyright © 2007 Pearson Education, Inc.
Publishing as Pearson Addison-Wesley, 75 Arlington Street, Boston, MA 02116

ISBN   0-321-47040-0

5 6 BB 08

TABLE OF CONTENTS

PREFACE

This manual contains the solutions to the odd-numbered exercises for each section of the textbook Elementary Statistics, Tenth Edition, by Mario Triola, and the solutions for all the end-of-chapter exercises described as Statistical Literacy and Critical Thinking, Review Exercises, and Cumulative Review Exercises. To aid in the comprehension of the calculations, worked problems typically include intermediate steps. When appropriate, additional hints and comments are included and prefaced by NOTE.

Many statistical problems are best solved using particular formats. Recognizing and following these patterns promote understanding and develop the capacity to apply the concepts to other problems. This manual identifies and employs such formats wherever practicable.

For best results, read the text carefully before attempting the exercises, and attempt the exercises before consulting the solutions. This manual has been prepared to provide a check and extra insights for exercises that have already been completed, and to provide guidance for solving exercises that have already been attempted but have not been successfully completed.

I would like to thank Mario Triola for writing an excellent elementary statistics book, and for inviting me to prepare this solutions manual.

Chapter 1

Introduction to Statistics

1-2 Types of Data

1. A parameter is a numerical value describing a population, while a statistic is a numerical value describing a sample.

3. Discrete data can take on only a countable number of specified values, while continuous data can take on an infinite number of values over some range.

5. Statistic, since the value 2.58 was obtained from a sample of households.

7. Parameter, since the value 706 was obtained from the population of all passengers.

9. Discrete, since the number of letters arriving at the target must be an integer.

11. Continuous, since weight can be any value on a continuum.

13. Nominal, since the numbers are for identification only. Numbers on jerseys are merely numerical names for the players.

15. Nominal, since the numbers are for identification only. Even though SS numbers are assigned chronologically within regions and can be placed in numerical order, there is no meaningful way to compare 208-34-3338 and 517-94-1438. If all the numbers had been assigned chronologically beginning with 000-00-0001, like the order of finishers in a race, then SS numbers would illustrate the ordinal level of measurement.

17. Interval, since differences are meaningful but ratios are not.

19. Ordinal, since the ratings are relative positions in a hierarchy.

IMPORTANT NOTE for exercises 21-24: The population and sample are determined by the intent of the researcher, which must be clearly defined at the outset of any project. Unfortunately these exercises state only what the researcher did, and do not specifically identify the intent of the researcher. Consequently there may be differences of interpretation in some of the exercises, but some general principles apply. (1) The sample is a subset of the population of interest and must have the same units as the population. If the population of interest is all households, for example, then the sample must be a selection of households and not a selection of adults – as households with more adults would have a higher chance of being included in the study, thus creating a bias. (2) The problem of nonresponse must be addressed. If 500 persons are randomly selected and asked a personal question, for example, but only 400 choose to answer the question, what is the sample? Depending on the situation, the sample could be either the 500 people randomly selected (and the 100 "no response" answers be reported as part of the sample data) or the just 400 people who actually gave data in the form of a specific answer.

21. a. The sample is the 25 Senators selected at random.
    b. The population is all 100 US Senators.
    Yes. If the selections are made at random there is no bias in the design of the experiment that would cause the sample not to be representative.

23. a. The sample is the 1059 randomly selected adults.

b. The population is all US adults.

Yes. If the adults were selected at random there is no reason why they would not be representative of the population.

NOTE: This is a sample of adults, not of homes. The appropriate inference is that about 39% of the adults live in a home with a gun, not that there is a gun in about 39% of the homes. Since adults and not homes were selected at random, homes with more adults are more likely to be represented in the survey – and that bias would prevent the survey from being representative of all homes.

25. Temperature ratios are not meaningful because a temperature of 0° does not represent the absence of temperature in the same sense that $0 represents the absence of money. The zero temperature in the exercise (whether Fahrenheit or Centigrade) was determined by a criterion other than "the absence of temperature."

27. This is an example of ordinal data. It is not interval data because differences are not meaningful – i.e., the difference between the ratings +4 and +5 does not necessarily represent the same differential in the quality of food as the difference between 0 and +1.

1-3 Critical Thinking

1. A voluntary response sample is one in which the participants choose whether or not to participate. This is generally unsuitable for statistical purposes because it is persons with strong feelings and/or a personal interest that tend to respond – and the opinions of such people are not necessarily representative of the entire population.

3. No. This is a voluntary response sample. Most of the data in a voluntary response sample comes from people with strong feelings or a personal interest in the topic, and there is usually a lack of data from "typical" people.

5. People who choose to play basketball are already taller than the general population when they choose to participate in the sport. It is not the participation that makes them taller.

7. If the population of Orange County includes significantly more minority drivers than white drivers, one would expect more speeding tickets to be issued to minorities than to whites – even if the percentage of white drivers who violated the speed limit was greater than the percentage of minority drivers who did so. It is also possible that police tend to target minority drivers – so that the numbers of tickets issued to the various racial/ethnic groups does not correspond to the actual amount of speed limit violations occurring. The fact that more speeding tickets are given to minorities does not warrant the conclusion that minority persons are more likely to speed.

9. The fact that the study was financed by a company with vested interests in the outcome might have influenced the conducting and/or reporting to concentrate on aspects of the study favorable to the company. The fact that a product contains one ingredient associated with positive health benefits may not outweigh other negative aspects of the product that were not reported.

11. There are at least two reasons why her conclusions should not be applied to the general population of all women. First, the target sample was biased because she sent the 100,000 questionnaires to members of various women's groups – and women who join such groups are not necessarily representative of the entire population, and certainly do not include the "non-

joiner" types. Secondly, the 4500 replies she received are a voluntary response sample (and a very small 4.5% response rate at that) – which means that the responses will include an over representation of those with strong feelings and/or a personal interest in the survey.

13. Using the telephone directory as a source of survey subjects eliminates persons without telephones and persons with unlisted numbers, and it also over-includes persons who live in a house with more than one telephone listing (as some households have different numbers and/or listings for each spouse as well as the children). A more subtle problem is that telephone listings are usually by household, not by individual – which means that a person in a one-adult household has a greater chance of being included in the survey than a person in a multi-adult household.

15. People who were killed in motorcycle crashes, when a helmet may have saved their lives, were not present to testify.

17. No, this is not likely to be a good estimate of the average of all wages earned in the United States. States with small populations but high wages (due to the cost of living) like Alaska and Hawaii, for example, contribute $2/50 = 4\%$ of the average for the 50 states – but since they do not contain 4% of the US wage earners, they inflate the estimate to make it higher than the true average of all US wage earners. In general, the average of averages does not give the overall average. Imagine, for example, two states: one where all 4 wage earners earned $100,000 each, and one where all 96 wage earners earned $50,000 each. The true average wage for the country would be $[(4)(100,000) + (96)(50,000)]/100 = 5,200,000/100 = \$52,000$, but the average of the state averages would be $[100,000 + 50,000]/2 = 150,000/2 = \$75,000$.

19. No, the results will not be representative of all households in the state. This approach violates the basic principle that the population and the sample have the same units – if the population of interest is all *households*, then you cannot survey *children*. This creates a bias because households with fewer children have a smaller chance of being included in the data, and households with no children will be missed entirely.

21. a. $3/20 = (3/20)(100\%) = 15\%$
    b. $56.7\% = 56.7/100 = 0.567$
    c. $(34\%)(500) = (34/100)(500) = 170$
    d. $0.789 = (0.789)(100\%) = 78.9\%$

23. a. $(52\%)(1038) = (52/100)(1038) = 540$
    b. $(52/1038)(100\%) = 5\%$

25. NOTE: The number who <u>actually committed</u> a campus crime while under the influence of alcohol or drugs cannot be determined from the survey, but the number who <u>say</u> that have committed such crimes can be determined as shown below. This distinction is an important one.
    $(8\%)(1875) = (8/100)(1875) = 150$ say they've committed a campus crime
    $(62\%)(150) = (62/100)(150) = 93$ say they've committed a campus crime while under the influence of alcohol or drugs

27. If the researcher started with 20 mice in each group, and none of the mice dropped out of the experiment for any reason, then the proportions of success could only be fractions like 0/20, 1/20, 2/20, etc. – and the success rates in percents could only be multiples of 5 like 0%, 5%, 10%, etc. A success rate of 53%, for example, would not be possible.

1-4 Design of Experiments

1. In a random sample, every individual member has an equal chance of being selected; while in a simple random sample of size n, every possible sample of size n has an equal chance of being selected.

3. Blinding occurs when the subject (or evaluator) does not know whether the real vaccine or a placebo has been given. Double-blinding occurs when both the subject and the evaluator don't know whether the real vaccine or a placebo has been given. Blinding is important in situations when the data involve subjective responses (e.g., whether there is less pain, more mobility, etc.) that may be influenced (deliberately or subconsciously) by knowing whether the real vaccine or the placebo has been given.

5. Observational study, since the researcher (Emily) merely measured whether the therapist could identify the chosen hand and did nothing to modify the therapist.

7. Observational study, since specific characteristics are measured on unmodified tablets.

9. Prospective, since the data are to be collected in the future from a group sharing a common factor.

11. Cross-sectional, since the data are collected for a specific point in time.

13. Systematic, since every 5th driver was stopped.

15. Stratified, since the population of interest was divided into activity subgroups from which the actual sampling was done.

17. Cluster, since all the wait staff at selected restaurants were surveyed.

19. Convenience, since the sample is those who happen to be in the author's classes.

21. Random, since each adult has an equal chance of being selected.
    NOTE: This is a complex situation. The above answer ignores the fact that there is not a 1 to 1 correspondence between adults and phone numbers. An adult with more than one number will have a higher chance of being selected. An adult who shares a phone number with other adults will have a lower chance of being selected. And, of course, adults with no phone numbers have no chance of being selected at all.

23. Cluster, since all the students in selected classes were interviewed.
    NOTE: Ideally the division of the population should place each subject in one and only one cluster, and every subject should have an equal chance to be included in the sample. In this scenario there are at least two problems: (1) students carrying more classes will have a greater chance of being included in the survey and (2) a student may be included in more than one cluster. While some techniques may eliminate those two problems, they could introduce others – for example, selecting the 10 classes from among all 1 pm MWF classes, means that a student not taking a class then has zero chance of being included in the survey.

25. Yes, this results in a random sample because each return has an equal chance of being included in the sample.
    No, this does not result in a simple random sample of size 48 because some groupings of 48 returns are not possible – for example, a grouping with 24 Monday returns and 24 Tuesday returns would not be possible.

27. Whether the sample is a random sample depends on how the first selection is made. If the engineer chooses the first one at random from 1 to 10,000 and every 10,000th one thereafter,

then every M&M has an equal chance of being selected (namely 1 in 10,000) and the sample is a random sample. If the engineer determines to start with #1 and choose every 10,000th one thereafter, then some M&M's have no chance of being selected (e.g., #2) and the sample is not a random sample.

No, no matter how the first selection is made the sample will not be a simple random sample of size n. All possible groupings of size n are not possible – any grouping containing #1 and #2, for example, could not occur.

29. Yes, this results in a random sample because every student has an equal chance of being selected.

Yes, this results in a simple random sample of size 6 because all possible groupings of size 6 have an equal chance to occur.

31. Assume that "students who use this book" refers to a particular semester, or other point in time. Before attempting any sampling procedure below, two preliminary steps are necessary: (1) obtain from the publisher a list of each school that uses the book and (2) obtain from each school the number of students enrolled in courses requiring the text. Note that using only the number of books sold to the school by the publisher for the semester of will miss those students purchasing used copies turned in from previous semesters, buying textbooks on the Internet, etc. "Students" in parts (a)-(e) below is understood to mean students using this book.

  a. Random: Conceptually (i.e., without actually doing so) make a list of all N students by assigning to each school one place on the list for each of its students. Pick 100 random numbers from 1 to N. Each number identifies a particular school and a conceptual student. For each particular school identified select a student at random (or more than one student, if that school was identified more than once). In this manner every student has the same chance of being selected. In fact every possible grouping of 100 students has the same chance of being selected, and the sample is also a simple random sample.

  b. Systematic. Place the n schools in a list. Pick a random number between 1 and (n/10) to determine a starting selection. Select that school and every (n/10)th school thereafter. At each of the 10 schools thus systematically selected, randomly (or systematically) select and survey 10 students.

  c. Convenience. Randomly (or conveniently) select and survey 100 students from your school.

  d. Stratified. Divide the schools into four categories: 2 year, 4 year public, 4 year private, proprietary. Randomly select one school from each category. For each school randomly (or in some stratified manner) select and survey the number of students equal to the percent of the N students that are in that category.

  e. Cluster. Select a school at random and survey all the students, or cluster sample again by selecting a class within that school and surveying all the students in that class. Repeat as necessary until the desired sample of 100 is reached.

33. The study may be described as

  a. prospective because the data are to be collected in the future, after the patients are assigned to different treatment groups.

  b. randomized because the patients were assigned to one of the three treatment groups at random.

  c. double-blind because neither the patient nor the evaluating physician knew which treatment had been administered.

  d. placebo-controlled because there was a treatment with no drug that allowed the researchers to distinguish between real effects of the drugs in question and (1) the psychological effects of being "treated" and/or (2) the effects (e.g., natural improvement over time) of not receiving any drug.

Statistical Literacy and Critical Thinking

1. A large sample is not necessarily a good sample. As in many other situations, quality is more important than quantity. More important than the size of the sample is whether it was selected in such as way as to be representative of the population of interest.

2. a. Height data are quantitative because height is an actual numerical measurement and not merely a categorization.
   b. Height data are continuous because they can take on any value over a specified range.
   c. The population is all runners who finish the New York Marathon.

3. This sample would not necessarily be representative of the general population. It is a voluntary response sample, which typically includes only those with certain motivation or characteristics – in this case that would probably be people who either (1) think they have a high IQ and /or enjoy such tests or (2) need the $50.

4. This would not be a good estimate for at least two reasons. First, she appears to be depending on values provided by the owners. An owner might not know a car's true value or might tend to exaggerate its value. Secondly, she is counting each state equally. Alaska has very few cars, but they probably include many relatively expensive (even considering that state's overall high cost of items) four-wheel drive utility vehicles. California has thousands of times as many cars as Alaska, and they would probably tend to be less expensive smaller vehicles. The average for 10 cars from each state would likely not be a good estimate for the average of all the cars in the nation.

Review Exercises

1. The results cannot be considered representative of the population of the United States for at least two reasons. First, only AOL Internet subscribers were polled – and they are not necessarily representative racially, socio-economically, geographically, etc. Secondly, the responders constitute a voluntary response sample – and such responders tend to be those with strong opinions and/or a personal stake in the issue.

2. Let N be the total number of full-time students and n be the desired sample size.
   a. Random. Obtain a list of all N full-time students, and number the students from 1 to N. Select n random numbers from 1 to N, and survey each student whose number on the list is one of the random numbers selected.
   b. Systematic. Obtain a list of all N full-time students, and number the students from 1 to N. Let m be the largest integer less than the fraction N/n, select a random number between 1 and m, and survey that student and every m^{th} student thereafter.   ·
   c. Convenience. Select a location by which most of the students usually pass and survey the first n full-time students that pass by.
   d. Stratified. Obtain a list of all N full-time students and the gender of each. Divide the list by gender and randomly survey n/2 students of each gender.

e. Cluster. Obtain a list of all the classes meeting at a popular time, select at random one of the classes (or as many of the classes as needed to obtain a sample of n students), and survey all the students in that class.

3. a. Ratio, since differences are valid and there is a meaningful 0.
   b. Ordinal, since there is a hierarchy but differences are not valid.
   c. Nominal, since the categories have no meaningful inherent order.
   d. Interval, since differences are valid but the 0 is arbitrary.

4. a. Discrete, since the number of shares held must be an integer.
      NOTE: Even if partial shares are allowed (e.g., 5½ shares), the number of shares must be some fractional value and cannot be any value on a continuum – e.g., a person cannot own π shares.
   b. Ratio, since differences between values are valid and there is a meaningful 0.
   c. Stratified, since the population of interest (all stockholders) was divided into subpopulations (by states) from which the actual sampling was done.
   d. Statistic, since it is calculated from a sample and not from the entire population.
   e. There is no unique correct answer, but the following are reasonable possibilities. (1) The proportion of stockholders holding above that certain number of shares that would make them "influential." (2) The proportion of stockholders holding below that certain number of shares that would make them "insignificant." (3) The numbers of shares held by the largest stockholders.
   f. There is no unique correct answer, but the following are reasonable possibilities. (1) The results would be from a voluntary response sample, and hence not necessarily be representative of all stockholders. (2) If the questionnaire did not include information on the number of shares owned, the views of small stockholders could not be distinguished from the views of large stockholders (whose views should carry more weight).

5. NOTE: The second part of the problem involves deciding "whether the sampling scheme is likely to result in a sample that is representative of the population." This is subjective, since the term "representative" has not been well-defined. Mathematically, any random sample is representative in the sense that there is no bias and a series of such samples can be expected to average out to the true population values – even though any one particular sample may not be representative. And so we expect a random sample of size n=50 to be representative. But how about a random sample of size n=2? It has all the mathematical properties of a random sample of size n=50, but common sense suggests that any one sample of size n=2 is not necessarily likely to be representative of the population.
   a. Systematic, since the selections were made at regular intervals. Yes, there is no reason why every 500th stockholder should have some characteristic that would introduce a bias.
   b. Convenience, since those selected were the ones who happened to attend. No, those who attend would tend to be the more interested and/or more well-off stockholders.
   c. Cluster, since the stockholders were organized into groups (by stockbroker) and all the stockholders in the selected groups were surveyed. Yes, since every stockholder has the same chance of being included in the sample (assuming each stockholder works through a single stockbroker), the sample will be a random sample – and considering the large number of stockbrokers involved (see the NOTE at the beginning of the problem), it is reasonable to expect the sample to be representative of the population.
   d. Random, since each stockholder has the same chance of being selected. Yes, since every stockholder has the same chance of being included in the sample, the scheme is likely to result in a representative sample.

e. Stratified, since the stockholders were divided into subpopulations (zip codes) from which the actual sampling was done. No, since all the zip codes were given equal weights (5 stockholders from each) – because "significant" zip codes (with large numbers of large stockholders) are counted equal with "insignificant" zip codes (with few numbers of small stockholders), the sample will probably be biased in favor of small stockholders.

6. a. Blinding occurs when those involved in an experiment (either as subjects or evaluators) do not know whether they are dealing with a treatment or a placebo. It might be used in this experiment by (1) not telling the subjects whether they are receiving Sleepeze or the placebo and/or (2) not telling any post-experiment interviewers or evaluators which subjects received Sleepeze and which ones received the placebo. Double-blinding occurs when neither the subjects nor the evaluators know whether they are dealing with a treatment or a placebo.

b. The data reported will probably involve subjective assessments (e.g., "on a scale of 1 to 10, how well did it work?") that may be influenced by whether the subject was known to have received Sleepeze or the placebo.

c. In a completely randomized design, subjects are assigned to the treatments (in this case Sleepeze or the placebo) at random.

d. In a rigorously controlled design, the subjects are carefully placed into treatment groups (in this case Sleepeze or the placebo) according to characteristics that might affect the outcome of the experiment. For example: if the Sleepeze group contains an athletic male under 20, then the placebo group should contain an athletic male under 20 – since gender, physical condition, and age are all characteristics that might reasonably affect the outcome of the experiment. This makes the treatment groups as similar as possible, so that any observed differences are likely to be due to the treatments and not to extraneous factors that might affect the experiment.

e. Replication involves performing the experiment on a sample of subjects large enough to ensure (1) that atypical responses of a few subjects will not give a distorted view of the true situation and (2) that repeating the experiment would likely produce similar results.

7. a. The figure 39.82% is a parameter, since it refers to the entire population.
b. Discrete, since the numbers of votes cast must be integers.
c. (39.82%)(1,865,908) = (39.82/100)(1,865,908) = 743,005

8. a. To contain 100% less fat would be to contain no fat at all – 100% of the fat has been removed. It is not physically possible to contain 125% less fat.
b. (19%)(237) = (19/100)(237) = 45
c. (30/186)(100%) = 16.1%

Cumulative Review Exercises

NOTE: Throughout the text intermediate mathematical steps will be shown to aid those who may be having difficulty with the calculations. In practice, most of the work can be done continuously on calculators and the intermediate values are unnecessary. Even when the calculations cannot be done continuously, DO NOT WRITE AN INTERMEDIATE VALUE ON YOUR PAPER AND THEN RE-ENTER IT IN THE CALCULATOR. That practice can introduce round-off and copying errors. Store any intermediate values in the calculator so that you can recall them with infinite accuracy and without copying errors. In general, the degree of accuracy appropriate depends upon the particular problem – and guidelines for this will be given as needed in subsequent chapters. Unless there is reason to otherwise, answers in this section are given with 3 decimal accuracy.

1. $\dfrac{3.0630+3.0487+2.9149+3.1358+2.9753}{5}=\dfrac{15.1377}{5}=3.02754$ grams

2. $\dfrac{98.20-98.60}{0.62}=\dfrac{-0.40}{0.62}=-0.645$

3. $\dfrac{98.20-98.60}{\dfrac{0.62}{\sqrt{106}}}=\dfrac{-0.40}{0.0602}=-6.642$

4. $\left[\dfrac{1.96\cdot0.25}{0.03}\right]^{2}=\left[16.333\right]^{2}=266.778$

5. $\dfrac{(50-45)^{2}}{45}=\dfrac{(5)^{2}}{45}=\dfrac{25}{45}=0.556$

6. $\dfrac{(2-4)^{2}+(3-4)^{2}+(7-4)^{2}}{3-1}=\dfrac{(-2)^{2}+(-1)^{2}+(3)^{2}}{2}=\dfrac{4+1+9}{2}=\dfrac{14}{2}=7$

7. $\sqrt{\dfrac{(2-4)^{2}+(3-4)^{2}+(7-4)^{2}}{3-1}}=\sqrt{\dfrac{(-2)^{2}+(-1)^{2}+(3)^{2}}{2}}=\sqrt{\dfrac{4+1+9}{2}}=\sqrt{\dfrac{14}{2}}=\sqrt{7}=2.646$

8. $\dfrac{8(151,879)-(516.5)(2176)}{\sqrt{8(34,525.75)-516.5^{2}}\sqrt{8(728,520)-2176^{2}}}=\dfrac{91128}{\sqrt{9433.75}\sqrt{1093184}}=0.89735$

9. $(0.5)^{10}=9.7656\cdot10^{-4}=0.00097656$

10. $2^{40}=1.099511628\cdot10^{12}=1,099,511,628,000$
    NOTE: The final 3 digits are not really zero, but most calculators will not be able to give all 13 digits of the exact answer.

11. $7^{12}=1.38412872\cdot10^{10}=13,841,287,200$
    NOTE: The final 2 digits are not really zero, but most calculators will not be able to give all 11 digits of the exact answer.

12. $0.8^{50}=1.4272\cdot10^{-5}=0.000014272$

Chapter 2

Summarizing and Graphing Data

2-2 Frequency Distributions

1. A frequency distribution is a table with two columns – one column that identifies data values either individually or in groups, and one column that tells how many data values meet that criterion. It is useful for summarizing data and giving information about the distribution of the data – the minimum, maximum, shape, and concentration of the data.

3. The given class intervals overlap. The classes of a frequency distribution should be defined so that each data value falls into one and only class. Because the value 10, for example, is in two different classes, the frequency distribution constructed using the given class intervals would not be well-defined.

5. Subtracting the first two lower class limits indicates that the class width is $40 - 35 = 5$. Since there is a gap of 1.0 between the upper class limit of one class and the lower class limit of the next, class boundaries are determined by increasing or decreasing the appropriate class limits by $(1.0)/2 = 0.5$ The class boundaries and class midpoints are given in the table below.

temperature (°F)	class boundaries	class midpoint	frequency
35 – 39	34.5 – 39.5	37	1
40 – 44	39.5 – 44.5	42	3
45 – 49	44.5 – 49.5	47	5
50 – 54	49.5 – 54.5	52	11
55 – 59	54.5 – 59.5	57	7
60 – 64	59.5 – 64.5	62	7
65 – 69	64.5 – 69.5	67	1
			35

NOTE: Although they often contain extra decimal places and may involve consideration of how the data were obtained, class boundaries are the key to tabular and pictorial data summaries. Once the class boundaries are obtained, everything else falls into place. In exercise #5 it is readily seen that the width of the first class is $39.5 - 34.5 = 5.0$ and the midpoint of the first class is $(34.5+39.5)/2 = 37.0$. In this manual class boundaries will typically be calculated first and that information used to determine other values. Because the sum of the frequencies is an informative number and is used in many subsequent calculations, it will be shown as an integral part of each frequency distribution.

7. Subtracting the first two lower class limits indicates that the class width is $65.0-60.0 = 5.0$ Since there is a gap of 0.1 between the upper class limit of one class and the lower class limit of the next, class boundaries are determined by increasing or decreasing the appropriate class limits by $(0.1)/2 = 0.05$ The class boundaries and class midpoints are given in the table on the following page.

height (in)	class boundaries	class midpoint	frequency
60.0 – 64.9	59.95 – 64.95	62.45	4
65.0 – 69.9	64.95 – 69.95	67.45	25
70.0 – 74.9	69.95 – 74.95	72.45	9
75.0 – 79.9	74.95 – 79.95	77.45	1
80.0 – 84.9	79.95 – 84.95	82.45	0
85.0 – 89.9	84.95 – 89.95	87.45	0
90.0 – 94.9	89.95 – 94.95	92.45	0
95.0 – 99.9	94.95 – 99.95	97.45	0
100.0 –104.9	99.95 –104.95	102.45	0
105.0 –109.9	104.95 –109.95	107.45	1
			40

9. Yes; in a relatively symmetric manner the frequencies start out low, increase to a maximum, and then decrease.

11. The height of the tallest man is stated to be between 105 and 110 inches – i.e., between 8'9" and 9'2". This is almost certainly an error. When this value is deleted the distribution of the heights is approximately normal.

NOTE: For elative frequency distributions, as with frequency distributions, the sum is included as an integral part of the table. For relative frequency distributions this should always be 1.000 (or 100.0%) and serves as a check for the calculations. [Because of rounding in the individual classes, the listed relative frequencies might actually sum to 0.999 or 1.001 – but the true sum with infinite accuracy is always 1.000, and deviations from that indicate that an error has been made.] Relative frequencies may be expressed as decimals or percents – while the choice is arbitrary, individual instructors may specify the use of one or the other.

13. The relative frequency for each class is found by dividing its frequency by 35, the sum of the frequencies.

temperature (°F)	relative frequency
35 – 39	2.9%
40 – 44	8.6%
45 – 49	14.3%
50 – 54	31.4%
55 – 59	20.0%
60 – 64	20.0%
65 – 69	2.9%
	100.0%

NOTE: For cumulative tables, this manual uses upper class boundaries in the "less than" column. Consider exercise #15, for example, to understand why this is done. Conceptually, temperatures occur on a continuum and the integer values reported are assumed to be the nearest whole number representation of the precise measure. An exact temperature of 39.7, for example, would be reported as 40 and fall in the second class. The values in the first class, therefore, are better described as "less than 39.5" (using the upper class boundary) than as "less than 40" (using the lower class limit of the next class). This distinction is crucial in the construction of pictorial representations in the next section.

In addition, that the final cumulative frequency must equal the total number (i.e., the sum of the frequency column in the frequency distribution) serves as a check for the calculations. The sum of the cumulative frequencies, however, has absolutely no meaning and is not included in the table.

15. The cumulative frequencies are the combined number in each class and all previous classes.

temperature (°F)	cumulative frequency
less than 39.5	1
less than 44.5	4
less than 49.5	9
less than 54.5	20
less than 59.5	27
less than 64.5	34
less than 69.5	35

17. The frequency distribution of the last digits shows unusually high numbers of 0's and 5's. This is typical for data that have been rounded off to "convenient" values. It appears that the heights were reported and not actually measured.

digit	frequency
0	9
1	2
2	1
3	3
4	1
5	15
6	2
7	0
8	3
9	1
	37

19. Since most of the days have no rainfall the data values accumulate in the lower end of the table. The frequency distribution does not appear to approximate a normal distribution.

rainfall (in)	frequency
0.00 – 0.19	44
0.20 – 0.39	6
0.40 – 0.59	1
0.60 – 0.79	0
0.80 – 0.99	0
1.00 – 1.19	0
1.20 – 1.39	1
	52

21. Most of the data values occur at the lower end of the table, and there are very high values that are not balanced by very low ones. The frequency distribution does not appear to approximate a normal distribution.

BMI	frequency
15.0 – 20.9	10
21.0 – 26.9	15
27.0 – 32.9	11
33.0 – 38.9	2
39.0 – 44.9	2
	40

23. The weights of the pennies appear to be normally distributed, but with an unusually heavy concentration of data values at the center and very few values at the extremes.

weight (lbs)	frequency
2.9500 – 2.9999	2
3.0000 – 3.0499	3
3.0500 – 3.0999	22
3.1000 – 3.1499	7
3.1500 – 3.1999	1
	35

25. The two relative frequency distributions, each constructed with 200 as the lower class limit of the first class and a class width of 20, are given below.

0.0109 CANS		0.0111 CANS	
	relative		relative
weight (lbs)	frequency	weight (lbs)	frequency
200 – 219	6.9%	200 – 219	3.4%
220 – 239	5.1%	220 – 239	2.9%
240 – 259	10.3%	240 – 259	6.9%
260 – 279	48.0%	260 – 279	20.6%
280 – 299	22.7%	280 – 299	49.7%
	100.0%	300 – 319	16.0%
		320 – 339	0.0%
		340 – 359	0.0%
		360 – 379	0.0%
		380 – 399	0.0%
		400 – 419	0.0%
		420 – 439	0.0%
		440 – 459	0.0%
		460 – 479	0.0%
		480 – 499	0.0%
		500 – 519	0.6%
			100.0%

For both distributions most of the data values occur near the center of the table, and the relative frequencies taper off toward each end. Both distributions have approximately the same "off-balance normal" shape – and as might be expected, the data values are slightly higher for the stronger 0.0111 cans.

27. Let n = the number of data values and let x = the number of classes.
Solve the given formula x = 1 + (log n)/(log 2) for n to get $n = 2^{x-1}$.

Use the values x = 5.5, 6.5, 7.5,… to get cut-off values for n shown below.

x	$n = 2^{x-1}$
5.5	22.63
6.5	45.25
7.5	90.51
8.5	181.02
9.5	362.04
10.5	724.04
11.5	1448.15
12.5	2896.31

Assuming n is at least16, use the cut-off values to complete the table as follows.

n	ideal # of classes
16 – 22	5
23 – 45	6
46 – 90	7
91 – 181	8
182 – 362	9
363 – 724	10
725 – 1448	11
1449 – 2896	12

NOTE: The cut-off values indicate that

For $n < 22.63$, x rounds to 5; for $n > 22.63$, x rounds to 6.

For $n < 45.25$, x rounds to 6; for $n > 45.25$, x rounds to 7.

etc.

2-3 Histograms

1. A histogram reveals the basic shape of the distribution. It also gives a preliminary indication of the typical values and the amount of spread of the values.

3. Histograms and other summaries are appropriate to organize the data when they represent too much information for the mind to assimilate as individual values – and data sets with only five values, for example, typically do not need to be summarized to get a feel for the data. In addition, five values would generally not be enough to provide an adequate visual or tabular representation of the underlying distribution.

5. The number of crew members is $2 + 10 + 5 + 1 = 18$. This represents 9 persons (1 coxswain and 8 rowers) from each school.

7. Rowers need to be muscular in order to power the boat. Since the coxswain directs the rowers and represents "unproductive" weight, it would be advantageous for a crew of rowers to have a light coxswain. The two lighter weights likely belong to the two coxswains.

9. The histogram is given below.

The digits 0 and 5 occur disproportionately more than the others. This is typical for data that have been rounded off to "convenient" values. It appears that the heights were reported and not actually measured.

11. The histogram is given below.

No, the data do not appear to have a distribution that is approximately normal.

13. The histogram is given below.

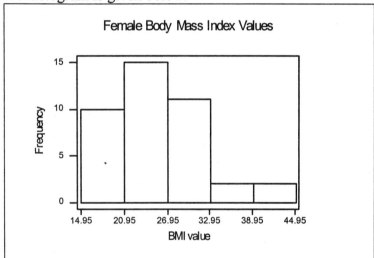

Yes, the data appear to have a distribution that is approximately normal.

15. The histogram is given below.

Yes, the data appear to have a distribution that is approximately normal.

17. The two relative frequency histograms are given below. For the sake of comparison, the same horizontal and vertical axes have been used for both histograms.

Each set of ages appears to have a distribution that is approximately normal, but there are two significant differences between the two sets: the ages for the actors are larger and more spread out than those for the actresses. Yes, the two genders seem to win Oscars at different ages – with males winning most often in their 40's, and females winning most often in their 30's.

19. The back-to-back relative frequency histogram is given below. Because it is an outlier and not helpful to the intended comparison, the 504 value for the 0.0111 thickness cans is not included.

2-4 Statistical Graphics

1. The main objective in graphing data is to better understand the information contained in the data. An appropriate graph can present the information efficiently and reveal nuances that otherwise might go unnoticed.

3. A time series graph requires data from a single variable that has been collected for different points in time. It can reveal trends or patterns that occur over time.

5. The dotplot is given below.

The dotplot suggests that the high temperatures during the period are fairly evenly distributed, with slightly fewer occurrences at the extremes.

7. The frequency polygon is given below.

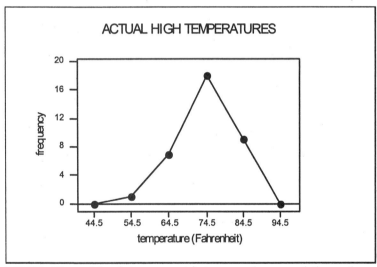

NOTE: The midpoints are not integer values, and the polygon must begin and end at zero at the midpoints of the adjoining classes that contain no data values.

9. The stemplot is given below.

```
 9 | 55
10 |
11 | 00055
12 | 000000005555
13 | 0000000066668
14 | 000088
15 | 00
```

The stemplot suggests that the distribution of the heights is approximately normal.
NOTE: The values suggest that the heights are being estimated and not measured precisely – hence the preponderance of heights ending in 0 and 5. This is to be expected, for how would it be possible to get an exact measurement of the height of a geyser? In the 130's and 140's, the only non-zero final digits are 6 and 8 – which may be a function of the algorithm used to estimate the height.

11. The ogive is given below.

Using the figure: move up from 119.5 on the horizontal scale to intersect the graph, then move left to intersect the vertical scale at 7. This indicates there were approximately 7 data values which would have been recorded as being below 120. This agrees with the actual data values. NOTE: Ogives always begin on the vertical axis at zero and end at n, the total number of data values. All cumulative values are plotted at the upper class boundaries.

13. See the figure below, with the bars arranged in order of magnitude.

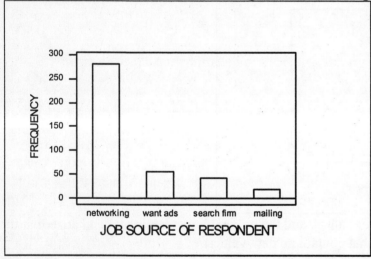

Networking appears to be the most effective job-seeking approach.

15. See the figure below. The corresponding central angles were determined as follows.
    vehicles: 2375/5524 = 43%, and 43% of 360° is 155°
    objects: 884/5524 = 16%, and 16% of 360° is 58°
    violence: 829/5524 = 15%, and 15% of 360° is 54°
    falls: 718/5524 = 13%, and 13% of 360° is 47°
    exposure: 522/5524 = 10%, and 10% of 360° is 36°
    fire: 166/5524 = 3%, and 3% of 360° is 11°

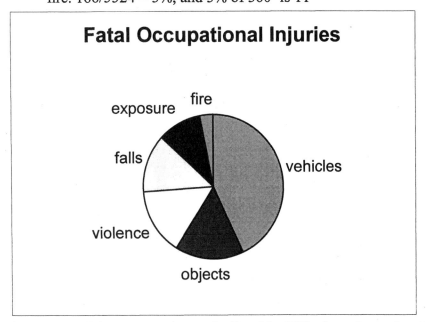

Fatal Occupational Injuries

Since some of the areas are very similar, it's difficult to determine the exact order of importance from the pie chart.

17. See the figure below. To be complete the figure should have a title, and each axis should be labeled both numerically and with the name of the variable.
    Note: When there are multiple occurrences of the same (or very similar) data pairs, some computer generated plots show only a single symbol. This does not give a true picture of the data or the relationship between the variables. In the figure below, the numbers 2 or 3 are superimposed on the computer generated circles to indicate multiple occurrences.

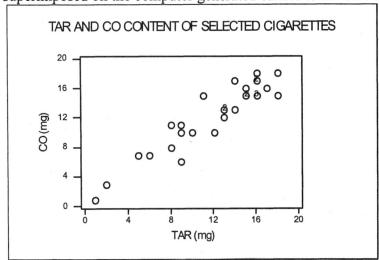

TAR AND CO CONTENT OF SELECTED CIGARETTES

Cigarettes high in tar tend to be high in CO. The points cluster around a straight line from (0,0) to (18,18), indicating that the mg of CO tends to be about equal to the mg of tar.

19. See the figure below.

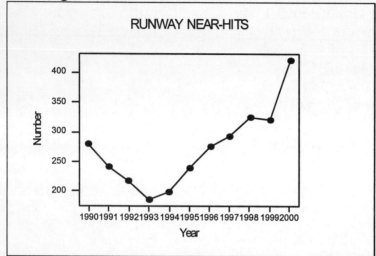

Yes, there appears to be a trend. The number of near-hits appears to have been decreasing prior to 1993 and increasing since 1993. Perhaps some change occurred about that time that may have contributed to the reversal in the trend.

IMPORTANT NOTE: Chapter 1 contained the warning that not starting the horizontal axis at zero "tends to produce a misleading subjective impression, causing readers to incorrectly believe that the difference is much greater than it really is" and that consequently "to correctly interpret a graph we should analyze the numerical information given in the graph." The visual impression above is that the occurrence of near-hits was almost eliminated in 1993 and that the numbers have risen dramatically since then. The time-series graph below which starts the horizontal axis at zero gives a very different impression – that except for a spike in 2000, the numbers seem relatively constant.

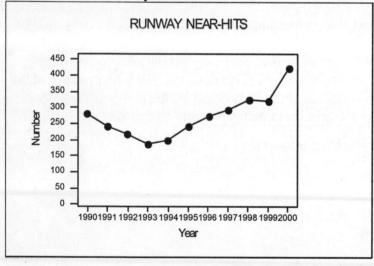

21. According to the figure, 422,000 started and 10,000 returned.
10,000/422,000 = 2.37%

23. The figure indicates that of the 400,000 – 80,000 = 320,000 men who started from Vilna toward Moscow, only 100,000 arrive in Moscow. The scale gives the length d on the chart that corresponds to 100 miles. The straight-line distance from Vilna to Moscow is about 4.9d = 490 miles, while the distance following the route is about 5.3d = 530 miles.

25. The final form of the back-to-back stemplot is given below, an example of adapting a standard visual form in order to better communicate the data. While such decisions are arbitrary, we chose to display "outward" from the central stem but to keep both ages in increasing order from left to right.

actresses' ages		actors' ages
124455556666777788889999999	2	9
0011122333334445555555677888899	3	001222444556667778888999
011111223569	4	0001111112222223333344455567778999
04	5	112222346677
013	6	00222
4	7	6
0	8	

Female Oscar winners tend to be younger than male Oscar winners. If one assumes that acting ability does not peak differently according to gender, the data may reveal a difference in the standard by which males and females are judged.

Statistical Literacy and Critical Thinking

1. Since it provides a visual image, a histogram is generally a more effective method for investigating the distribution of a data set.

2. If the data sets contain the same number of observations, the choice is arbitrary – but frequency distributions may be preferred since they give the added information of the actual counts. Since the actual number of observations per category is a function of the total number of observations as well as the nature of the distribution, relative frequency distributions should be used when comparing data sets with different total numbers of observations.

3. A time-series graph is designed to identify trends and patterns in the data over time. A histogram is designed to compare different categories. For the scenario suggested, the time-series graph would be more appropriate. In addition, if there are data from each of the 50 years, a line graph connecting 50 points would be much more readable than a series of 50 bars in a histogram.

4. A histogram constructed for data that are approximately normally distributed should have the following three features.
   (1) The tallest bar in the histogram should be near the center.
   (2) The heights of the other bars should decrease as their distance from the center increases.
   (3) The figure should be approximately symmetric.

Review Exercises

1. The requested frequency distribution is given below.

ages of actors	frequency
21-30	3
31-40	25
41-50	30
51-60	14
61-70	3
71-80	1
	76

The ages of the actors are greater than those of the actresses.

2. The requested histogram is given below, with the same class boundaries used in Figure 2-2.

The ages of the actors are greater than those of the actresses.

3. The requested dotplot is given below.

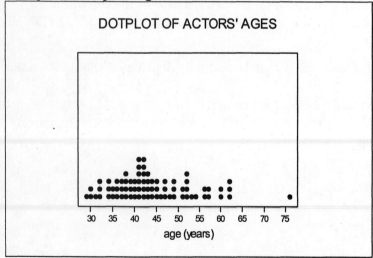

The ages of the actors are greater than those of the actresses.

4. The requested stemplot is given below.

Actors' Ages
2 | 9
3 | 00122244455666777788888999
4 | 00011111122222233333444455567778999
5 | 112222346677
6 | 00222
7 | 6

The ages of the actors are greater than those of the actresses.

5. The requested scatterplot is given below, using the first 10 entries by row for each gender and placing the ages of the actresses on the x axis.

There does not appear to be an association in any given year between the ages of the best actress and the best actor.

6. The requested time-series graph is given below, using the entries in order row by row.

There is no apparent trend or pattern over time.

Cumulative Review Exercises

1. No, these numbers do not measure or count anything.

2. The outcome results are at the nominal level of measurement – the numbers that result from the spins are merely names for the slots on the roulette wheel. The frequency results are at the ratio level of measurement – the numbers of occurrences are actual counts, and differences in counts are meaningful and there is a natural zero.

3. Since there were 380 spins and 38 slots on the roulette wheel, one would expect each slot to occur about 10 times. Each of the first 7 categories represents 5 slots and should have a frequency of about 50, while the last category represents 3 slots and should have a frequency of about 30. The frequencies reported agree with these expectations. Based on these results, there is no reason to suspect that the wheel is not fair.

4. Yes. On a fair wheel, each of the 38 slots is equally likely and the average of the resulting numbers should be close to $(1+2+3+...+35+36+0+00)/38 = 666/38 \approx 17.5$. If the average of 500 spins is close to 5.0, then the wheel is biased in favor of the lower numbers and betting on lower numbers would be a wise strategy.

5. No. These results cannot be used to make valid conclusions about the population of all car owners for at least two reasons. (1) The sample is a voluntary response sample, which is not necessarily representative of the original population. (2) The owners stated the amount they spent for their cars – which assumes that they correctly remember the prices and honestly report them, and neither of those assumptions can be counted on.

Chapter 3

Statistics for Describing, Exploring, and Comparing Data

3-2 Measures of Center

1. The mean, median, mode, and midrange are measures of "center" in the sense that they each attempt to determine (by various criteria – i.e., by using different approaches) what might be designated as a typical or representative value.

3. No; it makes no sense to calculate the mean of those numbers. The numbers are merely numerical names for the players and do not represent the measurement of any quantity or characteristic. If the individual numbers have no quantitative meaning, then neither would any value calculated from them.

NOTE: As it is common in mathematics and statistics to use symbols instead of words to represent quantities that are used often and/or that may appear in equations, this manual employs the following symbols for the various measures of center.

$$\text{mean} = \bar{x} \qquad \qquad \text{mode} = M$$
$$\text{median} = \tilde{x} \qquad \qquad \text{midrange} = \text{m.r.}$$

This manual will generally report means, medians and midranges accurate to one more decimal place than found in the original data. In addition, these two conventions will be employed.

(1) When there are an odd number of data, the median will be one of the original values. The manual follows the example given in this section (page 79) and reports the median as given in the original data. And so the median of 1,2,3,4,5 is reported as $\tilde{x} = 3$.

(2) When the mean falls <u>exactly</u> between two values accurate to one more decimal place than the original data, the round-off rule in this section (page 81) gives no specific direction. This manual follows the commonly accepted convention of always rounding up. And so the mean of 1,2,3,3 is reported as $\bar{x} = 9/4 = 2.3$ [i.e., $9/4 = 2.25$, rounded up to 2.3].

In addition, the median is the middle score <u>when the scores are arranged in order</u>, and the midrange is halfway between the first and last score <u>when the scores are arranged in order</u>. It is usually helpful to arrange the scores in order. This will not affect the mean, and it may also help in identifying the mode. Finally, no measure of center can have a value lower than the smallest score or higher than the largest score. Remembering this helps to protect against gross errors, which most commonly occur when calculating the mean.

5. Arranged in order, the n=8 scores are: 49 49 52 53 58 62 68 75
   a. $\bar{x} = (\Sigma x)/n = (466)/8 = 58.3$ sec   c. M = 49 sec
   b. $\tilde{x} = (53 + 58)/2 = 55.5$ sec   d. m.r. = (49 + 75)/2 = 62.0 sec
   A sample of statistics students would not be representative of all adults in terms of age, scientific background, etc.

7. Arranged in order, the n=25 scores are: 1 1 1 1 1 1 1 1 1 1 1 1 2 2 2 2 2 2 2 3 3 3 3 3 3 4
   a. $\bar{x} = (\Sigma x)/n = (47)/25 = 1.9$ c. M = 1
   b. $\tilde{x} = 2$ d. m.r. = (1 + 4)/2 = 2.5
   Yes, the measures of center can be calculated. No, the measures of center do not make sense – except for the mode of 1, which indicates that the smooth-yellow phenotype occurs more than any of the others

9. Arranged in order, the n=14 scores are:
   120 120 125 130 130 130 130 135 138 140 140 143 144 150
   a. $\bar{x} = (\Sigma x)/n = (1875)/14 = 133.9$ mm   c. M = 130 mm
   b. $\tilde{x} = (130 + 135)/2 = 132.5$ mm   d. m.r. = $(120 + 150)/2 = 135$ mm
   Given that the same person was measured, the data vary considerably. If these results are typical, it appears that blood pressure readings either (1) are dependent on the person taking the reading or (2) fluctuate widely from moment to moment or (3) are influenced by a lack of precision in the present equipment and technology.

11. Arranged in order, the n=11 scores are: 0.64 0.68 0.72 0.76 0.84 0.84 0.84 0.84 0.90 0.90 0.92
    a. $\bar{x} = (\Sigma x)/n = (8.88)/11 = 0.807$ mm   c. M = 0.84 mm
    b. $\tilde{x} = 0.84$ mm   d. m.r. = $(0.64 + 0.92)/2 = 0.780$ mm
    No, a sample from one location would not be representative of the genetic variety possible.

13. Arranged in order, the scores are as follows.
    Roof: 1 2 2 2 2 2 2 2 3 3 4 4 4 4 6 7 7 8 11
     Hat: 2 2 2 2 2 3 3 3 3 3 3 3 3 3 4 4 4 5 5

Roof	Hat
n = 20	n = 20
$\bar{x} = (\Sigma x)/n = 78/20 = 3.9$ letters	$\bar{x} = (\Sigma x)/n = 62/20 = 3.1$ letters
$\tilde{x} = (3 + 3)/2 = 3.0$ letters	$\tilde{x} = (3 + 3)/2 = 3.0$ letters

    Based on the means, the words appear to be longer in *Cat on a Hot Tin Roof*.

15. Arranged in order, the scores are as follows.
    One-day: -3 -2 -2 -1 0 0 0 0 1 1 1 2 2 8
    Five-day: -9 -6 -4 -3 -2 -2 -1 -1 0 2 4 5 6 6

One-day	Five-day
n = 14	n = 14
$\bar{x} = (\Sigma x)/n = 7/14 = 0.5$ °F	$\bar{x} = (\Sigma x)/n = -5/14 = -0.4$ °F
$\tilde{x} = (0 + 0)/2 = 0.0$ °F	$\tilde{x} = (-1 + -1)/2 = -1.0$ °F

    On the average both forecasts seem to have about the same degree of accuracy, but the errors in the five-day forecasts are understandably more variable.

17. Arranged in order, the scores are as follows.
    JVB: 6.5 6.6 6.7 6.8 7.1 7.3 7.4 7.7 7.7 7.7
    Prov: 4.2 5.4 5.8 6.2 6.7 7.7 7.7 8.5 9.3 10.0

JVB	Prov
n = 10	n = 10
$\bar{x} = (\Sigma x)/n = 71.5/10 = 7.15$ min	$\bar{x} = (\Sigma x)/n = 71.5/10 = 7.15$ min
$\tilde{x} = (7.1 + 7.3)/2 = 7.20$ min	$\tilde{x} = (6.7 + 7.7)/2 = 7.20$ min

    The two data sets have identical means and medians, but the Providence (individual lines) waiting times seem to be more variable.

19. Arranged in order, the scores are as follows.
    Before: 3.0398 3.0406 3.0586 3.0762 3.0775 3.1038 3.1043 3.1086 3.1274 3.1582
     After: 2.4823 2.4848 2.4907 2.4950 2.4998 2.5024 2.5113 2.5127 2.5163 2.5298

Before	After
n = 10	n = 10
$\bar{x} = (\Sigma x)/n = 30.8950/10 = 3.08950$ g	$\bar{x} = (\Sigma x)/n = 25.0251/10 = 2.50251$ g
$\tilde{x} = (3.0775 + 3.1038)/2 = 3.09065$ g	$\tilde{x} = (2.4998 + 2.5024)/2 = 2.50110$ g

    Yes, the weights after 1983 are considerably lower. Apparently zinc is lighter than copper.

21. Arrange the scores for each group in order (to aid in finding the median) and find the sum.

   One-day

   $n = 35$

   $\bar{x} = (\Sigma x)/n = -21/35 = -0.6$ °F

   $\tilde{x} = 0$ °F

   Five-day

   $n = 35$

   $\bar{x} = (\Sigma x)/n = -25/35 = -0.7$ °F

   $\tilde{x} = -1$ °F

   The results are very close to those obtained using only the first 14 days. No, the conclusions do not change when the larger data sets are used. Yes, we have more confidence in the results from the larger data set.

23. Arrange the scores for each group in order (to aid in finding the median) and find the sum.

   Before

   $n = 35$

   $\bar{x} = (\Sigma x)/n = 107.6174/35 = 3.07478$ g

   $\tilde{x} = 3.0763$ g

   After

   $n = 37$

   $\bar{x} = (\Sigma x)/n = 92.4668/37 = 2.49910$ g

   $\tilde{x} = 2.5004$ g

   The results are very close to those obtained using only the first 10 coins. No, the conclusions do not change when the larger data sets are used.

25. The x values below are the class midpoints from the given frequency table.

x	f	f·x
37	1	37
42	3	126
47	5	235
52	11	572
57	7	399
62	7	434
67	1	67
	35	1870

   $\bar{x} = \Sigma(f·x)/\Sigma f = 1870/35 = 53.4$ °F

   This is close to the 53.8 found using the original data.

27. The x values below are the class midpoints from the given frequency table, the class limits of which indicate the original data were integers.

x	f	f·x
43.5	25	1087.5
47.5	14	665.0
51.5	7	360.5
55.5	3	166.5
59.5	1	59.5
	50	2339.0

   $\bar{x} = \Sigma(f·x)/\Sigma f = 2339.0/50 = 46.8$ mph

   This is close to the 46.7 mph found using the original data. The mean speed of 46.8 mph of those ticketed by the police is more than 1.5 times the posted speed limit of 30 mi/hr. But this indicates nothing about the mean speed of all drivers, a figure which may or may not be higher than the posted limit.

29. a. Arranged in order, the original 54 scores are:

   26 29 34 40 46 48 60 62 64 65 76 79 80 86 90 94 105 114 116 120
   125 132 140 140 144 148 150 150 154 166 166 180 182 202 202 204 204 212 220 220
   236 262 270 316 332 344 348 356 360 365 416 436 446 514

   $\bar{x} = \Sigma x/n = 9876/54 = 182.9$ lbs

b. Trimming the highest and lowest 10% (or 5.4 = 5 scores), the remaining 44 scores are:

    48   60   62   64   65   76   79   80   86   90   94 105 114 116 120 125 132 140 140 144

    148 150 150 154 166 166 180 182 202 202 204 204 212 220 220 236 262 270 316 332

    344 348 356 360

    $\bar{x} = \Sigma x/n = 7524/44 = 171.0$ lbs

c. Trimming the highest and lowest 20% (or 10.8 = 11 scores), the remaining 32 scores are:

    79   80   86   90   94 105 114 116 120 125 132 140 140 144 148 150 150 154 166 166

    180 182 202 202 204 204 212 220 220 236 262 270

    $\bar{x} = \Sigma x/n = 5093/32 = 159.2$ lbs

Since physical considerations limit the lower values more than the upper values, increased trimming reduces the mean.

31. Treating the 5+ values as 5.0, $\bar{x} = \Sigma x/n = 17.1/5 = 3.42$. Since continuing the experiment would only increase Σx, one can conclude that the actual mean battery life in the sample is greater than 3.42 years.

33. Let $\bar{x}_h$ stand for the harmonic mean.

    $\bar{x}_h = n/[\Sigma(1/x)] = 2/[1/40 + 1/60] = 2/[0.0417] = 48.0$ mph

35. R.M.S. $= \sqrt{\Sigma x^2/n} = \sqrt{[(111.2)^2 + (108.7)^2 + (109.3)^2 + (104.8)^2 + (112.6)^2]/5} = 109.35$ volts

3-3 Measures of Variation

1. The standard deviation is considered a measure of variation because it gives an indication of how much spread there is in the data. Very loosely, the standard deviation measures the distance a typical score is from the mean.

3. Very loosely, the standard deviation measures the distance a typical score is from the mean. A student with a score of 85 would be 3.5 times farther away from the mean than is the typical student. Yes, in this context a score of 85 could be considered "unusual."

NOTE: Although not given in the text, the symbol R will be used for the range throughout this manual. Remember that the range is the difference between the highest and the lowest scores, and not necessarily the difference between the last and first scores as they are listed. Since calculating the range involves only the subtraction of 2 original pieces of data, the manual follows the example in this section (page 93) of reporting that measure of variation with the same accuracy as the original data. In general, the standard deviation and the variance will be reported with one more decimal place than the original data.

When finding the square root of the variance to obtain the standard deviation, use all the decimal places of the variance – and not the rounded value reported as the answer. To do this, keep the value on the calculator display or place it in the memory. Do not copy down all the decimal places and then re-enter them to find the square root – as that not only is a waste of time, but also could introduce copying and/or round-off errors.

5. preliminary values: n = 8 $\Sigma x = 466$ $\Sigma x^2 = 27772$

    R = 75 – 49 = 26 sec

    $s^2 = [n(\Sigma x^2) - (\Sigma x)^2]/[n(n-1)]$

       $= [8(27772) - (466)^2]/[8(7)] = 5020/56 = 89.6$ sec^2

    s = 9.5 sec

A sample of statistics students would not be representative of all adults in terms of age, scientific background, etc. Being a rather homogeneous group, they would likely exhibit less variability than is present in the general population of adults.

NOTE: The above format will be used in this manual for calculation of the variance and standard deviation. To reinforce some of the key symbols and concepts, the following detailed mathematical analysis is given for the data of the above problem.

x	x-$\bar{x}$	$(x-\bar{x})^2$	x^2
49	-9.25	85.5625	2401
49	-9.25	85.5625	2401
52	-6.25	39.0625	2704
53	-5.25	27.5625	2809
58	-0.25	0.0625	3364
62	3.75	14.0625	3844
68	9.75	95.0625	4624
75	16.75	280.5625	5625
466	0	627.5000	27772

$\bar{x} = (\Sigma x)/n = 466/8 = 58.25$ sec

by formula 3-4, $s^2 = \Sigma(x-\bar{x})^2/(n-1)$
$= 627.5000/7$
$= 89.6429$ sec^2

by formula 3-5, $s^2 = [n(\Sigma x^2) - (\Sigma x)^2]/[n(n-1)$
$= [8(27772) - (466)^2]/[8(7)]$
$= 5020/56$
$= 89.6429$ sec^2

NOTE: As shown in the preceding detailed mathematical analysis, formulas 3-4 and 3-5 give the same answer. The following three observations give hints for using these two formulas.
• When using formula 3-4, constructing a table having the first 3 columns shown above helps to organize the calculations and makes errors less likely. In addition, verify that $\Sigma(x-\bar{x}) = 0$ [except for a possible discrepancy at the last decimal due to rounding] before proceeding – if such is not the case, an error has been made and further calculation is fruitless.
• When using formula 3-5, the quantity $[n(\Sigma x^2) - (\Sigma x)^2]$ must be non-negative (i.e., greater than or equal to zero) – if such is not the case, an error has been made and further calculation is fruitless.
• In general, formula 3-5 is to be preferred over formula 3-4 because
  (1) it does not involve round-off error or messy calculations when the mean does not "come out even" and
  (2) the quantities Σx and Σx^2 can be found directly (on most calculators, with one pass through the data) without having to construct a table or make a second pass through the data after finding the mean.

7. preliminary values: n = 25 $\Sigma x = 47$ $\Sigma x^2 = 109$
   R = 4 – 1 = 3
   $s^2 = [n(\Sigma x^2) - (\Sigma x)^2]/[n(n-1)]$
   $= [25(109) - (47)^2]/[25(24)] = 516/600 = 0.9$
   s = 0.9
   Yes, the formulas can be applied to these values to obtain measures of variation – but no, the resulting values will have no meaning because the data are at the nominal level.

9. preliminary values: n = 14 $\Sigma x = 1875$ $\Sigma x^2 = 252179$
   R = 150 – 120 = 30 mm
   $s^2 = [n(\Sigma x^2) - (\Sigma x)^2]/[n(n-1)]$
   $= [14(252179) - (1875)^2]/[14(13)] = 14881/182 = 81.8$ mm^2
   s = 9.0 mm
   If the subject's blood pressure remains constant and is accurately measured, all the values would be the same and there would be no variability. Each value would equal the mean, and the standard deviation would be zero.

11. preliminary values: n = 11 $\Sigma x = 8.88$ $\Sigma x^2 = 7.2568$
    R = 0.92 – 0.64 = 0.28 mm
    $s^2 = [n(\Sigma x^2) - (\Sigma x)^2]/[n(n-1)]$
       $= [11(7.2568) - (8.88)^2]/[11(10)] = 0.9704/110 = 0.009 \text{ mm}^2$
    s = 0.094 mm
    No, a sample from one location would not be representative of the genetic variety possible.
    Being genetically homogeneous, the fruit flies in the sample would likely yield a smaller
    standard deviation than would result from a more representative group.

13. Cat on a Hot Tin Roof   Cat in the Hat
    n = 20, $\Sigma x = 78$, $\Sigma x^2 = 434$ n = 20, $\Sigma x = 62$, $\Sigma x^2 = 208$
    R = 11 – 1 = 10 letters   R = 5 – 2 = 3 letters
    $s^2 = [n(\Sigma x^2) - (\Sigma x)^2]/[n(n-1)]$ $s^2 = [n(\Sigma x^2) - (\Sigma x)^2]/[n(n-1)]$
       $= [20(434) - (78)^2]/[20(19)]$ $= [20(208) - (62)^2]/[20(19)]$
       $= 2596/380 = 6.8 \text{ letters}^2$ $= 316/380 = 0.8 \text{ letters}^2$
    s = 2.6 letters   s = 0.9 letters
    Yes, there is significant less variation in word length in the Dr. Seuss book.

15. One-day   Five-day
    n = 14, $\Sigma x = 7$, $\Sigma x^2 = 93$ n = 14, $\Sigma x = -5$, $\Sigma x^2 = 269$
    R = 8 – (-3) = 11 °F   R = 6 – (-9) = 15 °F
    $s^2 = [n(\Sigma x^2) - (\Sigma x)^2]/[n(n-1)]$ $s^2 = [n(\Sigma x^2) - (\Sigma x)^2]/[n(n-1)]$
       $= [14(93) - (7)^2]/[14(13)]$ $= [14(269) - (-5)^2]/[14(13)]$
       $= 1253/182 = 6.9 \text{ degrees}^2$ $= 3741/182 = 20.6 \text{ degrees}^2$
    s = 2.6 °F   s = 4.5 °F
    There appears to be a greater variation in the errors for the five-day forecasts. Yes, together
    with the fact that the forecasts do not differ significantly in mean error, this does suggest that
    the one-day forecasts are more accurate.

17. Jefferson Valley   Providence
    n = 10, $\Sigma x = 71.5$, $\Sigma x^2 = 513.27$ n = 10, $\Sigma x = 71.5$, $\Sigma x^2 = 541.09$
    R = 7.7 – 6.5 = 1.2 min   R = 10.0 – 4.2 = 5.8 min
    $s^2 = [n(\Sigma x^2) - (\Sigma x)^2]/[n(n-1)]$ $s^2 = [n(\Sigma x^2) - (\Sigma x)^2]/[n(n-1)]$
       $= [10(513.27) - (71.5)^2]/[10(9)]$ $= [10(541.09) - (71.5)^2]/[10(9)]$
       $= 20.45/90 = 0.23 \text{ min}^2$ $= 298.65/90 = 3.32 \text{ min}^2$
    s = 0.48 min   s = 1.82 min
    Exercise #17 of section 3-2 indicated that the mean waiting time was 7.15 minutes at each
    bank. The Jefferson Valley (single line) waiting times, however, are considerably less
    variable. The range measures the differences between the extremes. The longest and shortest
    waits at Jefferson Valley differ by a little over a minute (R=1.2), while the longest and shortest
    waits at Providence differ by almost 6 minutes (R=5.8). The standard deviation measures the
    typical deviation from the mean. A Jefferson Valley customer usually receives service within
    about ½ minute (s=0.48) of 7.15 minutes, while a Providence customer usually receives
    service within about 2 minutes (s=1.82) of the mean.

19. <u>Before 1983</u>

   n = 10, $\Sigma x = 30.8950$, $\Sigma x^2 = 95.46318394$

   R = 3.1582 – 3.0398 = 0.1184 g

   $s^2 = [n(\Sigma x^2) - (\Sigma x)^2]/[n(n-1)]$

   $\quad = [10(95.46318394) - (30.8950)^2]/[10(9)]$

   $\quad = 0.13081440/90 = 0.00145$ g^2

   s = 0.03812 g

<u>After 1983</u>

   n = 10, $\Sigma x = 25.0251$, $\Sigma x^2 = 62.62760433$

   R = 2.5298 – 2.4823 = 0.0475 g

   $s^2 = [n(\Sigma x^2) - (\Sigma x)^2]/[n(n-1)]$

   $\quad = [10(62.62760433) - (25.0251)^2]/[10(9)]$

   $\quad = 0.02041329/90 = 0.00023$ g^2

   s = 0.01506 g

   The weights of the pre-1983 pennies appear to vary about twice as much as the weights of the post-1983 pennies. The differences in variability could be caused by production factors – either zinc can be worked with more precisely than copper, or more precise penny-making machines have been developed since 1983. But it seems more likely that the differences in variability would be caused by wear – the older pennies have been through a wider variety of use, some being very worn from over 25 years of constant use and some being almost new because they have been out of heavy circulation for one reason or another.

NOTE: Exercises #21-#25 repeat earlier exercises, but use the full data sets instead of the first several values. In general, the variance and standard deviation of any sample size are good estimates of the population variance and standard deviation – although the larger the sample, the more confidence we have in the estimate. Increasing the sample size should not necessarily increase or decrease the variance or standard deviation. The range, however, is dependent upon the sample size because it is based only on the extremes and does not attempt to determine a "typical" spread based on all the data. In general, we expect larger samples to produce a larger range – in fact, it is not possible for the range to decrease as more data values are added to the sample.

21. NOTE: This exercise refers to differences between actual and predicted temperatures. Those differences must be calculated for each day, as they are not given directly in the data set.

   <u>One-day</u>

   n = 35, $\Sigma x = -21$, $\Sigma x^2 = 233$

   R = 8 – (-6) = 14 °F

   $s^2 = [n(\Sigma x^2) - (\Sigma x)^2]/[n(n-1)]$

   $\quad = [35(233) - (-21)^2]/[35(34)]$

   $\quad = 7714/1190 = 6.5$ degrees^2

   s = 2.5 °F

   <u>Five-day</u>

   n = 35, $\Sigma x = -25$, $\Sigma x^2 = 1077$

   R = 10 – (-14) = 24 °F

   $s^2 = [n(\Sigma x^2) - (\Sigma x)^2]/[n(n-1)]$

   $\quad = [35(1077) - (-25)^2]/[35(34)]$

   $\quad = 37070/1190 = 31.2$ degrees^2

   s = 5.6 °F

   There appears to be a grater variation in the errors for the five-day forecasts. Yes, together with the fact that the forecasts do not differ significantly in mean error, this does suggest that the one-day forecasts are more accurate. The conclusions are the same as in exercise #15.

23. <u>Before 1983</u>

   n = 35, $\Sigma x = 107.6174$, $\Sigma x^2 = 330.95211630$

   R = 3.1582 – 2.9593 = 0.1989 g

   $s^2 = [n(\Sigma x^2) - (\Sigma x)^2]/[n(n-1)]$

   $\quad = [35(330.95211630) - (107.6174)^2]/[35(34)]$

   $\quad = 1.81928774/1190 = 0.00153$ g^2

   s = 0.03910 g

<u>After 1983</u>

   n = 37, $\Sigma x = 92.4668$, $\Sigma x^2 = 231.09380700$

   R = 2.5298 – 2.4529 = 0.0769 g

   $s^2 = [n(\Sigma x^2) - (\Sigma x)^2]/[n(n-1)]$

   $\quad = [37(231.09380700) - (92.4668)^2]/[37(36)]$

   $\quad = 0.36175676/1332 = 0.00027$ g^2

   s = 0.01648 g

   The weights of the pre-1983 pennies appear to vary about twice as much as the weights of the post-1983 pennies. This is the same conclusion that was reached in exercise #19, and the individual results are almost identical.

25. The x values below are the class midpoints from the given frequency table.

x	f	f·x	f·x²
37	1	37	1369
42	3	126	5292
47	5	235	11045
52	11	572	29744
57	7	399	22743
62	7	434	26908
67	1	67	4489
	35	1870	101590

$$s^2 = [n\Sigma(f \cdot x^2) - (\Sigma f \cdot x)^2]/[n(n-1)]$$
$$= [35(101590) - (1870)^2]/[35(34)]$$
$$= 58750/1190$$
$$= 49.36$$
$$s = 7.0\ °F$$

This is close to the 6.9 °F found using the original data.

27. The x values below are the class midpoints from the given frequency table.

x	f	f·x	f·x²
43.5	25	1087.5	47306.25
47.5	14	665.0	31587.50
51.5	7	360.5	18565.75
55.5	3	166.5	9240.75
59.5	1	59.5	3540.25
	50	2339.0	110240.50

$$s^2 = [n\Sigma(f \cdot x^2) - (\Sigma f \cdot x)^2]/[n(n-1)]$$
$$= [50(110240.50) - (2339.0)^2]/[50(49)]$$
$$= 41104.00/2450$$
$$= 16.78$$
$$s = 4.1\ mph$$

This is close to the 4.0 mph found using the original data.

29. The Range Rule of Thumb suggests s ≈ range/4. Answers will vary, but one reasonable estimate for the range would be R = 70 − 26 = 44 years. This suggests s ≈ 44/4 = 11 years.

31. The Range Rule of Thumb suggests s ≈ range/4. Solving for the range, R ≈ 4s. If the distribution is approximately symmetric about the mean, then reasonable values for the upper and lower usual limits would be $\bar{x}$ ± 2s. In this instance,
    minimum usual value = $\bar{x}$ − 2s = 2838 − 2(504) = 1830 kwh
    maximum usual value = $\bar{x}$ + 2s = 2838 + 2(504) = 3846 kwh
Yes, in this context a consumption of 578 kwh would be considered unusual.

33. a. The range from 169 to 183 is $\bar{x}$ ± 1s. The empirical rule suggests that about 68% of the data values should fall within those limits.
    b. The range from 155 to 197 is $\bar{x}$ ± 3s. The empirical rule suggests that about 99.7% of the data values should fall within those limits.

35. The coefficient of variation for each sample is as follows.
    men: n=6 $\bar{x}$ = 69.167   s = 2.137   CV = s/$\bar{x}$ = 2.137/69.167 = 0.031 = 3.1%
    eggs: n=9 $\bar{x}$ = 22.144   s = 1.130   CV = s/$\bar{x}$ = 1.130/22.144 = 0.051 = 5.1%
The values are similar, but the lengths of the eggs are slightly more variable for their context.
NOTE: The coefficient of variation provides a direct comparison of the variability of two data sets even for objects that differ greatly in size and are measured in different units.

37. Even though the standard deviation measures the "typical" deviation and the deviation of the outlier will be "averaged" with the other deviations to get a "representative" value, the outlier will have a large effect and will increase the standard deviation.

3-4 Measures of Relative Standing

1. Since the sign of a z score is determined by subtracting $\bar{x}$ from the x value, a negative z score indicates that the x value is less than (i.e., below) the mean. Since the z score tells how many standard deviations a value is from the mean, an x value that has a z score of -2 is 2 standard deviations below the mean.

3. If 15 is the first quartile, then about 25% of the values fall below 15 – and about 75% of the values are above 15.

5. a. $|x - \bar{x}| = |182 - 176| = |6| = 6$ cm
   b. $6/7 = 0.86$ standard deviations
   c. $z = (x - \bar{x})/s$
      $= (182 - 176)/7 = 6/7 = 0.86$
   d. Since $-2 < 0.86 < 2$, Darwin's height is not considered unusual.

7. a. $|x - \bar{x}| = |67 - 71.5| = |-4.5| = 4.5$ in
   b. $4.5/2.1 = 2.14$ standard deviations
   c. $z = (x - \bar{x})/s$
      $= (67 - 71.5)/2.1 = -4.5/2.1 = -2.14$
   d. Since $-2.14 < -2$, McKinley's height is considered unusual.

9. In general $z = (x - \bar{x})/s$
   a. $z_{97.5} = (97.5 - 98.20)/0.62 = -0.7/0.62 = -1.13$
   b. $z_{98.60} = (98.60 - 98.20)/0.62 = 0.40/0.62 = 0.65$
   c. $z_{98.20} = (98.20 - 98.20)/0.62 = 0/0.62 = 0$

11. $z = (x - \bar{x})/s$
    $= (308 - 268)/15 = 40/15 = 2.67$
    Yes, since $2.67 > 2$, a pregnancy of 308 days is unusual. "Unusual" is just what the word implies – out of the ordinary. Other exercises in this section noted that there are unusual heights and IQ's, and so there can be unusual pregnancies.

13. Convert the two test scores to z scores.
    psychology: $z = (x - \bar{x})/s = (85 - 90)/10 = -5/10 = -0.50$
    economics: $z = (x - \bar{x})/s = (45 - 55)/5 = -10/5 = -2.00$
    Since $-0.50 > -2.00$, the psychology test score was relatively better.

For exercises #15-#29, refer to the list of ordered scores at the right side of the next page.

15. Let b = # of scores below x; n = total number of scores
    In general, the percentile of score x is $(b/n) \cdot 100$.
    The percentile score of 25 is $(4/76) \cdot 100 = 5$.

17. Let b = # of scores below x; n = total number of scores
    In general, the percentile of score x is $(b/n) \cdot 100$.
    The percentile score of 40 is $(57/76) \cdot 100 = 75$.

19. To find P_{10}, $L = (10/100) \cdot 76 = 7.6$, rounded up to 8.
    Since the 8th score is 25, $P_{10} = 25$.

21. To find P_{25}, $L = (25/100) \cdot 76 = 19$, a whole number.
    The mean of the 19th and 20th scores, $P_{25} = (28+28)/2 = 28.0$.

23. To find P_{33}, L = (33/100)·76 = 25.08, rounded up to 26.
Since the 26th score is 29, P_{33} = 29.

25. To find P_{01}, L = (1/100)·76 = 0.76, rounded up to 1.
Since the 1st score is 21, P_{01} = 21.

27. a. The interquartile range is $Q_3 - Q_1$.
   For $Q_3 = P_{75}$, L = (75/100)·76 = 57 – a whole number.
   The mean of the 57th and 58th scores, Q_3 = (39+40)/2 = 39.5.
   For $Q_1 = P_{25}$, L = (25/100)·76 = 19 – a whole number.
   The mean of the 19th and 20th scores, Q_1 = (28+28)/2 = 28.0.
   The interquartile range is 39.5 – 28.0 = 11.5

   b. The midquartile is $(Q_1+Q_3)/2$ = (28.8 + 39.5)/2 = 33.75.

   c. The 10-90 percentile range is $P_{90} - P_{10}$.
   For P_{90}, L = (90/100)·76 = 68.4, rounded up to 69.
   The 69th score, P_{90} = 49.
   For P_{10}, L = (10/100)·76 = 7.6, rounded up to 8.
   The 8th score, P_{10} = 25.
   The 10-90 percentile range is 49 – 25 = 24.

   d. Yes, $Q_2 = P_{50}$ by definition. Yes, they are always equal.

   e. For $Q_2 = P_{50}$, L = (50/100)·76 = 38 – a whole number.
   The mean of the 38th and 39th scores, Q_2 = (33 + 34)/2 = 33.5.
   No; in this case $Q_2 = 33.5 \neq 33.75 = (Q_1 + Q_3)/2$, which demonstrates
   that the median does not necessarily equal the midquartile.

29. a. $D_1 = P_{10}$, $D_5 = P_{50}$, $D_8 = P_{80}$

   b. For d = 1,2,3,4,5,6,7,8,9, L_d = (10d/100)·76 = 7.6d.
   For D_1, L = 7.6 and so $D_1 = x_8$ = 25
   For D_2, L = 15.2 and so $D_2 = x_{15}$ = 27
   For D_3, L = 22.8 and so $D_3 = x_{23}$ = 29
   For D_4, L = 30.4 and so $D_4 = x_{31}$ = 31
   For D_5, L = 38 and so $D_5 = (x_{38} + x_{39})/2$ = (33 + 34)/2 = 33.5
   For D_6, L = 45.6 and so $D_6 = x_{46}$ = 35
   For D_7, L = 53.2 and so $D_7 = x_{54}$ = 38
   For D_8, L = 60.8 and so $D_8 = x_{61}$ = 41
   For D_9, L = 68.4 and so $D_9 = x_{69}$ = 49

   c. For q = 1,2,3,4, L_q = (20q/100)·76 = 15.2q. These correspond to the even-numbered values
   computed in part (b) for the deciles.
   Quintile$_1 = P_{20} = D_2$ = 27
   Quintile$_2 = P_{40} = D_4$ = 31
   Quintile$_3 = P_{60} = D_6$ = 35
   Quintile$_4 = P_{80} = D_8$ = 41

#	age	#	age
1	21	39	34
2	22	40	34
3	24	41	34
4	24	42	35
5	25	43	35
6	25	44	35
7	25	45	35
8	25	46	35
9	26	47	35
10	26	48	35
11	26	49	36
12	26	50	37
13	27	51	37
14	27	52	38
15	27	53	38
16	27	54	38
17	28	55	38
18	28	56	39
19	28	57	39
20	28	58	40
21	29	59	41
22	29	60	41
23	29	61	41
24	29	62	41
25	29	63	41
26	29	64	42
27	30	65	42
28	30	66	43
29	31	67	45
30	31	68	46
31	31	69	49
32	32	70	50
33	32	71	54
34	33	72	60
35	33	73	61
36	33	74	63
37	33	75	74
38	33	76	80

3-5 Exploratory Data Analysis (EDA)

1. The boxplot gives the following information about the data set.

      minimum = 2 $Q_1 = 5$ median = 10

      maximum =20 $Q_3 = 12$

In addition, the data do not have a symmetric distribution about the mean – there is a concentration of scores from 10 to 12.

3. Sigma has less variation in repair times – and this is not merely a function of a few extreme values for Newport, as Newport also has a much larger the interquartile range Q_3-Q_1. Because the Sigma repair times have less variability, and costs are largely a function of the hourly rate, Sigma should have more predictable costs than Newport.

5. Order the n=11 scores to find the 5-number summary as follows.

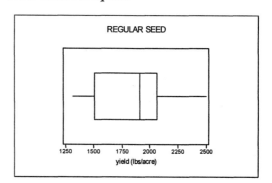

      minimum = x_1 = 1316

      $Q_1 = x_3$ = 1511

      median = x_6 = 1910

      $Q_3 = x_9$ = 2060

      maximum = x_{11} = 2496

The boxpolot is given at the right.

7. Order the n=40 scores to find the 5-number summary as follows.

      minimum = x_1 = 6.0002

      $Q_1 = (x_{10}+x_{11})/2$

          = (6.1309 + 6.1352)/2 = 6.13305

      median = $(x_{20}+x_{21})/2$

          = (6.1940 + 6.1947)/2 = 6.19435

      $Q_3 = (x_{30}+x_{31})/2$

          = (6.2647 + 6.2674)/2 = 6.26605

      maximum = x_{40} = 6.3669

The boxpolot is given at the right.

9. Order the n=54 scores to find the 5-number summary as follows.

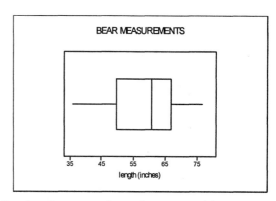

      minimum = x_1 = 36.0

      $Q_1 = x_{14}$ = 50.0

      median = $(x_{27}+x_{28})/2$

          = (60.5 + 61.0)/2 = 60.75

      $Q_3 = x_{41}$ = 66.5

      maximum = x_{54} = 76.5

The boxpolot is given at the right. While the median is toward the higher end of the main box (indicating a slight negative skew), the distribution is approximately symmetric.

11. Order the n=40 scores to find the 5-number
summary as follows.

$\text{minimum} = x_1 = 19.6$

$Q_1 = (x_{10}+x_{11})/2$
$= (23.5 + 23.8)/2 = 23.65$

$\text{median} = (x_{20}+x_{21})/2$
$= (26.2 + 26.2)/2 = 26.20$

$Q_3 = (x_{30}+x_{31})/2$
$= (27.4 + 27.8)/2 = 27.60$

$\text{maximum} = x_{40} = 33.2$

The boxpolot is given at the right.

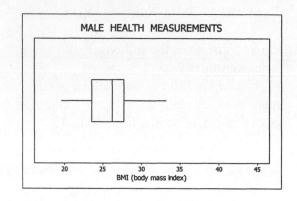

13. Choose Brand 1. Brand 1 and Brand 3 batteries both have a longer median life than Brand 2,
but the Brand 1 values have less variability than those for Brand 3. In addition to making it
easier to anticipate and schedule battery replacements, the small variability in values for
Brand 1 may also indicate better quality control and hence an overall better product.

15. Arrange the n=52 scores in order to determine

$Q_1 = (x_{13} + x_{14})/2 = (0.00 + 0.00)/2 = 0.00$

$Q_3 = (x_{39} + x_{40})/2 = (0.01 + 0.01)/2 = 0.01$

$IQR = Q_3 - Q_1 = 0.01 - 0.00 = 0.01$

$1.5(IQR) = 1.5(0.01) = 0.015$

$3(IQR) = 3(0.01) = 0.03$

There are no outliers at the lower end, since there are no values below $Q_1 = 0.00$.
The outliers at the upper end are as follows.

mild: $Q_3 + 1.5(IQR) < x \le Q_3 + 3(IQR)$ extreme: $x > Q_3 + 3(IQR)$

$0.01 + 0.015 < x \le 0.01 + 0.03$ $x > 0.01 + 0.03$

$0.025 < x \le 0.04$ $x > 0.04$

The only mild outlier is 0.03.
The extreme outliers are 0.05, 0.11, 0.12, 0.12, 0.41, 0.43, 0.47, 0.49, 0.59, 0.92, 1.41.

Statistical Literacy and Critical Thinking

1. No, reducing the standard deviation of the repair times means that there is less variability from
time to time – not that the times are smaller. Suppose, for example that the repair times were
previously about 2 hours each. The redesigned procedure could call for a minimum time of 2
hours spent on each repair, no matter how simple the problem. This would eliminate all the
very short times and the remaining times would be more tightly clustered at 2 hours and above
– a reduction in the standard deviation of the repair times, but a very inefficient increase in the
typical repair time.

2. No; even though the mean life is greater than 10 years as desired, the standard deviation cannot
be ignored. If the mean life is just barely greater than 10, about half the batteries would have a
life less than 10. And if the standard deviation is large, many of those lower values could be
considerably less than 10.

3. The additional value will have a considerable effect on the mean and the standard deviation.
The mean can be expected to increase by about ($1,000,000)/51 = $19,600. Because it
measures "typical deviation" from the mean, the standard deviation can also be expected to
increase – since scores less than the old mean will now be about $19,600 farther from the new

mean, and the individual deviations for scores above the old mean will also likely increase. The median will change very little – moving from the average of the 25th and 26th ordered scores to the 26th ordered score.

4. No, the results are not likely to yield a mean that is fairly close to the mean value of all cars owned by Americans. Some possible reasons for this are as follows.
    #1. The subscribers of any one particular ISP are not necessarily representative of the general population. Persons contracting with an ISP, for example, may tend to be more affluent than the general population.
    #2. The sample is a voluntary response sample, and those who choose to respond voluntarily are not necessarily representative of the group that was asked to respond.
    #3. The values were self-reported. People may not have an accurate or objective idea of the value of their cars – and some people may deliberately exaggerate the value of their cars to make themselves look good.

Review Exercises

1. The n=20 ordered circumferences are as follows.
        1.8 1.8 1.9 2.4 3.1 3.4 3.7 3.7 3.8 3.9 4.0 4.1 4.9 5.1 5.1 5.2 5.3 5.5 8.3 13.7

    a. $\bar{x} = (\Sigma x)/n = 90.7/20 = 4.54$ ft

    b. $\tilde{x} = (3.9 + 4.0)/2 = 3.95$ ft

    c. M = 1.8 ft, 3.7 ft, 5.1ft (tri-modal)

    d. m.r. = $(1.8 + 13.7)/2 = 7.75$ ft

    e. R = $13.7 - 1.8 = 11.9$ ft

    f. $s^2 = [n(\Sigma x^2) - (\Sigma x)^2]/[n(n-1)] = [20(544.85) - (90.7)^2]/[20(19)] = 2670.51/380 = 7.03$
        s = 2.65 ft

    g. $s^2 = 7.03$ ft^2 [from part (f)]

    h. $Q_1 = (x_5+x_6)/2 = (3.1 + 3.4)/2 = 3.25$ ft

    i. $Q_3 = (x_{15}+x_{16})/2 = (5.1 + 5.2)/2 = 5.15$ ft

    j. $P_{10} = (x_2 +x_3)/2 = (1.8 + 1.9)/2 = 1.85$ ft

2. a. $z = (x - \bar{x})/s$
        $z_{13.7} = (13.7 - 4.535)/2.65 = 9.165/2.65 = 3.46$

    b. Yes; since 13.7 is more than 2 standard deviations above the mean it is considered an unusual circumference.

    c. The Range Rule of Thumb suggests s ≈ range/4. Solving for the range, R ≈ 4s. If the distribution is approximately symmetric about the mean, then reasonable values for the upper and lower usual limits would be $\bar{x} \pm 2s$. In this instance,
        minimum usual value = $\bar{x} - 2s = 4.535 - 2(2.65) = -0.765$ ft
        maximum usual value = $\bar{x} + 2s = 4.535 + 2(2.65) = 9.835$ ft
    Other than the 13.7 already identified, no other listed circumferences are considered unusual.

3. The requested frequency distribution is given below.

circumference	frequency
1.0 – 2.9	4
3.0 – 4.9	9
5.0 – 6.9	5
7.0 – 8.9	1
9.0 – 10.9	0
11.0 – 12.9	0
13.0 – 14.9	1
	20

4. The requested histogram is given below.

The distribution is approximately bell-shaped, with a definite positive skew.

5. Order the n=20 scores to find the 5-number summary as follows.

minimum = x_1 = 1.8 ft

$Q_1 = (x_5 + x_6)/2$
$= (3.1 + 3.4)/2 = 3.25$ ft

median = $(x_{10} + x_{11})/2$
$= (3.9 + 4.0)/2 = 3.95$ ft

$Q_3 = (x_{15} + x_{16})/2$
$= (5.1 + 5.2)/2 = 5.15$ ft

maximum = x_{20} = 13.7 ft

The boxpolot is given at the right.

6. To make a comparison, convert each test score to a standard score (i.e., to a z score).

management: $z = (x - \overline{x})/s = (72 - 80)/12 = -8/12 = -0.67$

production: $z = (x - \overline{x})/s = (19 - 20)/5 = -1/5 = -0.20$

Since -0.20 > -0.67, the score of 19 on the production test is the better score.

7. Answers will vary, but the following are reasonable possibilities.
   a. The estimated mean age is 4 years.
   b. The estimated maximum and minimum ages are 20 years and 0 years, giving an estimated range of R = 20 – 0 = 20. The Range Rule of Thumb suggests s ≈ R/4. And so the estimated standard deviation is 20/4 = 5 years.

8. The Range Rule of Thumb suggests s ≈ range/4. Solving for the range, R ≈ 4s. If the distribution is approximately symmetric about the mean, then reasonable values for the upper and lower usual limits would be $\bar{x} \pm 2s$. In this instance,
   minimum usual value = $\bar{x} - 2s = 69.0 - 2(2.5) = 64$ in
   maximum usual value = $\bar{x} + 2s = 69.0 + 2(2.5) = 74$ in
Since 64 ≤ 72 ≤ 74, a height of 72 inches is within limits for usual heights and would not be considered unusual in this context.

Cumulative Review Exercises

1. a. The given values are continuous, since circumference can assume any value on a continuum.
   b. The given values are at the ratio level of measurement – since differences are meaningful and there is a natural zero.

2. a. The mode is the appropriate measure of "typicalness" for data at the nominal level. If the data is at the nominal level, numbers would be valid for identification only and could not be meaningfully ordered or otherwise mathematically manipulated. The mean, median and midrange all require ordering and/or other mathematical manipulations.
   b. This is convenience sampling, since he is contacting people who happen to be handy according to some criterion.
   c. This is cluster sampling, since all the subjects at randomly selected locations are sampled.
   d. Consistency is the lack of variability. Since she is dealing with variability, the standard deviation is the most relevant of the given statistics. It should be lowered. While the interquartile range (IQR = Q3 – Q1) also measures variability, it essentially ignores the extremely high and the extremely low scores – which are the very ones causing most of the problem. The standard deviation measures variability and considers all the scores.

Chapter 4

Probability

4-2 Fundamentals

1. "The probability of winning the grand prize in the Illinois lottery is 1/20,358,520" means that there are 20,358,520 equally likely possibilities (which may be the number of people who entered the lottery, the number of possible numeric outcomes, etc.) and that only one of them corresponds to winning. Yes, under these conditions such a win would be unusual – it has only 1 chance in 20,358,520 of occurring.

3. Yes, a probability of 0.001 = 1/1000 indicates that the event in question has only 1 chance in 1,000 of occurring.

5. a. "You will surely pass" indicates a probability of 1.00 – a certainty.
   b. "A 10% chance of rain" indicates a probability of 0.10.
   c. "A snowball's chance in hell" indicates a probability of 0.00 – an impossibility.

7. The probability of any event must be a value from 0 to 1 inclusive. The following listed values cannot be probabilities.
   -1, since it is less than 0
   2, since it is greater than 1
   5/3 = 1.667, since it is greater than 1
   $\sqrt{2}$ = 1.414, since it is greater than 1

9. a. Exactly one girl is the set with 3 simple events A = {BBG,BGB,GBB}.
   P(A) = 3/8 = 0.375
   b. Exactly two girls is the set with 3 simple events B = {BGG,GBG,GGB}.
   P(B) = 3/8 = 0.375
   c. All three girls is the set with 1 simple event C = {GGG}.
   P(C) = 1/8 = 0.125

11. There were 428 + 152 = 580 plants in the experiment.
    Let G = getting an offspring pea that is green.
    The estimate is P(G) = 428/580 = 0.738, which is very close to the expected value of 0.750.

13. There were 295 + 30 = 325 births in the experiment.
    Let G = getting a baby girl.
    The estimate is P(G) = 295/325 = 0.908.
    Yes, it appears that the technique is effective in increasing the likelihood of a girl.

15. a. There are 13R + 25O + 8Y + 8Br + 27Bl + 19G = 100 total M&M's, 27 of which are blue.
    The estimate is P(Blue) = 27/100 = 0.27.
    b. Yes, the estimate from part (a) agrees with the claim that 24% of all M&M's are blue. Even if precisely 24% of all M&M's are blue, we do not expect to get exactly 24 blue M&M's in every group of 100.

17. a. Let B = the birthdate is correctly identified. P(B) = 1/365 = 0.003.
    b. Yes. Since 0.003 ≤ 0.05, it would be unusual to guess correctly on the first try.
    c. Most people would probably believe that Mike's correct answer was the result of having inside information – and not making a lucky guess.
    d. Fifteen years is a big error. If the guess was made in all seriousness, Kelly would likely think that Mike was not a knowledgeable person and/or be personally insulted – and a second date would be unlikely. If the guess is perceived as being made in jest, Kelly might well appreciate his sense of humor and/or handling of a potentially awkward situation – and a second date would be likely.

19. Let A = a 20-24 year old has an accident this year. The estimate is P(A) = 136/400 = 0.340. If a driver who will be in that age bracket next year is selected, and the probabilities do not change from year to, the approximate probability that he or she will be in an accident next year is 0.340. Since 0.340 > 0.05, it would not be unusual for a driver that age to be involved in a crash next year. Yes, the result is high enough to be of concern.

21. There were 117 + 617 = 734 total patients in the experiment.
    Let H = a patient experiences a headache. The estimate is P(H) = 117/734 = 0.159.
    No. Since 0.159 > 0.05, experiencing a headache from taking the drug is not an unusual occurrence. Yes, the probability of experiencing a headache is high enough to be of concern – especially since a headache is one symptom that can interfere with the intended purpose for taking the drug in the first place.

23. a. The estimated probability of a wrong result for a person who does not use marijuana is 24/178 = 0.135.
    b. No. Since 0.135 > 0.05, it is not unusual to obtain a wrong test result for a person who does not use marijuana..

25. a. Listed by birth order, the sample space contains 4 simple events: BB BG GB GG.
    b. P(GG) = 1/4 = 0.25
    c. P(BG or GB) = 2/4 = 0.50

27. a. Listed with the brown/brown parent's contribution first, the sample space has 4 simple events: brown/brown brown/blue   brown/brown brown/blue
    b. P(blue/blue) = 0/4 = 0, an impossibility
    c. P(brown eyes) = P(blue/brown or brown/blue or brown/brown) = 4/4 = 1, a certainty

29. The odd against winning are P(not winning)/P(winning) = (423/500)/(77/500) = 423/77, or 423:77 – which is approximately 5.5:1, or 11:2.

31. Of the 38 slots, 18 are odd. Let W = winning because an odd number occurs
    a. P(W) = 18/38 = 9/19 = 0.474
    b. odds against W = P(not W)/P(W) = (20/38)/(18/38) = 20/18 = 10/9, or 10:9
    c. If the odds against winning are 1:1, a win gets back your bet plus $1 for every $1 bet. If you bet $18 and win, you get back $18 + $18 and your profit is $18.
    d. If the odds against winning are 10:9, a win gets get back your bet plus $10 for every $9 bet. If you bet $18 and win, you get back $18 + $20 and your profit is $20.

33. If the odds against A are a:b, the definition of odds indicates that P(not A)/P(A) = a/b. Since it is also true that P(not A) + P(A) = 1, algebra indicates that
    P(not A) = a/(a+b)
    P(A) = b/(a+b).
    If a=9 and b=1 as given in the problem, P(A) = b/(a+b) = 1/(9+1) = 1/10 = 0.100

35. No matter where the two flies land, it is possible to cut the orange in half to include flies on the same half. Since this is a certainty, the probability is 1. [Compare the orange to a globe. Suppose fly A lands on New York City, and consider all the circles of longitude. Wherever fly B lands, it is possible to slice the globe along some circle of longitude that places fly A and fly B on the same half.]

   NOTE: If the orange is marked into two predesignated halves before the flies land, the probability is different – once fly A lands, fly B has ½ a chance of landing on the same half. If both flies are to land on a specific one of the two predesignated halves, the probability is different still – fly A has ½ a chance of landing on the specified half, and only ½ the time will fly B pick the same half: the final answer would be ½ of ½, or ¼.

4-3 Addition Rule

1. Two events are disjoint if they cannot occur at the same time. The occurrence of one means that the other cannot occur – in other words, the occurrence of one prevents the occurrence of the other.

3. Close friends and relatives are a convenience sample and may not be representative of the general population.

5. a. No, it is possible to elect a female president.
   b. No, it is possible to select a male who smokes cigars.
   c. Yes, receiving Lipitor means that a person is not in the group that received no medication.

7. In general $P(\overline{A}) = 1 - P(A)$.
   a. $P(\overline{A}) = 1 - P(A)$
   $\qquad = 1 - 0.05$
   $\qquad = 0.95$
   b. Let R = a woman has red/green color blindness.
   $\quad P(\overline{R}) = 1 - P(R)$
   $\qquad\quad = 1 - 0.0025$
   $\qquad\quad = 0.9975$

NOTE: Throughout the manual we use the first letter (or other natural designation) of each category to represent that category. And so in exercises #9– #12,
   P(P) = P(selecting an accident in which the pedestrian was intoxicated)
   P(D) = P(selecting an accident in which the driver was intoxicated)
   P(PD) = P(selecting an accident in which pedestrian and driver were intoxicated)
   $P(P\overline{D})$ = P(selecting an accident in which pedestrian was intoxicated and driver was not)
   etc.
If there is ever cause for ambiguity, the notation will be clearly defined. Since mathematics and statistics use considerable notation and formulas, it is important to clearly define what various letters stand for.

ADDITIONAL NOTE: The following exercise, exercise #9, is worked two different ways. In general, using broad categories and correcting for double-counting is a more powerful technique that "lets the formula do the work" and requires less analysis by the solver. Except when more detailed analysis is instructive, this manual uses the first approach.

9. Make a table like the one at the right.
   Let P = the pedestrian was intoxicated.
   Let D = the driver was intoxicated.
   There are two approaches.

		PEDESTRIAN		
		yes	no	
DRIVER	yes	59	79	138
	no	266	581	847
		325	660	985

• Use broad categories and correct for
  double-counting – the "formal addition rule".
  P(P or D) = P(P) + P(D) – P(P and D)
   = 325/985 + 138/985 – 59/985
   = 404/985
   = 0.410

• Use individual mutually exclusive
  categories – the "intuitive addition rule".
  P(P or D) = P(PD or P$\overline{\text{D}}$ or $\overline{\text{P}}$D)
   = P(PD) + P(P$\overline{\text{D}}$) + P($\overline{\text{P}}$D)
   = 59/985 + 266/985 + 79/985
   = 404/985
   = 0.410

11. Refer to exercise #9.
   P(P or $\overline{\text{D}}$) = P(P) + P($\overline{\text{D}}$) – P(P and $\overline{\text{D}}$)
    = 325/985 + 847/985 – 266/985
    = 906/985 = 0.920

13. Make a table like the one at the right.
   P($\overline{\text{A}}$) = 1 – P(A)
    = 1 – 40/100
    = 60/100 = 0.60

	GROUP				
	O	A	B	AB	
TYPE Rh+	39	35	8	4	86
Rh-	6	5	2	1	14
	45	40	10	5	100

15. Refer to exercise #13.
   P(A or Rh-) = P(A) + P(Rh-) – P(A and Rh-)
    = 40/100 + 14/100 – 5/100
    = 49/100 = 0.49

17. Refer to exercise #13.
   P(not Rh+) = 1 – P(Rh+)
    = 1 – 86/100
    = 14/100 = 0.14

19. Refer to exercise #13.
   P(AB or Rh+) = P(AB) + P(Rh+) – P(AB and Rh+)
    = 5/100 + 86/100 – 4/100
    = 87/100 = 0.87

21. The table is given below.

	FLOWER		
	purple	white	
POD green	5	3	8
yellow	4	2	6
	9	5	14

23. Make a table like the one at the right.
    Let A = a person is age 18-21
        N = a person does not respond
    P(A or N) = P(A) + P(N) − P(A and N)
            = 84/359 + 31/359 − 11/359
            = 104/359 = 0.290

25. No, as illustrated by the following example.
    Let A = has prostate cancer.
    Let B = is a female.
    Let C = has testicular cancer.

27. The expression can be derived using algebra as follows.
    P(A or B or C)
        = P[(A or B) or C]
        = P(A or B)   + P(C) − P[(A or B) and C]
        = [P(A) + P(B) − P(A and B)] + P(C) − P[(A and C) or (B and C)]
        = P(A) + P(B) − P(A and B)   + P(C) − [P(A and C) + P(B and C) − P(A and B and C)]
        = P(A) + P(B) − P(A and B)   + P(C) − P(A and C) − P(B and C) + P(A and B and C)
        = P(A) + P(B) + P(C) − P(A and B) − P(A and C) − P(B and C) + P(A and B and C)

4-4 Multiplication Rule: Basics

1. Two events are independent if the occurrence of one does not affect the occurrence of the other. Mathematically, independence means that the probability that one event occurs is not affected by whether or not the other event has already occurred.

3. If only the 24 unselected students are available for the second selection, he is sampling without replacement. When sampling without replacement, the second outcome is not independent of the first – the first outcome affects the first in that whoever was chosen in the first selection cannot be chosen in the second selection.

5. a. Technically, dependent. The first outcome affects the second outcome in that whichever quarter was chosen in the first selection cannot be chosen in the second selection.
      NOTE: If selecting from a small set of pre-2001 quarters (e.g., the 8 that you have your possession), this dependence is significant. If selecting from "all" pre-2001 quarters, this dependence is negligible and for all practical purposes the events are independent.
   b. Technically, dependent. The first outcome affects the second outcome in that whichever viewer was chosen in the first selection cannot be chosen in the second selection.
      NOTE: If selecting from a small set of such viewers (e.g., the 8 that are in a given room), this dependence is significant. If selecting from "all" such viewers, this dependence is negligible and for all practical purposes the events are independent.
   c. In general these events are dependent, but the answer depends on the situation.
      • Dependent, if the invitation is made with visual contact – since physical appearance is one factor in deciding whether or not to accept a date with a person when there is visual contact.
      • Independent, if the invitation is made without visual contact (e.g., over the telephone) – in which case physical appearance has no opportunity to influence the decision on whether or not to accept the date.

7. Let T = getting the 2-choice true/false question correct by random guessing.
   Let M = getting the 4-choice multiple choice question correct by random guessing.
   P(T and M) = P(T)·P(M|T)
   $\qquad\qquad$ = (1/2)(1/4) = 1/8 = 0.125
   Random guessing does not appear to be a very good strategy.

9. Let O = selecting a hunter who was wearing orange.
   P(O) = 6/123 = 0.048, for the first selection from the sample
   a. P(O$_1$ and O$_2$) = P(O$_1$)·P(O$_2$|O$_1$)
   $\qquad\qquad$ = (6/123)(6/123) = 0.00238
   b. P(O$_1$ and O$_2$) = P(O$_1$)·P(O$_2$|O$_1$)
   $\qquad\qquad$ = (6/123)(5/122) = 0.00200
   c. Selecting without replacement makes more sense in this context. Selecting with
      replacement could lead to re-interviewing the same hunter, which would lead to no new
      information.

11. Let O = selecting a CD that is OK.
    P(O) = 4900/5000, for the first selection
    P(accepting the entire batch) = P(O$_1$ and O$_2$ and O$_3$ and O$_4$)
    $\qquad\qquad\qquad\qquad$ = P(O$_1$)·P(O$_2$|O$_1$)·P(O$_3$|O$_1$ and O$_2$)·P(O$_4$|O$_1$ and O$_2$ and O$_3$)
    $\qquad\qquad\qquad\qquad$ = (4900/5000)(4899/4999)(4898/4998)(4897/4997) = 0.922

13. Let G = a girl is born.
    P(G) = 0.50, for each couple
    P(all 12 girls) = P(G$_1$ and G$_2$ and…and G$_{12}$)
    $\qquad\qquad$ = P(G$_1$)·P(G$_2$)·…·P(G$_{12}$)
    $\qquad\qquad$ = (0.50)(0.50)…(0.50) = (0.50)^{12} = 0.000244
    If there are actually 12 girls born to the 12 couples in the experiment, then the gender-selection
    method appears to be effective. Either the method is effective or a very rare event (that has a 1
    in 4096 chance of occurrence) has taken place.

15. Let A = the alarm clock of that type functions properly.
    P(A) = 0.975
    a. P($\overline{A}$) = 1 − P(A)
    $\qquad$ = 1 − 0.975 = 0.025

    b. P($\overline{A}_1$ and $\overline{A}_2$) = P($\overline{A}_1$)·P($\overline{A}_2$)
    $\qquad\qquad$ = (0.025)(0.025) = 0.000625
    c. P("awakened with two clocks") = 1 − 0.000625 = 0.999375
       NOTE: This part could also be worked as follows.
       P(A$_1$ or A$_2$) = P(A$_1$) + P(A$_2$) − P(A$_1$ and A$_2$)
       $\qquad\qquad$ = 0.975 + 0.975 − (0.975)(0.975)
       $\qquad\qquad$ = 0.999375
    d. Yes, the second alarm clock results in greatly improved reliability.
       NOTE: Parts (c) and (d) actually give the probability that "at least one alarm clock functions
       properly" and not the probability that someone is "awakened with two clocks." If a person
       fails to arise because he sleeps through an alarm, then an additional clock may not increase
       the likelihood of awakening.

17. Make a table like the one at the right.
    Let D = the driver was intoxicated.
    Let P = the pedestrian was intoxicated.
    $P(D_1 \text{ and } D_2) = P(D_1) \cdot P(D_2|D_1)$
    $\phantom{P(D_1 \text{ and } D_2)} = (138/985)(137/984) = 0.0195$

		PEDESTRIAN			
		yes	no		
DRIVER	yes	59	79	138	
	no	266	581	847	
		325	660	985	

19. Refer to exercise #17.
    Let N = an accident in which neither the pedestrian nor the driver was intoxicated.
    a. $P(N) = P(\overline{P} \text{ and } \overline{D}) = 581/985 = 0.590$
    b. $P(N_1 \text{ and } N_2) = P(N_1) \cdot P(N_2|N_1)$
       $\phantom{P(N_1 \text{ and } N_2)} = (581/985)(580/984) = 0.34767$
    c. $P(N_1 \text{ and } N_2) = P(N_1) \cdot P(N_2|N_1)$
       $\phantom{P(N_1 \text{ and } N_2)} = (581/985)(581/985) = 0.34792$
    d. The answers in parts (c) and (d) are both 0.348 when rounded to 3 significant digits. The
       answer in part (d) is slightly larger because replacing the type-N accident after the first
       selection makes it more likely that another type-N accident will be chosen on the second
       selection.

21. Let D = a birthday is different from any yet selected.
    $P(D_1) = 366/366 = 1$ NOTE: With nothing to match, it <u>must</u> be different.
    $P(D_2) = P(D_2|D_1) = 365/366$
    $P(D_3) = P(D_3|D_1 \text{ and } D_2) = 364/366$
    ...
    $P(D_5) = P(D_5|D_1 \text{ and } D_2 \text{ and } D_3 \text{ and } D_4) = 362/366$
    ...
    $P(D_{25}) = P(D_{25}|D_1 \text{ and } D_2 \text{ and } D_3 \text{ and} \ldots \text{and } D_{24}) = 342/366$

    a. P(all different) = $P(D_1 \text{ and } D_2 \text{ and } D_3)$
       $ = P(D_1) \cdot P(D_2) \cdot P(D_3)$
       $ = (366/366)(365/366)(364/366) = 0.992$
    b. P(all different) = $P(D_1 \text{ and } D_2 \text{ and } D_3 \text{ and } D_4 \text{ and } D_5)$
       $ = P(D_1) \cdot P(D_2) \cdot P(D_3) \cdot P(D_4) \cdot P(D_5)$
       $ = (366/366)(365/366)(364/366)(363/366)(362/366) = 0.973$
    c. P(all different) = $P(D_1 \text{ and } D_2 \text{ and} \ldots \text{and } D_{25})$
       $ = P(D_1) \cdot P(D_2) \cdot \ldots \cdot P(D_{25})$
       $ = (366/366)(365/366) \ldots (342/366) = 0.432$

    NOTE: The above calculations assume that all birthdays (even February 29) are equally likely.
    If one omits February 29 and considers there to be only 365 assumed equally likely birthdays,
    the answers to (a) and (b) do not change – but the answer to (c) is 0.431. An algorithm to
    perform this calculation can be constructed using a programming language – or most
    spreadsheet or statistical software packages. In BASIC, for example, use

```
10  LET P=1
15  PRINT "HOW MANY PEOPLE?"
20  INPUT D
30  FOR K=1 TO D-1
40  LET P=P*(366-K)/366
50  NEXT K
60  PRINT "P(ALL DIFFERENT BIRTHDAYS) IS  ";P
70  END
```

23. This problem can be done by two different methods. In either case, let

> A = getting an ace
> S = getting a spade.

- Consider the sample space

The first card could be any of 52 cards, and for each first card there are 51 possible second cards. This makes a total of $52 \cdot 51 = 2652$ equally likely outcomes in the sample space. How many of them are A_1 and S_2?

The aces of hearts, diamond and clubs can be paired with any of the 13 spades for a total of $3 \cdot 13 = 39$ favorable possibilities. The ace of spades can only be paired with any of the remaining 12 members of that suit for a total of 12 favorable possibilities. Since there are $39 + 12 = 51$ total favorable possibilities among the equally likely outcomes, $P(A_1$ and $S_2) = 51/2652 = 0.0192$

- Use the formulas (and express the event as two mutually exclusive possibilities)

$$P(A_1 \text{ and } S_2) = P([\text{spade}A_1 \text{ and } S_2]\text{or } [\text{other}A_1 \text{ and } S_2])$$
$$= P(\text{spade}A_1 \text{ and } S_2) + P(\text{other}A_1 \text{ and } S_2)$$
$$= P(\text{spade}A_1) \cdot P(S_2|\text{spade}A_1) + P(\text{other}A_1) \cdot P(S_2|\text{other}A_1)$$
$$= (1/52)(12/51) + (3/52)(13/51)$$
$$= 12/2652 + 39/2652$$
$$= 51/2652 = 0.0192$$

4-5 Multiplication Rule: Complements and Conditional Probability

1. Knowing that "at least one" of ten items is defective does not indicate the exact number of defects – only that it is 1 or greater, and no more than 10.

3. Since about half the population is male, it may be reasonable to estimate that the probability a shopper in a gender-neutral situation is a male is ½. To estimate the probability such a shopper is a male given that a credit card was used for the purchase, one needs information about the proportion of all shoppers that use credit cards and the proportion of males that use credit cards.

5. If it is not true that at least one of them tests positive, then all six test negative – or, equivalently, none of the six tests positive.

7. If it is not true that none of them have the gene, then at least one of the 12 males have the particular X-linked recessive gene.

9. Let F = a credit card is being used fraudulently.
Let D = a credit card was used in several different countries in one day.
Answers will vary. One reasonable estimate is $P(F|D) \approx 0.999$.

11. $P(\text{at least one girl}) = 1 - P(\text{all boys})$
$$= 1 - P(B_1 \text{ and } B_2 \text{ and } B_3 \text{ and } B_4)$$
$$= 1 - P(B_1) \cdot P(B_2) \cdot P(B_3) \cdot P(B_4)$$
$$= 1 - (\tfrac{1}{2})(\tfrac{1}{2})(\tfrac{1}{2})(\tfrac{1}{2})$$
$$= 1 - 1/16 = 15/16 = 0.9375$$

Yes, this probability is high enough for the couple to be fairly confident that they will get at least one girl – but since $P(\text{all boys}) = 1/16 = 0.0625 > 0.05$, getting all boys would not be considered an unusual event according to the standard given in the previous chapter.

13. Since each birth is an independent event, $P(G_3|B_1 \text{ and } B_2) = P(G_3) = P(G) = \frac{1}{2}$.
    One can also consider the 8-outcome sample space S for 3 births and use the formula for conditional probability as follows.
    S = {BBB, BBG, BGB, GBB, BGG, GBG, GGB, GGG}
    Let F = the first two children are boys
    Let G_3 = the third child is a girl
    P(F) = 2/8
    $P(G_3)$ = 4/8
    P(F and G) = 1/8

    $$P(G_3|F) = \frac{P(G_3 \text{ and } F)}{P(F)} = \frac{1/8}{2/8} = 1/2$$

    Note: Using the sample space S also verifies directly that $P(G_3) = 4/8 = 1/2 = P(G)$, independent of what occurred during the first two births.
    No. This probability is not the same as P(BBB) = 1/8.

15. Make a table like the one at the right.
    P(neg|no) = 154/178 = 0.865
    This is the probability of correctly declaring
    that someone does not use marijuana.

		USE MARIJUANA?		
		yes	no	
TEST	pos	119	24	143
RESULTS	neg	3	154	157
		122	178	300

17. P(at least one clock works) = 1 – P(all clocks fail)
    $= 1 - P(F_1 \text{ and } F_2 \text{ and } F_3)$
    $= 1 - P(F_1) \cdot P(F_2) \cdot P(F_3)$
    = 1 – (0.05)(0.05)(0.05)
    = 1 – 0.000125 = 0.999875
    Yes. Her probability of having a functioning alarm goes up from 95% to 99.9875%.

19. P(HIV positive result) = P(at least one person is HIV positive)
    = 1 – P(all persons are HIV negative)
    $= 1 - P(N_1 \text{ and } N_2 \text{ and } N_3)$
    $= 1 - P(N_1) \cdot P(N_2) \cdot P(N_3)$
    = 1 – (0.9)(0.9)(0.9)
    = 1 – 0.729 = 0.271
    NOTE: This plan is very efficient. Suppose, for example, there were 3000 people to be tested.
    Only in 0.271 = 27.1% of the groups would a retest need to be done for each of the three individuals. Those (0.271)(1000) = 271 groups would generate (271)(3) = 813 retests. The total number of tests required would be 1813 (the 1000 original + the 813 retests), only 60% of the 3,000 tests that would have been required to test everyone individually.

21. Make a table like the one at the right.
    Let M = a Senator is a male.
    Let F = a Senator is a female.
    Let R = a Senator is a Republican.
    Let D = a Senator is a Democrat.
    Let I = a Senator is an Independent.
    P(R|M) = 46/86 = 0.535

		PARTY			
		Rep	Dem	Ind	
GENDER	male	46	39	1	86
	female	5	9	0	14
		51	48	1	100

23. Refer to exercise #21.
$P(F|I) = 0/1 = 0$

25. For 25 randomly selected people,
a. P(no two share the same birthday) = 0.432 [or .431, see exercise #21c of section 4-4]
b. P(at least two share the same birthday) = 1 – P(no two share the same birthday)
$$= 1 - 0.432 = 0.568$$

27. Let F = getting a seat in the front row.
$P(F) = 2/24 = 1/12 = 0.0833$
• formulate the problem in terms of "n" rides
P(at least one F in n rides) = 1 – P(no F in n rides)
$$= 1 - (11/12)^n$$
• try different values for n until P(at least one F) $\geq$ 0.95
P(at least one F in 32 rides) = $1 - (11/12)^{32} = 1 - 0.0616 = 0.9384$
P(at least one F in 33 rides) = $1 - (11/12)^{33} = 1 - 0.0566 = 0.9434$
P(at least one F in 34 rides) = $1 - (11/12)^{34} = 1 - 0.0519 = 0.9481$
P(at least one F in 35 rides) = $1 - (11/12)^{35} = 1 - 0.0476 = 0.9524$
You must ride 35 times in order to have at least a 95% chance of getting a first row seat at least once.

4-6 Probabilities Through Simulations

1. A simulation is an approximation of some phenomenon that behaves in the same way as the phenomenon it represents. It often takes the form of a mathematical model for some real-world phenomenon that would be difficult or dangerous to observe first-hand. If a simulation method is used for a probability problem, the result will not necessarily be exactly correct – but it should be close to the true value.

3. No. People have subconscious biases and notions about what is random that prevent them from generating sequences that are truly random.

NOTE: In exercises #5-8 we suggest using random two-digit numbers from 00 to 99 inclusive. For convenience of association with the natural counting numbers, consider 00 to represent 100 so that the conceptual "order" of the digits is 01,02,03,…98,99,00. Even though other schemes using fewer than 100 possibilities could be used, this has the advantage of being able to read through a printed list of random digits and associating them into pairs – without having to discard any values as being outside the range of desired random numbers.

5. Generate 20 random two-digit integers as suggested in the NOTE before exercise #5.
Let 01-95 represent a male and 96-00 represent a female.

7. Generate 500 random two-digit integers as suggested in the NOTE before exercise #5.
Let 01-02 represent a defective cell phone and 03-00 represent a good cell phone.

9. a. Refer to the table at th right. The simulated percentage of males is
      19/20 = 95%, which is exactly the desired value.
   b. The numbers of males in the 10 simulations prepared for this manual
      were as follows: 19 19 20 20 20 20 20 20 20 20.
      These numbers are very consistent. Based on these results it would be
      very unusual to find 10 females among 20 randomly selected cyclists.

11. a. The simulation prepared for this manual produced 18/500 =
       3.6% defective cell phones, which is fairly close to 2%.
    b. The numbers of defective cell phones in the 5 simulations
       prepared for this manual were as follows: 18 18 18 11 17.
       Based on these results, it would be very unusual to select 500
       such cell phones and find that 0 of them are defective.

exercise #9	
#	gender
80	male
86	male
8	male
69	male
81	male
42	male
35	male
96	female
1	male
53	male
8	male
7	male
91	male
45	male
49	male
87	male
30	male
79	male
2	male
52	male

13. When the simulation was performed 20 times for this manual, the numbers
    of smokers who quit (assuming the usual 20% success rate) out of 50 were
    as follows: 9 14 10 8 13 9 16 13 11 10 13 5 6 13 6 9 14 12 9 6
    Yes. It appears that 12 successes can easily occur by chance when a drug is
    ineffective and the usual 20% success rate is still in effect. In fact, the
    estimate is P(12 or more successes with an ineffective drug) $\approx 8/20 = 40\%$.

15. The probability of originally selecting the door with the car is 1/3.
    You can either switch or not switch.
    Let W = winning the car.
    • If you make a selection and do not switch, P(W) = 1/3.
    • If you make a selection and switch, there are only 2 possibilities.
        A = you originally had the winning door and switched to a losing door.
        B = you originally had a losing door and switched to a winning door.
      NOTE: You cannot switch from a losing door to another losing door. Monty Hall always
      opens a losing door. If you have a losing door and Monty Hall opens a losing door, then
      the only possible door to switch to is the winning door. Since you win if B occurs,
            P(W) = P(B) = P(originally had a losing door) = 2/3.
    • Conclusion: the better strategy is to switch.
    While the preceding analysis makes it unnecessary, a simulation of n trails can be done as
    follows.
        1. Randomly select door 1,2 or 3 to hold the prize.
        2. Randomly select door 1,2 or 3 as your original choice.
        3. Determine m = the number of times the doors match.
        4. Estimated P(winning if you don't switch) = m/n. [m/n should be about 1/3]
        5. Estimated P(winning if you switch) = (n-m)/n. [(n-m)/n should be about 2/3]
        NOTE: If you switch, you lose m times and win (n-m) times – if you originally selected a
        losing door, remember, a switch always wins the car.
    The following Minitab commands produce the desired simulation, where column 1 indicates
    the winning door and column 2 indicates your choice.
        MTB > RANDOM 100 C1 C2;   MTB > LET C3= C1-C2
        SUBC> INTEGER 1 3.   MTB > TABLE C3
    A zero in C3 indicates a match and that the original choice would win the car. The simulation
    produced for this manual yielded 37 0's in C3.
        Estimated P(winning if you don't switch) = 37/100 = 37%
        Estimated P(winning if you switch) = 63/100 = 63%

17. No, his reasoning is not correct. No, the proportion of girls will not increase. Each family would consist of zero to several consecutive girls, followed by a boy. The types of possible families and their associated probabilities would be as follows.

$P(B) = \frac{1}{2} = 1/2$ $= 16/32$
$P(GB) = (\frac{1}{2})(\frac{1}{2}) = 1/4$ $= 8/32$
$P(GGB) = (\frac{1}{2})(\frac{1}{2})(\frac{1}{2}) = 1/8$ $= 4/32$
$P(GGGB) = (\frac{1}{2})(\frac{1}{2})(\frac{1}{2})(\frac{1}{2}) = 1/16$ $= 2/32$
$P(GGGGB) = (\frac{1}{2})(\frac{1}{2})(\frac{1}{2})(\frac{1}{2})(\frac{1}{2}) = 1/32$ $= 1/32$
etc.

Each collection of 32 families would be expected to look like this, where * represents one family with 5 or more girls and a boy.

B	B	B	B	B	B	B	B
B	B	B	B	B	B	B	B
GB	GB	GB	GB	GB	GB	GB	GB
GGB	GGB	GGB	GGB	GGGB	GGGB	GGGGB *	

A gender count reveals 32 boys and 31 or more girls, an approximately equal distribution of genders. In practice, however, families would likely stop after some maximum number of children – whether they had a boy or not. If that number was 5 for all families, then * = GGGGG and the expected gender count for the 32 families would be an exactly equal distribution of 31 boys and 31 girls.

NOTE: Since the "stopping point" would likely vary from family to family, the problem is actually more complicated. Suppose, for example, that 25% of the families decided in advance to stop at two children regardless of gender. We expect those $1/4 = 8/32$ of the families to be distributed approximately evenly among the four rows in the above listing – 2 in each row. The 6 such families in the first 3 rows would be stopped by the king's decree and remain unchanged, but the 2 such families in the last row would both become GG families. Counting the * as GGGGG, this reduces the expected number of boys by $2 \cdot 7/8 = 1.75$ and the expected number of girls by $2 \cdot 7/8 = 1.75$ – and there would still be an equal number of each gender.

4-7 Counting

1. The permutations rule applies when different orderings of the same collection of objects represent a different solution. The combinations rule applies when all orderings of the same collection of objects represent the same solution.

3. No. The methods of this section are used when objects are selected at random from various populations. She could find the probability that the letters z, i and p are chosen at random (in any order, or in a particular order) in 3 selections from the alphabet – but this is in no way related to the relative frequency of word usage.

5. $5! = 5 \cdot 4 \cdot 3 \cdot 2 \cdot 1 = 120$

7. $_{24}C_4 = 24!/(20!4!) = (24 \cdot 23 \cdot 22 \cdot 21 \cdot 20!)/(20!4!) = (24 \cdot 23 \cdot 22 \cdot 21)/4! = 10,626$
NOTE: This technique of reducing the problem by "canceling out" the 20! from both the numerator and denominator can be used in most combinations and permutations problems. In general, a smaller factorial in the denominator can be completely divided into a larger factorial in the numerator to leave only the "excess" factors not appearing in the denominator. Furthermore, $_nC_r$ and $_nP_r$ will always be integers. More generally, a non-integer answer to <u>any counting problem</u> (but not a probability problem) indicates that an error has been made.

9. $_{52}P_2 = 52!/50! = 52 \cdot 51 = 2652$

11. $_{30}C_3 = 30!/(27!3!) = 30 \cdot 29 \cdot 28/3! = 4060$

13. Let W = winning the described lottery with a single selection.
The total number of possible combinations is
$$_{35}C_5 = 35!/(30!5!) = 35 \cdot 34 \cdot 33 \cdot 32 \cdot 31/5! = 324,632.$$
Since only one combination wins, P(W) = 1/324,632.

15. Let W = winning the described lottery with a single selection.
The total number of possible combinations is
$$_{38}C_6 = 38!/(32!6!) = 38 \cdot 37 \cdot 36 \cdot 35 \cdot 34 \cdot 33/6! = 2,760,681.$$
Since only one combination wins, P(W) = 1/2,760,681.

17. Let W = winning the described lottery with a single selection.
The total number of possible permutations is
$$_{39}P_5 = 39!/34! = 39 \cdot 38 \cdot 37 \cdot 36 \cdot 35 = 69,090,840.$$
Since only one permutation wins, P(W) = 1/69,090,840.

19. Let O = selecting at random and getting the 5 oldest managers.
The total number of possible combinations is
$$_{15}C_5 = 15!/(10!5!) = 15 \cdot 14 \cdot 13 \cdot 12 \cdot 11/5! = 3003.$$
Since only one combination includes the oldest 5, P(O) = 1/3003 = 0.000333.
Yes, that probability is low enough to charge that the dismissals resulted from age discrimination instead of random selection.

21. By the fundamental counting rule, there are $4 \cdot 4 \cdot 4 \cdot 4 \cdot 4 \cdot 4 \cdot 4 \cdot 4 \cdot 4 \cdot 4 = 4^{10} = 1,048,576$ different possible treatment arrangements.

23. a. "Permutations" implies that order makes a difference, but the only distinguishing feature of each pill is whether it is good (G) or defective (D). The problem reduces to how many ways can you arrange 7G's and 3 D's, which further reduces to how many ways can you pick 3 of the 10 positions to be D's – and the order in which the 3 positions are picked is irrelevant. And so the solution is $_{10}C_3 = 10!/(7!3!) = 10 \cdot 9 \cdot 8/3! = 120$.
    b. Since the order in which the pills are picked does not affect the sample, use combinations.
       The total number of possible samples is $_{10}C_3 = 10!/(7!3!) = 10 \cdot 9 \cdot 8/3! = 120$.
       Let A = the sample consists of precisely all 3 defective pills. Since only one of the possible samples includes all 3 defective pills, P(A) = 1/120.

25. Let S = rolling 5 consecutive 6's.
By the fundamental counting rule, there are $6 \cdot 6 \cdot 6 \cdot 6 \cdot 6 = 6^5 = 7776$ possible outcomes.
Since only one of them is 5 consecutive 6's, P(S) = 1/7776 = 0.000129.
Yes, that probability is low enough to rule out chance as an explanation for Mike's results.

27. a. Since each order is a different slate of officers, use permutations.
       The total number of possible slates is $_{12}P_4 = 12!/8! = 12 \cdot 11 \cdot 10 \cdot 9 = 11,880$.
    b. Since the order does not affect the composition of the committee, use combinations.
       The total number of possible committees is $_{12}C_4 = 12!/(8!4!) = 12 \cdot 11 \cdot 10 \cdot 9/4! = 495$.

29. The first note is given by *. There are 3 possibilities for each of the next 15 notes.
By the fundamental counting rule, there are
$$3 \cdot 3 \cdot 3 \cdot 3 \cdot 3 \cdot 3 \cdot 3 \cdot 3 \cdot 3 \cdot 3 \cdot 3 \cdot 3 \cdot 3 \cdot 3 \cdot 3 = 3^{15} = 14,348,907 \text{ possible sequences.}$$
NOTE: This assumes that each song has at least 16 notes, and it does not guarantee that two different melodies will not have the same representation – if one goes up two steps every time the other goes up one step, for example, they both will show a U in that position.

31. There are 10 digits: 0,1,2,3,4,5,6,7,8,9.
    The first digit could be any of 8 digits: 2,3,4,5,6,7,8,9.
    The middle digit could be any of 2 digits: 0,1.
    The last digit could be any of 9 digits: any digit except the middle one.
    By the fundamental counting rule there are $8 \cdot 2 \cdot 9 = 144$ possible area codes under the old system.

33. a. Since there are 64 teams at the start and only 1 left at the end, 63 teams must be eliminated.
       Since each game eliminates 1 team, it takes 63 games to eliminate 63 teams.
    b. Let W = picking all 63 winners by random guessing.
       Since there are 2 possible outcomes for each game, there are $2^{63} = 9.223 \times 10^{18}$ possible sets of results. Since only one such result gives all the correct winners,
       $$P(W) = 1/[9.223 \times 10^{18}] = 1.084 \times 10^{-19}.$$
       One could also reason: since each guess has a 50% chance of being correct,
       $$P(W) = (0.5)^{63} = 1.084 \times 10^{-19}.$$
    c. Let E = picking all 63 winners by expert guessing.
       Since each guess has a 70% chance of being correct, $P(W) = (0.7)^{63} = 1.743 \times 10^{-10}$.
       Since $1.743 \times 10^{-10} = 1/5,738,831,575$, the expert has a 1 in 5,738,831,575 chance [or about a 1 in 5.7 billion chance] of selecting all 63 winners.

35. There are $_{47}C_5 = 47!/(42!5!) = 1,533,939$ ways to select the five regular numbers.
    There are 27 ways to select the special number.
    By the fundamental counting rule, there are $1,533,939 \cdot 27 = 41,416,353$ total possible entry selections. Since only one of them is a winner, the probability of winning the jackpot with a single entry selection is $1/41,416,353 = 0.0000000241$.

37. There are 26 possible first characters, and 36 possible characters for the other positions.
    Find the number of possible names using 1,2,3,...,8 characters and then add to get the total.

characters	possible names		
1	26	=	26
2	26·36	=	936
3	26·36·36	=	33,696
4	26·36·36·36	=	1,213,056
5	26·36·36·36·36	=	43,670,016
6	26·36·36·36·36·36	=	1,572,120,576
7	26·36·36·36·36·36·36	=	56,596,340,736
8	26·36·36·36·36·36·36·36	=	2,037,468,266,496
		total =	2,095,681,645,538

39. a. The calculator factorial key gives $50! = 3.04140932 \times 10^{64}$
       Using the approximation, $K = (50.5) \cdot \log(50) + 0.39908993 - 0.43429448(50)$
       $$= 85.79798522 + 0.39908993 - 21.71472400 = 64.48235225$$
       and then $50! = 10^K = 10^{64.48235225} = 3.036345215 \times 10^{64}$
       NOTE: The two answers differ by 0.0051×10^{64} – i.e., 5.1×10^{61}, or 51 followed by 60 zeros. While such an error of "zillions and zillions" may seem quite large, it is only an error of $(5.1 \times 10^{61})/(3.04 \times 10^{64}) = 1.7\%$.

    b. The number of possible routes is 300!
       Using the approximation, $K = (300.5) \cdot \log(300) + 0.39908993 - 0.43429448(300)$
       $$= 744.374937 + 0.39908993 - 130.288344 = 614.485683$$
       and then $300! = 10^K = 10^{614.485683}$
       Since the number of digits in 10^x is the next whole number above x, 300! has 615 digits.

41. This problem cannot be solved directly using permutations, combinations or other techniques presented in this chapter. It is best solved by listing all the possible solutions in an orderly fashion and then counting the number of solutions. Often this is the most reasonable approach to a counting problem.

   While you are encouraged to develop your own systematic approach to the problem, the following table represents one way to organize the solution. The table is organized by rows, according to the numbers of pennies in each way that change can be made. The numbers in each row give the numbers of the other coins as explained in the footnotes below the table.

   The three numbers in the 80 row, for example, indicate that there are three ways to make change using 80 pennies. The 4 in the "only 5¢" column represents one way (80 pennies, 4 nickels). The 2 in the 10¢ column of the "nickels and one other coin" columns represents another way (80 pennies, 2 dimes). The 1 in the 10¢ column of the "nickels and one other coin" columns represents a third way (80 pennies, 1 dime, and [by default] 2 nickels).

   The bold-face **3** in the 20 row and the 10¢/25¢ column of the "nickels and two other coins" columns represents 20 pennies, 3 dimes, one quarter, and [by default] 5 nickels.

   Using the following table, or the system of your own design, you should be able to
(1) take a table entry and determine what way to make change it represents and
(2) take a known way to make change and find
   its representation in the table.

1¢	only 5¢	nickels and one other coin[a] 10¢	25¢	50¢	nickels and two other coins[b] 10¢/25¢	10¢/50¢	25¢/50¢	all[c] 10¢/25¢/50¢
100	0	.			.		.	.
95	1	.			.		.	.
90	2	1			.		.	.
85	3	1			.		.	.
80	4	2,1			.		.	.
75	5	2,1	1		.		.	.
70	6	3,2,1	1		.		.	.
65	7	3,2,1	1	.	1		.	.
60	8	4,3,2,1	1	.	1		.	.
55	9	4,3,2,1	1	.	2,1		.	.
50	10	5,4,3,2,1	2,1	1	2,1		.	.
45	11	5,4,3,2,1	2,1	1	3,2,1		.	.
40	12	6,5,4,3,2,1	2,1	1	3,2,1	1	.	.
35	13	6,5,4,3,2,1	2,1	1	4,3,2,1	1	.	.
30	14	7,6,5,4,3,2,1	2,1	1	4,3,2,1	2,1	.	.
25	15	7,6,5,4,3,2,1	3,2,1	1	5,4,3,2,1	2,1	1	.
20	16	8,7,6,5,4,3,2,1	3,2,1	1	5,4,**3**,2,1	3,2,1	1	.
15	17	8,7,6,5,4,3,2,1	3,2,1	1	6,5,4,3,2,1	3,2,1	1	1
10	18	9,8,7,6,5,4,3,2,1	3,2,1	1	6,5,4,3,2,1	4,3,2,1	1	1
5	19	9,8,7,6,5,4,3,2,1	3,2,1	1	7,6,5,4,3,2,1	4,3,2,1	1	2,1
0	20	10,9,8,7,6,5,4,3,2,1	4,3,2,1	2,1	7,6,5,4,3,2,1	5,4,3,2,1	2,1	2,1
ways	21	100	34	12	56	25[d]	7	6[e]

[a]The possible numbers of the non-nickel coin are given. The numbers of nickels are found by default.
[b]This is for a single occurrence of the second coin listed. The possible numbers of the leading non-nickel coin are given. The numbers of nickels are found by default.
[c]The is for a single occurrence of the second and third coin listed. The possible numbers of the leading non-nickel coin are given. The numbers of nickels are found by default.
[d]There are another 25 ways when the 50¢ is in the form of 2 quarters.
[e]There are another 6 ways when the 50¢ is in the form of 2 quarters.

   The total number of ways identified is 21+100+34+12+56+25+25+7+6+6 = 292.
   If a one-dollar coin is also considered as change for a dollar, there are 293 ways.

Statistical Literacy and Critical Thinking

1. A probability must be a value between 0 and 1 inclusive. The term "50-50" is not a valid probability. The correct statement is that the probability a fair coin turns up heads is 0.5.

2. No, chance cannot be ruled out as a reasonable explanation. The probability of an event occurring by chance alone must be less than 0.05 in order for that event to be considered unusual enough to attribute it to factors other than chance.

3. No, his reasoning is not correct. Using the classical approach to calculate probabilities requires that all the outcomes are equally likely. These two outcomes (life exists, life does not exist) are not equally likely.

4. When two events are disjoint, they cannot both occur at the same time. When two events are independent, the occurrence of one does not affect the occurrence of the other. Disjoint events cannot be independent – since they cannot both occur at the same time, the occurrence of one definitely affects (i.e., it actually prevents) the occurrence of the other.

Review Exercises

1. Make a chart like the one on the right.
   Let H = a person had a headache
   N = a person had no headache
   D = a person used the drug
   P = a person used the placebo
   $P(H) = 80/100 = 0.80$

		TREATMENT		
		drug	placebo	
RESULT	headache	15	65	80
	no headache	17	3	20
		32	68	100

2. Refer to exercise #1.
   $P(D) = 32/100 = 0.32$

3. Refer to exercise #1.
   $P(H \text{ or } H) = P(H) + P(D) - P(H \text{ and } D)$
   $= 80/100 + 32/1000 - 15/100$
   $= 97/100 = 0.97$

4. Refer to exercise #1.
   $P(P \text{ or } N) = P(P) + P(N) - P(P \text{ and } N)$
   $= 68/100 + 20/100 - 3/100$
   $= 85/100 = 0.85$

5. Refer to exercise #1.
   $P(P_1 \text{ and } P_2) = P(P_1) \cdot P(P_2|P_1)$
   $= (68/100)(67/99)$
   $= 1139/2475 = 0.460$

6. Refer to exercise #1.
   $P(H_1 \text{ and } H_2) = P(H_1) \cdot P(H_2|H_1)$
   $= (80/100)(79/99)$
   $= 316/495 = 0.638$

7. Refer to exercise #1.
   $P(H|D) = 15/32 = 0.469$

8. Refer to exercise #1.
   $P(D|H) = 15/80 = 0.1875$

9. a. P(birthday of randomly selected person is October 18) = 1/365.
   b. P(birthday of randomly selected person is in October) = 31/365.
   c. Answers will vary, but a reasonable answer for America might be
      P(randomly selected adult knows October 18 is NSD in Japan) $\approx 1/1,000,000$.
   d. Yes. Since $1/1,000,000 = 0.000001 < 0.05$, such a selection would be unusual.

10. There were $132 + 880 = 1012$ people in the survey.
Let D = the randomly selected person thinks it should be used as a doorstop.
$P(D) = 132/1012 = 0.130$

11. $P(G_1 \text{ and } G_2 \text{ and} \dots \text{and } G_{12}) = P(G_1) \cdot P(G_2) \cdot \dots \cdot P(G_{12})$
$$= (\tfrac{1}{2})(\tfrac{1}{2}) \dots (\tfrac{1}{2})$$
$$= (\tfrac{1}{2})^{12} = 0.000244$$
Yes. The probability of getting 12 girls in 12 births by chance alone is so small that the results appear to support the company's claim.

12. Let S = a 27-year-old man survives for the year.
$P(S) = 0.9982$, for each such man
$P(S_1 \text{ and } S_2 \text{ and} \dots \text{and } S_{12}) = P(S_1) \cdot P(S_2) \cdot \dots \cdot P(S_{12})$
$$= (0.9982)(0.9982) \dots (0.9982)$$
$$= (0.9982)^{12} = 0.9786$$

13. Let L = selecting the two wires that are live
There are $_5C_2 = 5!/(3!2!) = 10$ ways to select two of the five wires.
Since only one of those selections is the two wires that are live, $P(L) = 1/10$.

14. Let G = selecting a good aspirin.
Since the population is so large, assume all probabilities remain constant as tablets are selected for examination.
$P(G) = 0.98$, for each aspirin selected.
P(at least one defective) = 1 – P(all are good)
$$= 1 - P(G_1 \text{ and } G_2 \text{ and } G_3 \text{ and } G_4)$$
$$= 1 - P(G_1) \cdot P(G_2) \cdot P(G_3) \cdot P(G_4)$$
$$= 1 - (0.98)^4$$
$$= 1 - 0.9224 = 0.0776$$
NOTE: The exact answer, not assuming the probabilities remain constant, is 0.0777.
P(at least one defective) = 1 – P(all are good)
$$= 1 - P(G_1 \text{ and } G_2 \text{ and } G_3 \text{ and } G_4)$$
$$= 1 - P(G_1) \cdot P(G_2) \cdot P(G_3) \cdot P(G_4)$$
$$= 1 - (2450/2500)(2449/1499)(2448/2498)(2447/2497)$$
$$= 1 - 0.9223 = 0.0777$$

15. Let C = a person has chlamydia.
Since the population is so large, assume all probabilities remain constant as persons are selected.
a. $P(C) = 278.32/100,000 = 0.0027832$, for each person selected.
b. $P(C_1 \text{ and } C_2) = P(C_1) \cdot P(C_2)$
$$= (0.0027832)(0.0027832)$$
$$= 0.00000775$$
c. $P(\overline{C}) = 1 - P(C) = 1 - 0.0027832 = 0.9972168$, for each person selected.
$P(\overline{C}_1 \text{ and } \overline{C}_2) = P(\overline{C}_1) \cdot P(\overline{C}_2)$
$$= (0.9972168)(0.9972168)$$
$$= 0.9944413$$

16. There are 10 possibilities (0,1,2,3,4,5,6,7,8,9) for each digit.
By the fundamental counting rule, the number of possibilities is now
$$10 \cdot 10 \cdot \dots \cdot 10 = 10^{13} = 10,000,000,000,000 = 10 \text{ trillion}$$

Cumulative Review Exercises

1. The values in order are: 0 0 0 2 3 3 3 3 4 4 4 5 5 5 5 6 6 6 6 7 7
   The summary statistics are: $n = 21$ $\Sigma x = 84$ $\Sigma x^2 = 430$

   a. $\bar{x} = (\Sigma x)/n = 84/21 = 4.0$

   b. $\tilde{x} = 4$

   c. $s^2 = [n(\Sigma x^2) - (\Sigma x)^2]/[n(n-1)]$
      $= [21(430) - (84)^2]/[21(20)]$
      $= 1974/420 = 4.70$
      $s = 2.2$

   d. $s^2 = 4.7$ [from part (c) above]

   e. Yes. Since there were no negative scores at all, it appears that the treatment was effective.

   f. Since 18 of the 21 values are positive,
      P(selecting a positive value at random) $= 18/21 = 6/7 = 0.857$.

   g. $P(P_1 \text{ and } P_2) = P(P_1) \cdot P(P_2|P_1)$
      $= (18/21)(17/20)$
      $= 51/70 = 0.729$

   h. $P(P_1 \text{ and } P_2 \text{ and} \ldots \text{and } P_{18}) = P(P_1) \cdot P(P_2) \cdot \ldots \cdot P(P_{18})$
      $= (\frac{1}{2})(\frac{1}{2}) \ldots (\frac{1}{2})$
      $= (\frac{1}{2})^{18} = 0.00000381$
   Yes. The probability of getting 18 positive values in 18 subjects by chance alone is so small as to justify rejecting the assumption that the treatment is ineffective and to conclude that the treatment appears to be effective.

2. From the given five-number summary,
   minimum $= 62$ 25th percentile $= 76$ median $= 76$ 75th percentile $= 80$ maximum $= 85$
   a. $\tilde{x} = 76 \,°F$

   b. $P(72 \leq t \leq 76) = 0.50 - 0.25 = 0.25$

   c. $P(t < 72 \text{ or } t > 76) = 1 - P(72 \leq t \leq 76)$
      $= 1 - 0.25$
      $= 0.75$

   d. Let B $=$ selecting a day for which $72 \leq t \leq 76$
      Since the population is relatively large compared to the sample, assume all probabilities
         remain constant as subsequent days are selected.
      $P(B) = 0.25$, for each day
      $P(B_1 \text{ and } B_2) = P(B_1) \cdot P(B_2)$
         $= (0.25)(0.25) = 0.0625$

   e. No. Hot and cold days tend to occur in cycles. If the temperature is above $80\,°F$ one day, it is more likely than usual to be above $80\,°F$ on the next day.

Chapter 5

Discrete Probability Distributions

5-2 Random Variables

1. The requested probability distribution is given below.

x	P(x)
1	1/6
2	1/6
3	1/6
4	1/6
5	1/6
6	1/6
	6/6 = 1

NOTE: Officially a probability distribution for a random variable is a list or other representation of the possible values for the variable and their associated probabilities. Just as we included the sum of the frequencies as an integral part of a frequency distribution in Chapter 2, we now choose to include the sum of the probabilities as an integral part of a probability distribution. This sum should always be 1.00 and serves as a check to the set-up of the problem.

3. The "probability distribution" described would be as follows.

x	P(x)
1	0.1
2	0.2
3	0.3
4	0.4
5	0.5
6	0.6
	2.1

No, the gambler cannot load the die as described.
Since $\Sigma P(x) = 2.1 \neq 1.0$, this is not a possible probability distribution.

5. a. Continuous, since height can be any value on a continuum.
   b. Discrete, since the number of bald eagles must be an integer.
   c. Continuous, since time can be any value on a continuum.
   d. Discrete, since the number of authors so occupied must be an integer.
   e. Discrete, since the number of students so occupied must be an integer.

NOTE: In exercises #7-#12, we use the decimal values for P(x) given in the text. In truth these are decimal approximations for fractions or other exact mathematical values. This sometimes produces slight round-off errors. In exercise #11, for example, exact calculations would produce $\Sigma P(x) = 1$ exactly. In general, this manual ignores such discrepancies.

7. The given table is a probability distribution since $0 \leq P(x) \leq 1$ for each x and $\Sigma P(x)=1$.

x	P(x)	x·P(x)	x²	x²·P(x)	$\mu = \Sigma[x \cdot P(x)]$
0	0.4219	0	0	0	$= 0.7499$, rounded to 0.7
1	0.4219	0.4219	1	0.4219	$\sigma^2 = \Sigma[x^2 \cdot P(x)] - \mu^2$
2	0.1406	0.2812	4	0.5624	$= 1.1247 - (0.7499)^2$
3	0.0156	0.0468	9	0.1404	$= 0.56235$
	1.0000	0.7499		1.1247	$\sigma = 0.7499$, rounded to 0.7

NOTE: If one of the conditions for a probability distribution does not hold, the formulas do not apply – and they produce numbers that have no meaning. When working with probability distributions and formulas in the exercises that follow, always keep these important facts in mind.
- $\Sigma P(x)$ must always equal 1.000
- $\Sigma[x \cdot P(x)]$ gives the mean of the x values and must be a number between the highest and lowest x values.
- $\Sigma[x^2 \cdot P(x)]$ gives the mean of the x^2 values and must be a number between the highest and lowest x^2 values.
- Σx and Σx^2 have no meaning and should not be calculated
- The quantity "$\Sigma[x^2 \cdot P(x)] - \mu^2$" cannot possibly be negative – if it is, there is a mistake.
- Always be careful to use the unrounded mean in the calculation of the variance, and to take the square root of the unrounded variance to find the standard deviation.

9. The given table is not a probability distribution since $\Sigma P(x) = 1.16 \neq 1$.

11. The given table is a probability distribution since $0 \leq P(x) \leq 1$ for each x and $\Sigma P(x)=1$.

x	P(x)	x·P(x)	x^2	x^2·P(x)	
4	0.1818	0.7272	16	2.9088	$\mu = \Sigma[x \cdot P(x)]$
5	0.2121	1.0605	25	5.3025	$= 5.7974$, rounded to 5.8
6	0.2323	1.3938	36	8.3628	$\sigma^2 = \Sigma[x^2 \cdot P(x)] - \mu^2$
7	0.3737	2.6159	49	18.3113	$= 34.8854 - (5.7974)^2$
	0.9999	5.7974		34.8854	$= 1.2756$
					$\sigma = 1.1294$, rounded to 1.1

Since $P(x=4) = 0.1818 > 0.05$, winning in 4 games is not an unusual event.

13. a. $P(x=5) = 0.003$
   b. $P(x \leq 5) = 0^+ + 0^+ + 0^+ + 0^+ + 0.001 + 0.003 = 0.004$
   c. The probability in part (b) is relevant. When there are many possible outcomes, the probability that any one of them occurs will typically be small – whether or not that outcome is really unusual. Determining whether an outcome is rare should consider that outcome and "more unusual" outcomes, as this places the outcome in a proper context.
   d. Since $0.004 \leq 0.05$, selecting five or fewer Mexican-Americans by chance alone would be an unusual event. Either an unusual event has occurred or factors other than chance are at work in the selection process. Yes, conclude that there is discrimination.

15. a. $P(x \leq 8) = 0^+ + 0^+ + 0^+ + 0^+ + 0.001 + 0.003 + 0.016 + 0.053 + 0.133 = 0.206$
   b. Since $0.206 > 0.05$, selecting 8 or fewer Mexican-Americans by chance alone would not be an unusual event. No, there is no evidence of discrimination.

17. Consider the following tables, where x represents the net financial result.

   a. betting on 7

x	P(x)	x·P(x)
-5	37/38	-185/38
175	1/38	175/38
	38/38	-10/38

   $E = \Sigma[x \cdot P(x)]$
   $= -10/38$
   $= -26.3¢$

   b. betting on an odd number

x	P(x)	x·P(x)
-5	20/38	-100/38
5	18/38	90/38
	38/38	-10/38

   $E = \Sigma[x \cdot P(x)]$
   $= -10/38$
   $= -26.3¢$

   c. not betting

x	P(x)	x·P(x)
0	1.00	0
	1.00	0

   $E = \Sigma[x \cdot P(x)]$
   $= 0$
   $= 0¢$

   The expected value for (a) and (b) is -26.3¢. This means that on the average you lose 26.3 cents for every time you play. The expected value for (c) is 0. This means you expect to break even if you don't play – which makes perfect sense. The best option is (c).

19. a. He "wins" $100,000 − $250 = $99,750 if he dies.
     He "loses" $250 if he lives.

   b. The following table describes the situation.

x	P(x)	x·P(x)
-250	0.9985	-249.625
99750	0.0015	149.625
	1.0000	-100.000

   $E = \Sigma[x \cdot P(x)] = -\$100.000$. The expected value is a loss of $100.

   c. Since the company is making a $100 profit at this price, it would break even if it sold the policy for $100 less – i.e., for $250 - $100 = $150. NOTE: This oversimplified analysis ignores the cost of doing business. If the costs (printing, salaries, advertising, etc.) associated with offering the policy are $25, for example, then the company's profit is only $75 and selling the policy for $75 less (i.e., for $175) would represent the break-even point.

   d. Buying life insurance is similar to purchasing other services. You let someone make a profit by shoveling the snow in your driveway, for example, because the cost is worth the frustration you save. You let someone make a profit by selling you life insurance because the cost is worth the peace you have from providing for the financial security of your heirs.

21. There are eight equally likely possible outcomes: GGG GGB GBG BGG GBB BGB BBG BBB.
   The following table describes the situation, where x is the number of girls per family of 3.

x	P(x)	x·P(x)	x^2	x^2·P(x)	
0	0.125	0	0	0	$\mu = \Sigma[x \cdot P(x)]$
1	0.375	0.375	1	0.375	$= 1.500$, rounded to 1.5
2	0.375	0.750	4	1.500	$\sigma^2 = \Sigma[x^2 \cdot P(x)] - \mu^2$
3	0.125	0.375	9	1.125	$= 3.000 - (1.500)^2$
	1.000	1.500		3.000	$= 0.75$
					$\sigma = 0.8660$, rounded to 0.9

   No. Since P(x=3) = 0.125 > 0.05, it is not unusual for a family of 3 children to have all girls.

23. The following table describes the situation.

x	P(x)	x·P(x)	x^2	x^2·P(x)	
0	0.1	0	0	0	$\mu = \Sigma[x \cdot P(x)]$
1	0.1	0.1	1	0.1	$= 4.5$
2	0.1	0.2	4	0.4	$\sigma^2 = \Sigma[x^2 \cdot P(x)] - \mu^2$
3	0.1	0.3	9	0.9	$= 28.5 - (4.5)^2$
4	0.1	0.4	16	1.6	$= 8.25$
5	0.1	0.5	25	2.5	$\sigma = 2.8723$, rounded to 2.9
6	0.1	0.6	36	3.6	
7	0.1	0.7	49	4.9	
8	0.1	0.8	64	6.4	
9	0.1	0.9	81	8.1	
	1.0	4.5		28.5	

The probability histogram for this distribution would be flat, with each bar having the same height.

25. Both distributions are tables with two columns – the first column stating the possible outcomes, and the second column giving some information about the outcomes. In a frequency distribution, the second column consists of whole numbers that give the actual observed counts. In a probability distribution, the second column consists of values between 0 and 1 that give the theoretical proportion of times the outcomes are expected to occur in the long run.

27. Let C and N represent correctly and not correctly calibrated altimeters respectively. For 8 C's and 2 N's, there are 7 possible samples of size n=3. These are given, along with their associated probabilities, in the box at the right. Letting x be the number of N's selected yields the following table.

$$P(CCC) = (8/10)(7/9)(6/8) = 336/720$$
$$P(CCN) = (8/10)(7/9)(2/8) = 112/720$$
$$P(CNC) = (8/10)(2/9)(7/8) = 112/720$$
$$P(NCC) = (2/10)(8/9)(7/8) = 112/720$$
$$P(CNN) = (8/10)(2/9)(1/8) = 16/720$$
$$P(NCN) = (2/10)(8/9)(1/8) = 16/720$$
$$P(NNC) = (2/10)(1/9)(8/8) = \underline{16/720}$$
$$720/720$$

x	P(x)	x·P(x)	x^2	x^2·P(x)
0	336/720	0	0	0
1	336/720	336/720	1	336/720
2	48/720	96/720	4	192/720
	720/720	432/720		528/720

$\mu = \Sigma[x \cdot P(x)]$
  $= 432/720 = 0.600$, rounded to 0.6
$\sigma^2 = \Sigma[x^2 \cdot P(x)] - \mu^2$
  $= (528/720) - (432/720)^2$
  $= 0.3733$
$\sigma = 0.611$, rounded to 0.6

5-3 Binomial Probability Distributions

1. The symbols p and x are defined so as to refer to the same event. If p is the probability of getting a correct answer, then x must be the number of correct answers.

3. Table A-1 is limited because it includes only selected values of p and stops at n=15.

NOTE: For exercises #5-#12, the four requirements for a binomial experiment are
  (1) There is a fixed number of trials.
  (2) The trials are independent.
  (3) Each trial has exactly two possible named outcomes.
  (4) The probabilities remain constant for each trial.

5. No. Requirement (3) is not met. There are more than two possible outcomes.

7. No. Requirement (3) is not met. There are more than two possible outcomes.

9. Yes. All four requirements are met.

11. Yes. All four requirements are met.

13. a. P(WWC) = P(W)·P(W)·P(C)
              = (4/5)(4/5)(1/5) = 16/125 = 0.128
    b. There are three possible arrangements: WWC, WCW, CWW
          P(WWC) = P(W)·P(W)·P(C) = (4/5)(4/5)(1/5) = 16/125
          P(WCW) = P(W)·P(C)·P(W) = (4/5)(1/5)(4/5) = 16/125
          P(CWW) = P(C)·P(W)·P(W) = (1/5)(4/5)(4/5) = 16/125
    c. P(exactly one correct) = P(WWC or WCW or CWW)
              = P(WWC) + P(WCW) + P(CWW)
              = 16/125 + 16/125 + 16/125 = 48/125 = 0.384

15. From Table A-1 in the .05 column and the 3-0 row, $P(x=0) = 0.857$.

17. From Table A-1 in the .05 column and the 8-4 row, $P(x=4) = 0^+$ [rounds to 0 to 3 decimals].

19. From Table A-1 in the .30 column and the 14-2 row, $P(x=2) = 0.113$.

21. $P(x) = \{n!/[(n-x)!x!]\}p^x q^{n-x}$
    $P(x=2) = [5!/(3!2!)](0.25)^2(0.75)^3$
    $= [10](0.0625)(0.421875) = 0.264$

NOTE: The intermediate values 10, 0.0625, and 0.421875 are given in exercise #21 to help those with an incorrect answer to identify the portion of the problem in which the mistake was made. In the future, only the value $n!/[(n-x)!x!]$ will be given separately. In practice, all calculations can be done in one step on a calculator. You may choose (or be asked to) write down intermediate values for you own (or the instructor's) benefit, but…
 • Never round off in the middle of a problem.
 • Do not write the values down on paper and then re-enter them in the calculator.
 • Use the memory to let the calculator remember with complete accuracy any intermediate values
      that will be used in subsequent calculations.
In addition, always make certain that $n!/[(n-x)!x!]$ is a whole number and that the final answer is between 0 and 1.

23. $P(x) = \{n!/[(n-x)!x!]\}p^x q^{n-x}$
    $P(x=3) = [9!/(6!3!)](0.25)^3(0.75)^6$
        $= [84](0.25)^3(0.75)^6$
        $= 0.234$

25. $P(x \geq 5) = P(x=5) + P(x=6)$
        $= 0.0006 + 0.0000$
        $= 0.0006$
    Yes. Since $0.0006 \leq 0.05$, it is unusual for at least 5 subjects to experience headaches..

27. $P(x>1) = P(x=2) + P(x=3) + P(x=4) + P(x=5) + P(x=6)$
        $= 0.2014 + 0.0538 + 0.0081 + 0.0006 + 0.0000$
        $= 0.2639$
    No. Since $0.2639 > 0.05$, it is not unusual for more than 1 subject to experience headaches.

29. Let x = the number of households tuned to *60 Minutes*.
    binomial problem: n=10 and p=0.20, use Table A-1
    a. $P(x=0) = 0.107$
    b. $P(x \geq 1) = 1 - P(x=0)$
            $= 1 - 0.107 = 0.893$
    c. $P(x \leq 1) = P(x=0) + P(x=1)$
            $= 0.107 + 0.268 = 0.375$
    d. No. Since $0.375 > 0.05$, it would not be unusual to find at most one household tuned to
        60 Minutes when that show had a 20% share of the market.

31. Let x = the number of defects found in the sample.
    binomial problem: n=24 and p=0.04, use the binomial formula $P(x) = \{n!/[(n-x)!x!]\}p^x q^{n-x}$
    P(shipment accepted) = $P(x \leq 1)$
                        $= P(x=0) + P(x=1)$
                        $= [24!/(24!0!)](0.04)^0(0.96)^{24} + [24!/(23!1!)](0.04)^1(0.96)^{23}$
                        $= [1](1)(0.96)^{24} + [24](0.04)(0.96)^{23}$
                        $= 0.3754 + 0.3754 = 0.7508$, rounded to 0.751

33. Let x = the number of booked passengers that actually arrive.
   binomial problem: n=15 and p=0.85, use the binomial formula $P(x) = \{n!/[(n-x)!x!]\}p^x q^{n-x}$
   $P(x=15) = [15!/(0!15!)](0.85)^{15}(0.15)^0$
   $= [1](0.85)^{15}(1) = 0.0874$
   No. Since 0.0874 > 0.05, not having enough seats is not an unusual event. The probability is not low enough to be of no concern.

35. Let x = the number of women hired.
   binomial problem: n=20 and p=0.5, use the binomial formula $P(x) = \{n!/[(n-x)!x!]\}p^x q^{n-x}$
   $P(x \leq 2) = P(x=0) + P(x=1) + P(x=2)$
   $= [20!/(20!0!)](0.5)^0(0.5)^{20} + [20!/(19!1!)](0.5)^1(0.5)^{19} + [20!/(18!2!)](0.5)^2(0.5)^{18}$
   $= [1](1)(0.5)^{20} + [20](0.5)(0.5)^{19} + [190](0.5)^2(0.5)^{18}$
   $= 0.000000954 + 0.000019073 + 0.000181198 = 0.0002012$, rounded to 0.000201
   Yes. Since $0.0002012 \leq 0.05$, it would be unusual to have only 2 or fewer female hires by chance alone. All other factors being equal, the probability does support a charge of gender discrimination.

37. Let x = the number of components tested to find the first defect.
   geometric problem: p=0.2, use the geometric formula $P(x) = p(1-p)^{x-1}$
   $P(x=7) = (0.2)(0.8)^6 = 0.0524$

39. Extending the patter to cover six types of outcomes, where $\Sigma x = n$ and $\Sigma p = 1$,
   n=20 and $p_1=p_2=p_3=p_4=p_5=p_6=1/6$ and $x_1=5, x_2=4, x_3=3, x_4=2, x_5=3, x_6=3$
   use the multinomial formula
   $P(x_1,x_2,x_3,x_4,x_5,x_6) = [n!/(x_1!x_2!x_3!x_4!x_5!x_6!)](p_1)^{x_1}(p_2)^{x_2}(p_3)^{x_3}(p_4)^{x_4}(p_5)^{x_5}(p_6)^{x_6}$
   $= [20!/(5!4!3!2!3!3!)](1/6)^5(1/6)^4(1/6)^3(1/6)^2(1/6)^3(1/6)^3$
   $= [1,955,457,504,030] \cdot (1/6)^{20}$
   $= 0.000535$

5-4 Mean, Variance, and Standard Deviation for the Binomial Distribution

1. Yes. Since 70 is more than two standard deviations from the mean, it would be unusual to get 70 girls in 100 births. Any number higher than $50 + 2(5) = 60$ would be considered unusual.

3. If the standard deviation is 1.2 females, the variance is the square of that – i.e., 1.44 females^2.

5. $\mu = np = (200)(0.4) = 80.0$
   $\sigma^2 = npq = (200)(0.4)(0.6) = 48$
   $\sigma = 6.928$, rounded to 6.9
   minimum usual value $= \mu - 2\sigma = 80 - 2(6.928) = 66.1$
   maximum usual value $= \mu + 2\sigma = 80 + 2(6.928) = 93.9$

7. $\mu = np = (1492)(0.25) = 373.0$
   $\sigma^2 = npq = (1492)(0.25)(0.75) = 279.75$
   $\sigma = 16.726$, rounded to 16.7
   minimum usual value $= \mu - 2\sigma = 373.0 - 2(16.726) = 339.5$
   maximum usual value $= \mu + 2\sigma = 373.0 + 2(16.726) = 406.5$

9. Let x = the number of correct answers.
   binomial problem: n=16 and p=0.5
   a. $\mu = np = (16)(0.5) = 8.0$
   $\sigma^2 = npq = (16)(0.5)(0.5) = 4$
   $\sigma = 2.0$
   b. Unusual values are those outside $\mu \pm 2\sigma$
   $$8.0 \pm 2(2.0)$$
   $$8.0 \pm 4.0$$
   $$4.0 \text{ to } 12.0$$
   No. Since 10 is within the above limits, it would not be unusual for a student to pass by getting at least 10 correct answers.

11. Let x = the number of orange M&M's.
    binomial problem: n=100 and p=0.2
    a. $\mu = np = (100)(0.2) = 20.0$
    $\sigma^2 = npq = (100)(0.2)(0.8) = 16$
    $\sigma = 4.0$
    b. Unusual values are those outside $\mu \pm 2\sigma$
    $$20.0 \pm 2(4.0)$$
    $$20.0 \pm 8.0$$
    $$12.0 \text{ to } 28.0$$
    No. Since 25 is within the above limits, it would not be unusual to find 25 orange M&M's. There is no reason to conclude that the 20% rate is wrong.

13. Let x = the number of girl births.
    binomial problem: n=325 and p=0.5
    a. $\mu = np = (325)(0.5) = 162.5$
    $\sigma^2 = npq = (325)(0.5)(0.5) = 81.25$
    $\sigma = 9.0$
    b. Unusual values are those outside $\mu \pm 2\sigma$
    $$162.5 \pm 2(9.0)$$
    $$162.5 \pm 18.0$$
    $$144.5 \text{ to } 180.5$$
    Yes. Since 295 is not within the above limits, it would be an unusual number of girl births. The gender-selection method appears to be effective.

15. Let x = the number of letter r's per page.
    binomial problem: n=2600 and p=0.077
    a. $\mu = np = (2600)(0.077) = 200.2$
    $\sigma^2 = npq = (2600)(0.077)(0.923) = 184.78$
    $\sigma = 13.6$
    b. Unusual values are those outside $\mu \pm 2\sigma$
    $$200.2 \pm 2(13.6)$$
    $$200.2 \pm 27.2$$
    $$173.0 \text{ to } 227.4$$
    No. Since 175 is within the above limits, it would not be an unusual result.

17. Let x = the number of persons who actually voted.
    binomial problem: n=1002 and p=0.61
    a. $\mu = np = (1002)(0.61) = 611.2$
       $\sigma^2 = npq = (1002)(0.61)(0.39) = 238.3758$
       $\sigma = 15.439$, rounded to 15.4
    b. Unusual values are those outside $\mu \pm 2\sigma$

$$611.2 \pm 2(15.439)$$
$$611.2 \pm 30.9$$
$$580.3 \text{ to } 642.1 \text{ [using rounded } \sigma, 580.4 \text{ to } 642.0]$$

No. Since 701 is not within the above limits, it is not consistent with the actual voter
turnout.
    c. No. Based on these results, it appears that accurate voting data cannot be obtained by asking
       possible voters how they acted.

19. Let x = the number of persons receiving the drug who experience flu symptoms.
    binomial problem: n=863 and p=0.019
    a. $\mu = np = (863)(0.019) = 16.4$
       $\sigma^2 = npq = (863)(0.019)(0.981) = 16.09$
       $\sigma = 4.0$
    b. Unusual values are those outside $\mu \pm 2\sigma$

$$16.4 \pm 2(4.0)$$
$$16.4 \pm 8.0$$
$$8.4 \text{ to } 24.4$$

No. Since 19 is within the above limits, it is not an unusual result.
    c. No. These results do not suggest that persons receiving the drug experience flu symptoms
       any differently from persons not receiving the drug.

21. Let x = the number of girl births.
    binomial problem: n=100 and p=0.5
    a. $\mu = np = (100)(0.5) = 50.0$
       $\sigma^2 = npq = (100)(0.5)(0.5) = 25.0$
       $\sigma = 5.0$
       minimum usual value $= \mu - 2\sigma = 50.0 - 2(5.0) = 40.0$
       maximum usual value $= \mu + 2\sigma = 50.0 + 2(5.0) = 60.0$
    b. Yes, the probability distribution is bell-shaped. Values near 50.0 are most likely – and the
       farther a value is from 50.0 in either direction, the less likely it is to occur.
    c. According to the empirical rule for bell-shaped distributions, the number of girls will fall
       between 40 and 60 (i.e., within 2σ of the mean) about 95% of the time.

5-5 Poisson Probability Distributions

1. The Poisson distribution describes the numbers of occurrences, over some defined interval, of
   independent random events that are uniformly distributed over that interval.

3. The formula to generate the entire distribution needs only the value of μ, the mean. The
   standard deviation, variance and shape of the distribution are all determined by the mean.

5. $P(x) = \mu^x e^{-\mu}/x!$ with $\mu=5$
   $P(x=4) = (5)^4(e^{-5})/4!$
   $\qquad = (625)(0.0067)/24$
   $\qquad = 0.175$

NOTE: The intermediate values of 625, 0.0067 and 24 are given in exercise #5 to help students who are unable to get the correct answer and need to identify where they are going wrong. In practice, the answer can be obtained on a calculator in one step without writing down intermediate values. If you write down intermediate values (for your sake or the instructor's), DO NOT USE ROUNDED INTERMEDIATE VALUES IN SUBSEQUENT CALCULATIONS. IF NECESSARY, USE THE MEMORY TO STORE INTERMEDIATE VALUES WITH INFINITE ACCURACY. Intermediate values given in subsequent Poisson problems are for instructional purposes only.

7. $P(x) = \mu^x e^{-\mu}/x!$ with $\mu=0.5$
   $P(x=3) = (0.5)^3(e^{-0.5})/3!$
   $\qquad = (0.125)(0.6065)/6$
   $\qquad = 0.0126$

9. Let x = the number of dandelions per square meter.
   Poisson problem: $\mu = 7.0$, use $P(x) = \mu^x e^{-\mu}/x!$
   a. $P(x=0) = (7.0)^0(e^{-7.0})/0!$
   $\qquad = (1)(0.0009)/1 = 0.0009$
   b. $P(x \geq 1) = 1 - P(x=0)$
   $\qquad = 1 - 0.0009 = 0.9991$
   c. $P(x \leq 2) = P(x=0) + P(x=1) + P(x=2)$
   $\qquad = (7.0)^0(e^{-7.0})/0! + (7.0)^1(e^{-7.0})/1! + (7.0)^2(e^{-7.0})/2!$
   $\qquad = 0.0009 + 0.0064 + 0.0223$
   $\qquad = 0.0296$

IMPORTANT NOTE: In the problems that follow, store the <u>unrounded</u> value for μ for use in all Poisson calculations.

11. Let x = the number of atoms lost per day.
    Poisson problem: $\mu = 62.2$, use $P(x) = \mu^x e^{-\mu}/x!$
    a. The number of atoms lost is $1,000,000 - 997,287 = 22,713$.
    $\qquad \mu = 22,713/365 = 62.2$
    b. $P(x=50) = (62.2)^{50}(e^{-62.2})/50!$
    $\qquad = 0.0155$

NOTE: The following exercises compare actual results to predictions by the Poisson probability distribution. The comparison can be made either by changing the actual frequencies to relative frequencies (by dividing by $n = \Sigma f$) and comparing them to the predicted Poisson probabilities, or by changing the Poisson probabilities to predicted frequencies (by multiplying by $n = \Sigma f$) and comparing them to the actual frequencies. We use the first approach, but the choice is arbitrary.

13. Let x = the number of homicides per day.
    Poisson problem: $\mu = 116/365 = 0.3178$, use $P(x) = \mu^x e^{-\mu}/x!$
    a. $P(x=0) = (0.3178)^0(e^{-0.3178})/0! = \qquad (1)(0.7277)/1 \quad = 0.7277$
    b. $P(x=1) = (0.3178)^1(e^{-0.3178})/1! = (0.3178)(0.7277)/1 \quad = 0.2313$
    c. $P(x=2) = (0.3178)^2(e^{-0.3178})/2! = (0.1010)(0.7277)/2 \quad = 0.0368$
    d. $P(x=3) = (0.3178)^3(e^{-0.3178})/3! = (0.0321)(0.7277)/6 \quad = 0.00389$
    e. $P(x=4) = (0.3178)^4(e^{-0.3178})/4! = (0.0102)(0.7277)/24 = 0.000309$
    The following table compares the actual relative frequencies to the Poisson probabilities.

x	f	r.f.	P(x)	
0	268	0.7342	0.7277	
1	79	0.2164	0.2313	
2	17	0.0466	0.0368	
3	1	0.0027	0.0038	
4 or more	0	0.0000	0.0004	(by subtraction)
	365	1.0000	1.0000	

NOTE: r.f. = f/Σf = f/365

The agreement is very good.

15. Let x = the number of successes in 100 trials.
   binomial problem: n=100 and p=0.1
   Poisson approximation appropriate since n = 100 $\geq$ 100 and np = (100)(0.1) = 10 $\leq$ 10.
   use $P(x) = \mu^x e^{-\mu}/x!$ with μ = np = (100)(0.1) = 10
   $P(x=101) = (10)^{101}(e^{-10})/101!$

This cannot be evaluated directly on most calculators because the numbers are too large. Even many statistical software packages cannot evaluate this.

   Using Excel, however, Poisson(10,101,0) = 4.1 x 10^{-31}
   An approximation is possible using logarithms and exercise #39 of section 4-7.
   $n! \approx 10^k$, where k = (n+.5)(log n) + 0.39908993 – 0.43429448n
   for 101, k = (101.5)(log 101) + 0.39908993 – 0.43429448(101) = 159.9739669
   $$P(x=101) \approx (10)^{101}(e^{-10})/10^{159.9739669}$$
   $$\log[P(x=101)] \approx 101(\log 10) - 10(\log e) - 159.9739669(\log 10)$$
   $$\approx 101(1) - 10(0.434298882) - 159.9739669(1)$$
   $$\approx -63.31691172$$
   $$P(x=101) \approx 10^{-63.31691172}$$
   $$\approx 4.82 \times 10^{-64}$$

Both this value and the Excel value are so small that they can be considered 0, agreeing with the fact that 101 is impossible in such a binomial distribution.

Statistical Literacy and Critical Thinking

1. A probability distribution is a statement (as a graph, table, or formula) of the values a random variable can assume and the probability associated with each of those values.

2. The two main requirements of a probability distribution are as follows.
   (1) Each individual probability must be between 0 and 1 inclusive.
   (2) The sum of all the individual probabilities must be 1.000.

3. A discrete probability distribution is one for which all the individual possible values of the random variable can be listed in some logical order so that each one can be identified by a counting number – even if there is an infinite number of them. There are also continuous probability distributions – for which there are an infinite number of possible values for the random variable that occur over some continuum.

4. No. The discrete probability distribution at the right, for example, describing the possible outcomes for tossing a single die, is neither a binominal probability distribution nor a Poisson probability distribution.

x	P(x)
1	1/6
2	1/6
3	1/6
4	1/6
5	1/6
6	1/6
	6/6

Review Exercises

1. The following table summarizes the calculations for this exercise.

x	P(x)	x·P(x)	x^2	x^2·P(x)	
0	0.107	0	0	0	$\mu = \Sigma[x \cdot P(x)]$
1	0.268	0.268	1	0.268	$= 2.000$, rounded to 2.0
2	0.302	0.604	4	1.208	$\sigma^2 = \Sigma[x^2 \cdot P(x)] - \mu^2$
3	0.201	0.603	9	1.089	$= 5.608 - (2.000)^2$
4	0.088	0.352	16	1.408	$= 1.608$
5	0.026	0.130	25	0.650	$\sigma = 1.268$, rounded to 1.3
6	0.006	0.036	36	0.216	
7	0.001	0.007	49	0.049	
8	0.000^+	0.000^+	64	0.000^+	
9	0.000^+	0.000^+	81	0.000^+	
10	0.000^+	0.000^+	100	0.000^+	
	0.999	2.000		5.608	

a. The table satisfies the requirements for a probability distribution since
(1) $0 \leq P(x) \leq 1$ for each x
(2) $\Sigma P(x) = 1$
NOTE: The "missing" 0.001 in the table is distributed among the three 0.000^+ categories.

b. Since this is a binomial distribution with n=10 and p=0.2, $\mu = np = (10)(0.2) = 2$. This agrees with the 2.000 in the above table.

c. Since this is a binomial distribution with n=10 and p=0.2, $\sigma^2 = npq = (10)(0.2)(0.8) = 1.6$ and $\sigma = 1.265$. This agrees (within round-off error) with the 1.268 in the table above.

d. $P(x \geq 5) = P(x=5) + P(x=6) + P(x=7) + P(x=8) + P(x=9) + P(x=10)$
   $= 0.026 + 0.006 + 0.001 + 0.000^+ + 0.000^+ + 0.000^+$
   $= 0.033$

e. $E = \Sigma[x \cdot P(x)] = 2.000$

f. $P(x \geq 1) = 1 - P(x=0)$
   $= 1 - 0.107$
   $= 0.893$

g. No. Since $0.893 > 0.05$, getting at least one answer correct by chance alone is not unusual.

2. Let x = the number of TV sets tuned to *Cold Case*.
   binomial problem: n=12 and p=0.15

a. $E = \mu = np = (12)(0.15) = 1.8$
b. $\mu = np = (12)(0.15) = 1.8$

c. $\sigma^2 = npq = (12)(0.15)(0.85) = 1.53$
   $\sigma = 1.237$, rounded to 1.2.

d. $P(x) = \{n!/[(n-x)!x!]\}p^x q^{n-x}$
   $P(x=3) = [12!/(9!3!)](0.15)^3(0.85)^9$
   $= [220](0.15)^3(0.85)^9$
   $= 0.1720$

e. $P(x) = \{n!/[(n-x)!x!]\}p^x q^{n-x}$
$\quad P(x=0) = [12!/(12!0!)](0.15)^0(0.85)^{12}$
$\quad\quad\quad\quad = [1](0.15)^0(0.85)^{12}$
$\quad\quad\quad\quad = 0.1422$

No. Since $0.1422 > 0.05$, it would not be unusual to find in a sample of 12 sets that 0 sets were tuned to *Cold Case*.

3. Let x = the number of workers fired for inability to get along with others.
binomial problem: n=5 and p=0.17, use $P(x) = \{n!/[(n-x)!x!]\}p^x q^{n-x}$

a. $P(x=4) = [5!/(1!4!)](0.17)^4(0.83)^1$
$\quad\quad\quad\quad = [5](0.17)^4(0.83)$
$\quad\quad\quad\quad = 0.00347$
$\quad P(x=5) = [5!/(0!5!)](0.17)^5(0.83)^0$
$\quad\quad\quad\quad = [1](0.17)^5(1)$
$\quad\quad\quad\quad = 0.00014$
$\quad P(x \geq 4) = P(x=4) + P(x=5)$
$\quad\quad\quad\quad = 0.00347 + 0.00014$
$\quad\quad\quad\quad = 0.00361$

b. Yes. Since $0.00361 \leq 0.05$, having at least 4 of the 5 firings due to an inability to get along with others would be unusual for a company with the standard rate of 17%.

4. Let x = the number of deaths per year.
Poisson problem: $\mu = 0.019$, use $P(x) = \mu^x e^{-\mu}/x!$
a. $\mu = 7/365 = 0.019$

b. $P(x=0) = (0.019)^0(e^{-0.019})/0!$
$\quad\quad\quad\quad = (1)(0.9810)/1$
$\quad\quad\quad\quad = 0.9810$

c. $P(x=1) = (0.019)^1(e^{-0.019})/1!$
$\quad\quad\quad\quad = (0.019)(0.9810)/1$
$\quad\quad\quad\quad = 0.0188$

d. $P(x \geq 1) = 1 - [P(x=0) + P(x=1)]$
$\quad\quad\quad\quad = 1 - [.9810 + 0.0188]$
$\quad\quad\quad\quad = 1 - 0.9998$
$\quad\quad\quad\quad = 0.0002$

e. No. Since $0.0002 \leq 0.05$ by such a wide margin, having more than one death on a single day is such a rare event that no such contingency plans are necessary.
NOTE: This exercise specifically stated that the information concerned the deaths of village residents, not deaths occurring within the village limits or the nearby area. If there are other circumstances to consider (e.g., a nearby highway prone to fatal accidents), then having such contingency plans would be advisable.

Cumulative Review Exercises

1. The following tables summarize the calculations for this exercise.

original digits as reported				relative frequency distribution		theoretical distribution of digits			
x	f	f·x	f·x^2	x	r.f.	x	P(x)	x·P(x)	x^2·P(x)
0	7	0	0	0	0.0987	0	0.1	0	0
1	14	14	14	1	0.1750	1	0.1	0.1	0.1
2	5	10	20	2	0.0625	2	0.1	0.2	0.4
3	11	33	99	3	0.1375	3	0.1	0.3	0.9
4	8	32	128	4	0.1000	4	0.1	0.4	1.6
5	4	20	100	5	0.0500	5	0.1	0.5	2.5
6	5	30	180	6	0.0625	6	0.1	0.6	3.6
7	6	42	294	7	0.0750	7	0.1	0.7	4.9
8	12	96	768	8	0.1500	8	0.1	0.8	6.4
9	8	72	648	9	0.1000	9	0.1	0.9	8.1
	80	349	2251		1.0000		1.0	4.5	28.5

a. Note that $n = \Sigma f = 80$

$\bar{x} = \Sigma(f \cdot x)/n = 349/80 = 4.36$

$s^2 = [n\Sigma(f \cdot x^2) - (\Sigma f \cdot x)^2]/[n(n-1)]$

  $= [80(2251) - (349)^2]/[80/79]$

  $= 58279/6320 = 9.221$

$s = 3.04$

b. The relative frequency distribution is given among the tables above.

c. $\mu = \Sigma[xP(x)] = 4.5$

   $\sigma^2 = \Sigma[x^2P(x)] - \mu^2 = 28.5 - (4.5)^2 = 8.25$

   $\sigma = 2.87$

d. Yes. It appears that the given last digits roughly agree with the distribution we expect with random selection.

2. Let x = the number of HIV cases.

   binomial problem: n=150 and p=0.10, use the binomial formula $P(x) = \{n!/[(n-x)!x!]\}p^x q^{n-x}$

   a. $\mu = np = (150)(0.10) = 15.0$

      $\sigma^2 = npq = (150)(0.10)(0.90) = 13.5$

      $\sigma = 3.674$, rounded to 3.7

   b. minimum usual value $= \mu - 2\sigma = 15.0 - 2(3.67) = 7.7$ [using rounded σ, 7.6]

      maximum usual value $= \mu + 2\sigma = 15.0 + 2(3.67) - 22.3$ [using rounded σ, 22.4]

      No. Since 12 is within the above limits, it is not an unusually low result. There is not sufficient evidence to suggest that the program is effective in lowering the 10% rate.

3. The table at the right summarizes the calculations.

   a. Yes. Since $0 \le P(x) \le 1$ for each x, and $\Sigma P(x) = 1$, it is a valid probability distribution.

   b. The population represented is all of the student's friends. No, the population is not all credit card holders in the United States.

   c. Yes. Conclusions from the data cannot be generalized to a general population.

x	P(x)	x·P(x)	x^2·P(x)
0	0.16	0	0
1	0.24	0.24	0.24
2	0.40	0.80	1.60
3	0.16	0.48	1.44
4	0.04	0.16	0.64
	1.00	1.68	3.92

   d. $\mu = \Sigma[x \cdot P(x)] = 1.68$, rounded to 1.7

   e. $\sigma^2 = \Sigma[x^2 \cdot P(x)] - \mu^2 = 3.92 - (1.68)^2 = 1.0976$

      $\sigma = 1.048$, rounded to 1.0

Chapter 6

Normal Probability Distributions

6-2 The Standard Normal Distribution

1. The word "normal" as used when referring to a normal distribution does carry with it some of the meaning the word has in ordinary language. Normal distributions occur in nature and describe the normal, or natural, state of many common phenomena. But in statistics the term "normal" has a specific and well-defined meaning in addition to its generic connotations of being "typical" – it refers to a specific bell-shaped distribution generated by a particular mathematical formula.

3. A normal distribution can be centered about any value and have any level of spread. A *standard* normal distribution has a center (as measured by the mean) of 0 and has a spread (as measured by the standard deviation) of 1.

5. The height of the rectangle is 0.5.
   Probability corresponds to area, and the area of a rectangle is (width)·(height).
   $$P(x<51.5) = (width)·(height)$$
   $$= (51.5 - 50.0)(0.5)$$
   $$= (1.5)(0.5)$$
   $$= 0.75$$

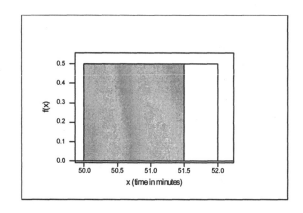

7. The height of the rectangle is 0.5.
   Probability corresponds to area, and the area of a rectangle is (width)·(height).
   $$P(50.5<x<51.5) = (width)·(height)$$
   $$= (51.5 - 50.5)(0.5)$$
   $$= (1.0)(0.5)$$
   $$= 0.50$$

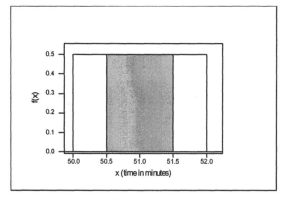

NOTE: The sketch is the key to exercises #9-#28. It tells whether to subtract two Table A-2 probabilities, to subtract a Table A-2 probability from 1, etc. For the remainder of chapter 6, THE ACCOMPANYING SKETCHES ARE NOT TO SCALE and are intended only as aids to help the reader understand how to use the table values to answer the questions. In addition, the probability of any single point in a continuous distribution is zero – i.e., P(x=a) = 0 for any single point a. For normal distributions, therefore, this manual ignores P(x=a) and uses P(x>a) = 1 – P(x<a).

9. $P(z<-1.00) = 0.1587$

11. $P(z<1.00) = 0.8413$

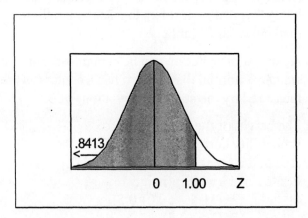

13. $P(z>1.25)$
    $= 1 - 0.8944$
    $= 0.1056$

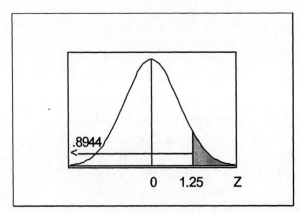

15. $P(z>-1.75)$
    $= 1 - 0.0401$
    $= 0.9599$

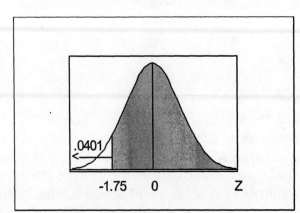

17. P(1.00<z<2.00)
    = 0.9772 − 0.8413
    = 0.1359

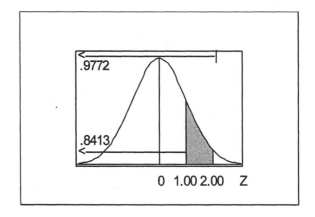

19. P(-2.45<z<-2.00)
    = 0.0228 − 0.0071
    = 0.0157

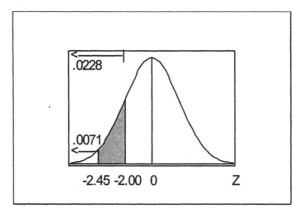

21. P(-2.11<z<1.55)
    = 0.9394 − 0.0174
    = 0.9220

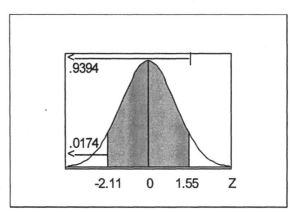

23. P(-1.00<z<4.00)
    = 0.9999 − 0.1587
    = 0.8412

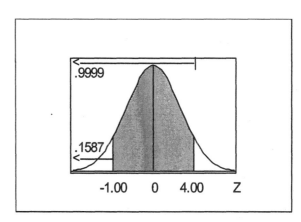

25. P(z>3.52)

       = 1 − 0.9999

       = 0.0001

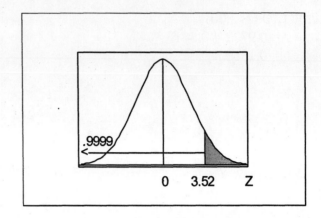

27. P(z>0.00)

       = 1 − 0.5000

       = 0.5000

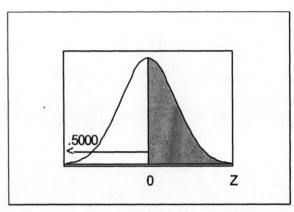

29. P(−1.00<z<1.00)

       = P(z<1.00) − P(z<−1.00)

       = 0.8413 − 0.1587

       = 0.6826 or 68.26%

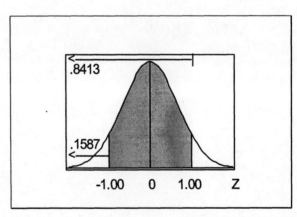

31. P(−3.00<z<3.00)

       = P(z<3.00) − P(z<−3.00)

       = 0.9987 − 0.0013

       = 0.9974 or 99.74%

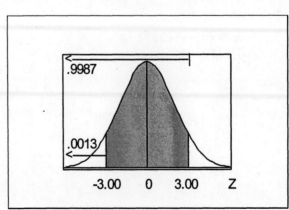

33. P(-1.96<z<1.96)
   = P(z<1.96) – P(z<-1.96)
   = 0.9750 – 0.0250
   = 0.9500

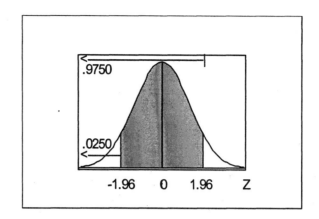

35. P(z<-2.575 or z>2.575)
   = P(z<-2.575) + P(z>2.575)
   = 0.0050 + [1 – 0.9950]
   = 0.0050 + 0.0050
   = 0.0100

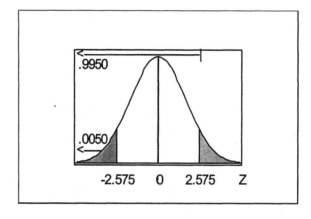

NOTE: The sketch is the key to exercises #37-#40. It tells what probability (i.e., cumulative area) to look up when reading Table A-2 "backwards." It also provides a check against gross errors by indicating at a glance whether a z score is above or below 0.

37. For P_{90}, the cumulative area is 0.9000.
    The closest entry is 0.8997,
       for which z = 1.28

39. For the lowest 2.5%, the cumulative
    area is 0.0250 (exact entry in the table),
    indicated by z = -1.96.
    By symmetry, the highest 2.5%
    [= 1–.9750] are above z = 1.96.

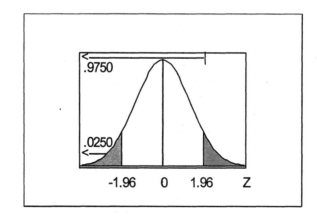

41. Rewrite each statement in terms of z, recalling that z is the number of standard deviations a score is from the mean.

a. $P(-1.00 < z < 1.00)$
   $= P(z < 1.00) - P(z < -1.00)$
   $= 0.8413 - 0.1587$
   $= 0.6826$ or 68.26%

b. $P(-1.96 < z < 1.96)$
   $= P(z < 1.96) - P(z < -1.96)$
   $= 0.9750 - 0.0250$
   $= 0.9500$ or 95.00%

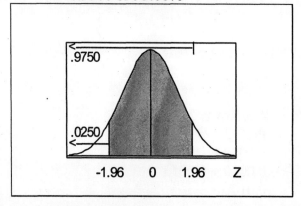

c. $P(-3.00 < z < 3.00)$
   $= P(z < 3.00) - P(z < -3.00)$
   $= 0.9987 - 0.0013$
   $= 0.9974$ or 99.74%

d. $P(-1.00 < z < 2.00)$
   $= P(z < 2.00) - P(z < -1.00)$
   $= 0.9772 - 0.1587$
   $= 0.8185$ or 81.85%

e. $P(z < -2.00$ or $z > 2.00)$
   $= P(z < -2.00) + P(z > 2.00)$
   $= P(z < -2.00) + [1 - P(z < 2.00)]$
   $= 0.0228 + [1 - 0.9772]$
   $= 0.0228 + 0.0228$
   $= 0.0456$ or 4.56%

43. The sketches are the key. They tell what probability (i.e., cumulative area) to look up when reading Table A-2 "backwards." They also provide a check against gross errors by indicating whether a score is above or below zero.

a. $P(0<z<a) = P(z<a) - P(z<0)$
  $0.3907 = P(z<a) - 0.5000$
  $0.8907 = P(z<a)$
    $a = 1.23$

b. Since $P(z<-b) = P(z>b)$ by symmetry & $\Sigma P(z) = 1$,
  $P(z<-b) + P(-b<z<b) + P(z>b) = 1$
  $P(z<-b) + 0.8664 + P(z<-b) = 1$
    $2 \cdot P(z<-b) + 0.8664 = 1$
      $2 \cdot P(z<-b) = 0.1336$
       $P(z<-b) = 0.0668$
         $-b = -1.50$
          $b = 1.50$

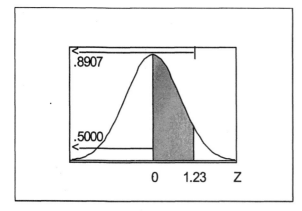

Note that $0.9332 - 0.0668 = 0.8664$, as given.

c. $P(z>c) = 0.0643)$
 $P(z<c) = 1 - 0.0643$
    $= 0.9357$
   $c = 1.52$

d. $P(z>d) = 0.9922$
 $P(z<d) = 1 - 0.9922$
    $= 0.0078$
   $d = -2.42$

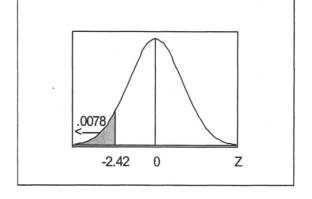

e. $P(z<e) = 0.4500$ [closest entry is 0.4483]
  $e = -0.13$

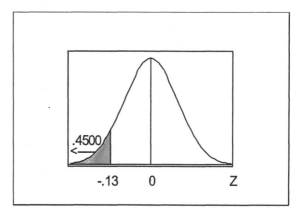

45. a. Moving across the x values from the minimum to the maximum will accumulate probability
    at a constant rate from 0 to 1. The result will be a straight line (i.e., with a constant slope)
    from (minimum x, 0) to (maximum x, 1). The sketch is given below.

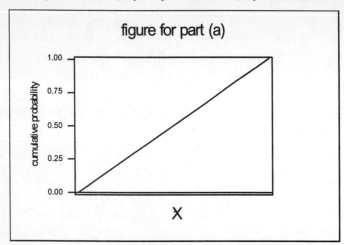

b. Moving across the x values from the minimum to the maximum will accumulate probability
   slowly at first, at a larger rate near the middle, and more slowly again at the upper end. The
   result will be a curve that has a variable slope from (minimum x, 0) to (maximum x, 1) –
   starting with a slope near 0 at the lower end, gradually increasing until reaching a maximum
   slope at μ, and then gradually decreasing again toward a slope near 0 at the upper end. The
   sketch is given below.

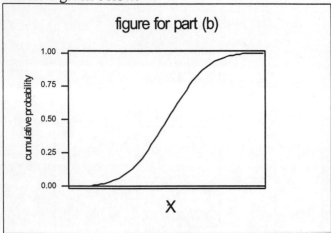

6-3 Applications of Normal Distributions

1. A normal distribution can have any mean and any non-negative standard deviation. A standard
   normal distribution has mean 0 and standard deviation 1 – and it follows a "nice" bell-shaped
   curve. Non-standard normal distributions can follow bell-shaped curves that are tall and thin,
   or short and fat.

3. No. Since the distribution of random digits does not follow a normal distribution, the methods
   of this section are not relevant. Because random digits follow a discrete uniform distribution,
   in which each digit is equally likely, probability questions can be answered using the methods
   of chapter 4. When selecting a random digit, there are 5 possibilities [0,1,2,3,4] less than 5 and
   $P(x<5) = 5/10 = 0.5$.

NOTE: In each nonstandard normal distribution, x scores are converted to z scores using the formula $z = (x-\mu)/\sigma$ and rounded to two decimal places. For "backwards" problems, solving the formula for x yields $x = \mu + z\sigma$. The area (cumulative probability) given in Table A-2 will be designated by A. As in the previous section, drawing and labeling the sketch is the key to successful completion of the exercises.

5. normal distribution: $\mu = 100$ and $\sigma = 15$
   P(x<130)
       = P(z<2.00)
       = 0.9772

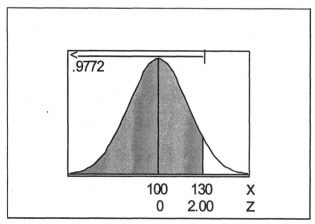

7. normal distribution: $\mu = 100$ and $\sigma = 15$
   P(90<x<110)
       = P(-0.67<z<0.67)
       = P(z<0.67) − P(z<-0.67)
       = 0.7486 − 0.2514
       = 0.4972

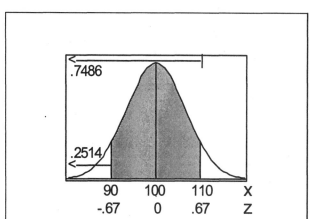

9. normal distribution: $\mu = 100$ and $\sigma = 15$
   For P_{10}, A = 0.1000 [0.1003]
   and z = -1.28.
   x = μ + zσ
       = 100 + (-1.28)(15)
       = 100 − 19.2
       = 80.8

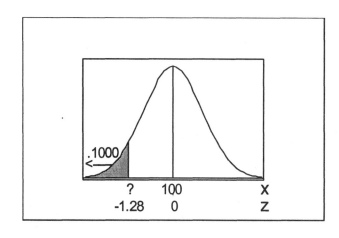

11. normal distribution: $\mu = 100$ and $\sigma = 15$
For the top 35%,
    $A = 0.6500 \ [0.6517]$ and $z = 0.39$.
$x = \mu + z\sigma$
    $= 100 + (0.39)(15)$
    $= 100 + 5.85$
    $= 105.85$

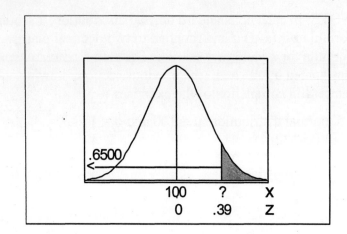

13. normal distribution: $\mu = 63.6$ and $\sigma = 2.5$
$P(x > 70)$
    $= 1 - P(x < 70)$
    $= 1 - P(z < 2.56)$
    $= 1 - 0.9948$
    $= 0.0052$ or 0.52%

15. normal distributions

a. men: $\mu = 69.0$ and $\sigma = 2.8$ women: $\mu = 63.6$ and $\sigma = 2.5$
$P(x > 80)$ $P(x > 80)$
    $= 1 - P(x < 80)$ $= 1 - P(x < 80)$
    $= 1 - P(z < 3.93)$ $= 1 - P(z < 6.44)$
    $= 1 - 0.9999$ $= 1 - 0.9999$
    $= 0.0001$ $= 0.0001$

Yes. It appears that the current doorway design is OK.

b. normal distribution: $\mu = 69.0$ and $\sigma = 2.8$
   For the tallest 5%,
   $\qquad$ A = 0.9500 and z = 1.645.
   $x = \mu + z\sigma$
   $\qquad = 69.0 + (1.645)(2.8)$
   $\qquad = 69.0 + 4.6$
   $\qquad = 73.6$ in

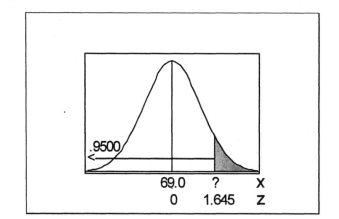

17. normal distribution: $\mu = 3420$ and $\sigma = 495$
    For the lightest 2%,
    $\qquad$ A = 0.0200 [0.0202] and z = -2.05.
    $x = \mu + z\sigma$
    $\qquad = 3420 + (-2.05)(495)$
    $\qquad = 3420 - 1015$
    $\qquad = 2405$ g

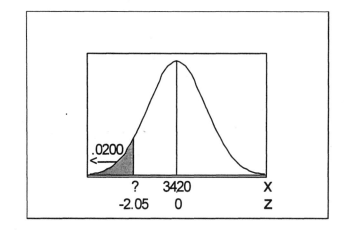

19. normal distribution: $\mu = 184.0$ and $\sigma = 55.0$
    P(x>230)
    $\qquad = 1 - P(x<230)$
    $\qquad = 1 - P(z<0.84)$
    $\qquad = 1 - 0.7995$
    $\qquad = 0.2005$

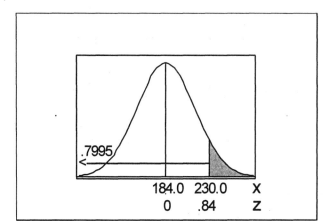

21. normal distribution: $\mu = 268$ and $\sigma = 15$
    a. P(x>308)
    $\qquad = 1 - P(x<308)$
    $\qquad = 1 - P(z<2.67)$
    $\qquad = 1 - 0.9962$
    $\qquad = 0.0038$
    The result suggests that an unusual
    event has occurred – but certainly not
    an impossible one, as about 38 of every
    10,000 pregnancies can be expected to
    last as long.

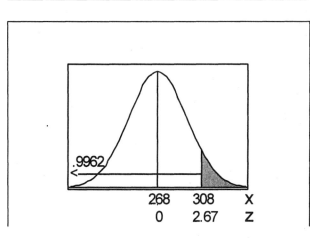

b. For the lowest 4%,
  A = 0.0400 [0.0401]and z = -1.75.
  $x = \mu + z\sigma$
    = 268 + (-1.75)(15)
    = 268 – 26
    = 242 days

23. normal distribution: $\mu = 6.0$ and $\sigma = 1.0$
  For the smallest 2.5%,
    A = 0.0250 and z = -1.96.
  For the largest 2.5%,
    A = 0.9750 and z = 1.96
  $x_S = \mu + z\sigma$
    = 6.0 + (-1.96)(1.0)
    = 6.0 – 1.96
    = 4.04, rounded to 4.0 in
  $x_L = \mu + z\sigma$
    = 6.0 + (1.96)(1.0)
    = 6.0 + 1.96
    = 7.96, rounded to 8.0 in

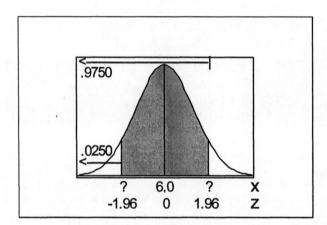

25. a. The distribution is given at the right.
  The summary statistics are as follows.
      n = 40
      $\bar{x}$ = 118.9
      s = 10.46
  The data appear to have a distribution that
  is approximately normal.

pressure	frequency
90- 99	1
100-109	4
110-119	17
120-129	12
130-139	5
140-149	0
150-159	1
	40

b. normal distribution:
      $\mu = 118.9$ and $\sigma = 10.46$
  For P_5, A = 0.0500 and z = -1.645.
  For P_{95}, A = 0.9500 and z = 1.645.
  $x_5 = \mu + z\sigma$
      = 118.9 + (-1.645)(10.46)
      = 118.9 – 17.2
      = 101.7 mm [using rounded σ, 101.6]
  $x_{95} = \mu + z\sigma$
      = 118.9 + (1.645)(10.46)
      = 118.9 + 17.2
      = 136.1 mm [using rounded σ, 136.2]

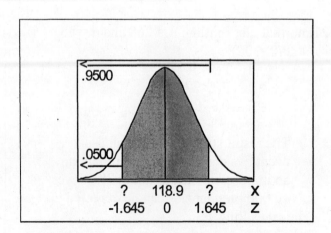

27. a. The z scores are always unit free. Because the numerator and the denominator of the fraction $z = (x-\mu)/\sigma$ have the same units, the units will divide out.
    b. For a population of size N, $\mu = \Sigma x/N$ and $\sigma^2 = \Sigma(x-\mu)^2/N$.
       As shown below, $\mu = 0$ and $\sigma = 1$ will be true for *any* set of z scores – regardless of the shape of the original distribution.

$$\Sigma z = \Sigma(x-\mu)/\sigma = (1/\sigma)[\Sigma(x-\mu)]$$
$$= (1/\sigma)[\Sigma x - \Sigma\mu]$$
$$= (1/\sigma)[N\mu - N\mu] = (1/\sigma)[0] = 0$$
$$\Sigma z^2 = \Sigma[(x-\mu)/\sigma\,]^2 = (1/\sigma)^2[\Sigma(x-\mu)^2]$$
$$= (1/\sigma^2)[N\sigma^2] = N$$
$$\mu_Z = (\Sigma z)/N = 0/N = 0$$
$$\sigma^2_Z = \Sigma(z-\mu_Z)^2/N = \Sigma(z-0)^2/N = \Sigma z^2/N = N/N = 1 \text{ and } \sigma_Z = 1$$

   The re-scaling from x scores to z scores will not affect the basic shape of the distribution – and so in this case the z scores will be normal, as was the original distribution.

29. normal distribution: $\mu = 25$ and $\sigma = 5$
    a. For a population of size N, $\mu = \Sigma x/N$ and $\sigma^2 = \Sigma(x-\mu)^2/N$.
       As shown below, adding a constant to each score increases the mean by that amount but does not affect the standard deviation. In non-statistical terms, shifting everything by k units does not affect the spread of the scores. This is true for *any* set of scores – regardless of the shape of the original distribution. Let $y = x + k$.

$$\mu_Y = [\Sigma(x+k)]/N$$
$$= [\Sigma x + \Sigma k]/N$$
$$= [\Sigma x + Nk]/N$$
$$= (\Sigma x)/N + (Nk)/N$$
$$= \mu_X + k$$
$$\sigma^2_Y = \Sigma[y - \mu_Y]^2/N$$
$$= \Sigma[(x+k) - (\mu_X + k)]^2/N$$
$$= \Sigma[x - \mu_X]^2/N$$
$$= \sigma^2_X \text{ and so } \sigma_Y = \sigma_X$$

   If the teacher adds 50 to each grade,
   new mean $= 25 + 50 = 75$
   new standard deviation $= 5$ (same as before).

    b. No. Curving should consider the variation. Had the test been more appropriately constructed, it is not likely that every student would score exactly 50 points higher. If the typical student score increased by 50, we would expect the better students to increase by more than 50 and the poorer students to increase by less than 50. This would make the scores more spread out and would increase the standard deviation.

    c. For the top 10%, A = 0.9000 [0.8997] and z = 1.28.
$$x = \mu + z\sigma = 25 + (1.28)(5) = 25 + 6.4 = 31.4$$
    For the bottom 70%, A = 0.7000 [0.6985] and z = 0.52.
$$x = \mu + z\sigma = 25 + (0.52)(5) = 25 + 2.6 = 27.6$$
    For the bottom 30%, A = 0.3000 [0.3015] and z = -0.52.
$$x = \mu + z\sigma = 25 + (-0.52)(5) = 25 - 2.6 = 22.4$$
    For the bottom 10%, A = 0.1000 [0.1003] and z = -1.28.
$$x = \mu + z\sigma = 25 + (-1.28)(5) = 25 - 6.4 = 18.6$$
    This produces the grading scheme given at the right.

A: higher than 31.4
B: 27.6 to 31.4
C: 22.4 4o 27.6
D: 18.6 to 22.4
E: less than 18.6

d. The curving scheme in part (c) is fairer because it takes into account the variation as discussed in part (b). Assuming the usual 90-80-70-60 letter grade cut-offs, for example, the percentage of A's under the scheme in part (a) with $\mu = 75$ and $\sigma = 5$ is

$$P(x>90) = 1 - P(x<90)$$
$$= 1 - P(z<3.00)$$
$$= 1 - 0.9987$$
$$= 0.0013 \text{ or } 0.13\%$$

This is considerably less than the 10% A's under the scheme in part (c) and reflects the fact that the variation in part (a) is unrealistically small.

6-4 Sampling Distributions and Estimators

1. Given a population distribution of scores, one can take a sample of size n and calculate any of several statistics. A sampling distribution is the distribution of all possible values of such a particular statistic.

3. Table 6-7 indicates that the sample median, range and standard deviation are not necessarily unbiased estimators of the population median, range and standard deviation. Table 6-7 indicates that the sample mean, variance and proportion were unbiased estimators of the population mean, variance and proportion for the given example – and the text states that such is always the case.

5. No. The sample proportion is an unbiased estimator of the population proportion. This means that it gives the correct value on the average, but not necessarily for any one particular sample.

NOTE: Section 5-2 defined the mean of a probability distribution of x's as $\mu_x = \Sigma[x \cdot P(x)]$. If the variable is designated by the symbol y, then the mean of a probability distribution of y's is $\mu_y = \Sigma[y \cdot P(y)]$. In this section, the variables are statistics – like $\bar{x}$ and $\hat{p}$. In such cases, the formula for the mean may be adjusted – to $\mu_{\bar{x}} = \Sigma[\bar{x} \cdot P(\bar{x})]$ and $\mu_{\hat{p}} = \Sigma[\hat{p} \cdot P(\hat{p})]$. In a similar manner, the formula for the variance of a probability distribution may also be adjusted to match the variable being considered.

7. For the original population,
   $\mu = \Sigma x/N$
   $= (10+6+5)/3$
   $= 21/3 = 7.0$ calls
   a. The 9 equally likely samples and their means are given at the immediate right.
   b. Each sample has a probability of 1/9. The sampling distribution of the means, a list of each possible sample mean and its total probability, is given at the far right. The distribution appears to "bunch up" in the middle and taper off at the ends.
   c. $\mu_{\bar{x}} = \Sigma[\bar{x} \cdot P(\bar{x})] = 63.0/9 = 7.0$ calls
   d. Yes. The sample mean is always an unbiased estimator of the population mean.

sample	$\bar{x}$	$\bar{x}$	$P(\bar{x})$	$\bar{x} \cdot P(\bar{x})$
5,5	5.0	5.0	1/9	5.0/9
5,6	5.5	5.5	2/9	11.0/9
5,10	7.5	6.0	1/9	6.0/9
6,5	5.5	7.5	2/9	15.0/9
6,6	6.0	8.0	2/9	16.0/9
6,10	8.0	10.0	1/9	10.0/9
10,5	7.5		9/9	63.0/9
10,6	8.0			
10,10	10.0			
	63.0			

9. For the original population,

   $\mu = \Sigma x/N$

     $= (47+43+21+20+20)/5$

     $= 151/5 = 30.2$ billion \$

   a. The 25 equally likely samples and their means are given at the immediate right.
   b. Each sample has a probability of 1/25. The sampling distribution of the means, a list of each possible sample mean and its total probability, is given at the far right. The distribution appears to "bunch up" toward the lower end.
   c. $\mu_{\bar{x}} = \Sigma[\bar{x} \cdot P(\bar{x})] = 755.0/25 = 30.2$ billion \$
   d. Yes. The sample mean is always an unbiased estimator of the population mean.

sample	$\bar{x}$	$\bar{x}$	$P(\bar{x})$	$\bar{x} \cdot P(\bar{x})$
20,20	20.0	20.0	4/25	80.0/25
20,20	20.0	20.5	4/25	82.0/25
20,21	20.5	21.0	1/25	21.0/25
20,43	31.5	31.5	4/25	126.0/25
20,47	33.5	32.0	2/25	64.0/25
20,20	20.0	33.5	4/25	134.0/25
20,20	20.0	34.0	2/25	68.0/25
20,21	20.5	43.0	1/25	43.0/25
20,43	31.5	45.0	2/25	90.0/25
20,47	33.5	47.0	1/25	47.0/25
21,20	20.5		25/25	755.0/25
21,20	20.5			
21,21	21.0			
21,43	32.0			
21,47	34.0			
43,20	31.5			
43,20	31.5			
43,21	32.0			
43,43	43.0			
43,47	45.0			
47,20	33.5			
47,20	33.5			
47,21	34.0			
47,43	45.0			
47,47	47.0			
	755.0			

11. For the original population, $p = 3/4 = 0.75$.
   a. The 16 equally likely samples and their proportions are given at the immediate right. Each sample has a probability of 1/16. The sampling distribution of the proportions, a list of each possible sample proportion and its total probability, is given at the far right. The distribution appears to "bunch up" toward the upper end.
   b. $\mu_{\hat{p}} = \Sigma[\hat{p} \cdot P(\hat{p})] = 12.0/16 = 0.75$
   c. Yes. The sample proportion is always an unbiased estimator of the population proportion.

sample	$\hat{p}$	$\hat{p}$	$P(\hat{p})$	$\hat{p} \cdot P(\hat{p})$
MM	0.0	0.0	1/16	0.0/16
MA	0.5	0.5	6/16	3.0/16
MB	0.5	1.0	9/16	9.0/16
MC	0.5		16/16	12.0/16
AM	0.5			
AA	1.0			
AB	1.0			
AC	1.0			
BM	0.5			
BA	1.0			
BB	1.0			
BC	1.0			
CM	0.5			
CA	1.0			
CB	1.0			
CC	1.0			
	12.0			

13. For the original population,

$\mu = \Sigma x/N$

$= (3+9+23)/3$

$= 35/3 = 11.7$

sample	$\overline{x}$	$\overline{x}$	$P(\overline{x})$	$\overline{x} \cdot P(\overline{x})$
3,3	3.0	3.0	1/9	3.0/9
3,9	6.0	6.0	2/9	12.0/9
3,23	13.0	9.0	1/9	9.0/9
9,3	6.0	13.0	2/9	26.0/9
9,9	9.0	16.0	2/9	32.0/9
9,23	16.0	23.0	1/9	23.0/9
23,3	13.0		9/9	105.0/9
23,9	16.0			
23,23	23.0			
	105.0			

a. The 9 equally likely samples and their means are given in the table above. Each sample has a probability of 1/9. The sampling distribution of the means, a list of each possible sample mean and its total probability, is given at the far right. The distribution appears to "bunch up" toward the middle and taper off at the ends. As expected,

$\mu_{\overline{x}} = \Sigma[\overline{x} \cdot P(\overline{x})] = 105.0/9 = 11.7$

b. Rank data is at the ordinal level – and the difference between ranks 1 and 2 is not necessarily the same, in actual performance, as the difference between ranks 25 and 26. And since the values of the ranks are predetermined (1,2,…,n), the data are not independent. But one can still average ranks to make valid comparisons and make sample selections from a population of ranks. In that sense ranks are just as meaningful as other scores and it makes sense to identify sampling distributions for statistics calculated using ranks.

15. Use of the formula for the given values is shown below. The resulting distribution agrees with the sampling distribution for the proportion of girls in a family of size 2.

$P(x) = 1/[2(2-2x)!(2x)!]$

$P(x=0) = 1/[2(2)!(0)!] = 1/[2\cdot2\cdot1] = 1/4$

$P(x=0.5) = 1/[2(1)!(1)!] = 1/[2\cdot1\cdot1] = 1/2$

$P(x=1) = 1/[2(0)!(2)!] = 1/[2\cdot1\cdot2] = 1/4$

6-5 The Central Limit Theorem

1. The standard error of a statistic is the standard deviation of its sampling distribution. For any given sample size n, the standard error of the mean is the standard deviation of the distribution of all possible sample means of size n. Its symbol and numerical value are $\sigma_{\overline{x}} = \sigma/\sqrt{n}$.

3. The subscript $\overline{x}$ is used to distinguish the mean and standard deviation of the sample means from the mean and standard deviation of the original population. This produces the notation

$\mu_{\overline{x}} = \mu$ $\qquad$ $\sigma_{\overline{x}} = \sigma/\sqrt{n}$

5. a. normal distribution

$\mu = 63.6$

$\sigma = 2.5$

$P(x<64)$

$= P(z<0.16)$

$= 0.5636$

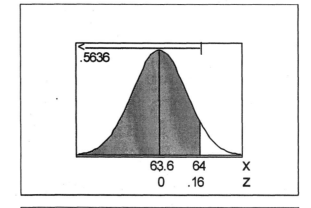

b. normal distribution,
  since the original distribution is so

$\mu_{\bar{x}} = \mu = 63.6$

$\sigma_{\bar{x}} = \sigma/\sqrt{n} = 2.5/\sqrt{36} = 0.417$

$P(\bar{x}<64)$

$= P(z<0.96)$

$= 0.8315$

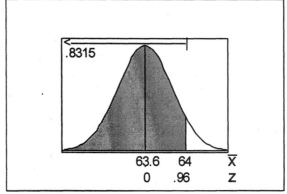

7. a. normal distribution

$\mu = 63.6$

$\sigma = 2.5$

$P(63.5<x<64.5)$

$= P(-0.04<z<0.36)$

$= P(z<0.36) - P(z<-0.04)$

$= 0.6406 - 0.4840$

$= 0.1566$

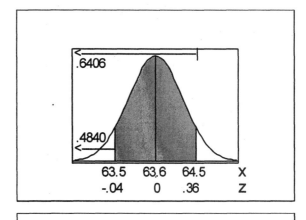

b. normal distribution,
  since the original distribution is so

$\mu_{\bar{x}} = \mu = 63.6$

$\sigma_{\bar{x}} = \sigma/\sqrt{n} = 2.5/\sqrt{9} = 0.833$

$P(63.5<\bar{x}<64.5)$

$= P(-0.12<z<0.1.08)$

$= 0.8599 - 0.4522$

$= 0.4077$

c. Since the original distribution is normal, the Central Limit Theorem can be used in part (b) even though the sample size does not exceed 30.

9. a. normal distribution

$\mu = 172$

$\sigma = 29$

$P(x>167)$

$\qquad = 1 - P(x<167)$

$\qquad = 1 - P(z<-0.17)$

$\qquad = 1 - .4325$

$\qquad = 0.5675$

b. normal distribution,

since the original distribution is so

$\mu_{\bar{x}} = \mu = 172$

$\sigma_{\bar{x}} = \sigma/\sqrt{n} = 29/\sqrt{12} = 8.372$

$P(\bar{x}>167)$

$\qquad = 1 - P(\bar{x}<167)$

$\qquad = 1 - P(z<-0.60)$

$\qquad = 1 - 0.2743$

$\qquad = 0.7257$

c. No. It appears that the 12 person capacity could easily exceed the 2004 lbs – especially when the weight of clothes and other equipment is considered. On the other hand, skiers may not be representative of the general population – they may be lighter than the general population, as skiing may not be an activity that attracts heavier persons.

11. a. normal distribution,

by the Central Limit Theorem

$\mu_{\bar{x}} = \mu = 12.00$

$\sigma_{\bar{x}} = \sigma/\sqrt{n} = 0.11/\sqrt{36} = 0.0183$

$P(\bar{x}>12.19)$

$\qquad = 1 - P(\bar{x}<12.19)$

$\qquad = 1 - P(z<10.36)$

$\qquad = 1 - 0.9999$

$\qquad = 0.0001$

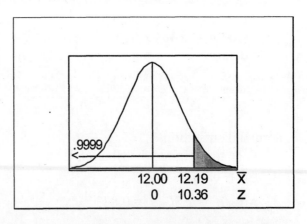

b. No; if $\mu = 12.00$, then a very rare event has occurred. No; there is no cheating, since the actual amount is more than the advertised amount. If the cans were actually filled with $\mu = 12.00$, then about one-half of the cans would contain less than that – and that would not be acceptable.

13. a. normal distribution

$\mu = 3.00$

$\sigma = 0.40$

$P(x > 4.00)$

$= 1 - P(x < 4.00)$

$= 1 - P(z < 2.50)$

$= 1 - 0.9938$

$= 0.0062$

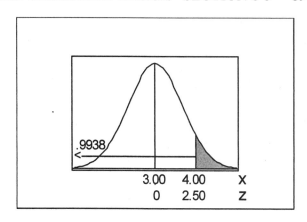

b. normal distribution,

since the original distribution is so

$\mu_{\bar{x}} = \mu = 4.00$

$\sigma_{\bar{x}} = \sigma/\sqrt{n} = 0.40/\sqrt{60} = 0.0516$

$P(\bar{x} > 4.00)$

$= 1 - P(\bar{x} < 4.00)$

$= 1 - P(z < 19.36)$

$= 1 - 0.9999$

$= 0.0001$

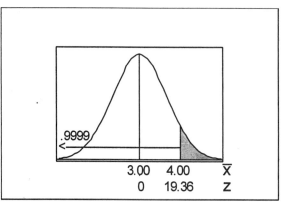

c. Part (a). It is the individual flashes that guide pilots, and not the mean of several flashes.

15. a. normal distribution

$\mu = 114.8$

$\sigma = 13.1$

$P(x > 140)$

$= 1 - P(x < 140)$

$= 1 - P(z < 1.92)$

$= 1 - 0.9726$

$= 0.0274$

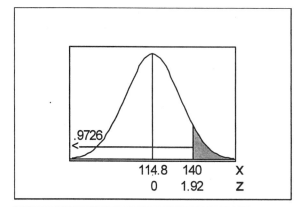

b. normal distribution,

since the original distribution is so

$\mu_{\bar{x}} = \mu = 114.8$

$\sigma_{\bar{x}} = \sigma/\sqrt{n} = 13.1/\sqrt{4} = 6.55$

$P(\bar{x} > 140)$

$= 1 - P(\bar{x} < 140)$

$= 1 - P(z < 3.85)$

$= 1 - 0.9999$

$= 0.0001$

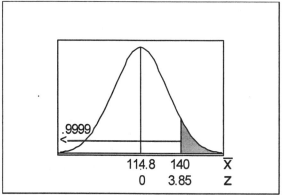

c. Since the original distribution is normal, the Central Limit Theorem can be used in part (b) even though the sample size does not exceed 30.

d. No. The mean can be less than 140 when one or more of the values is greater than 140.

17. a. normal distribution

$\mu = 143$

$\sigma = 29$

$P(140 < x < 211)$

$= P(-0.10 < z < 2.34)$

$= P(z < 2.34) - P(z < -0.10)$

$= 0.9904 - 0.4602$

$= 0.5302$

b. normal distribution,

since the original distribution is so

$\mu_{\bar{x}} = \mu = 143$

$\sigma_{\bar{x}} = \sigma/\sqrt{n} = 29/\sqrt{36} = 4.833$

$P(140 < \bar{x} < 211)$

$= P(-0.62 < z < 14.07)$

$= P(z < 14.07) - P(z < -0.62)$

$= 0.9999 - 0.2676$

$= 0.7323$

c. The information from part (a) is more relevant, since the seats will be occupied by one woman at a time.

19. a. normal distribution

$\mu = 5.670$

$\sigma = 0.062$

$P(5.550 < x < 5.790)$

$= P(-1.94 < z < 1.94)$

$= P(z < 1.94) - P(z < -1.94)$

$= 0.9738 - 0.0262$

$= 0.9476$

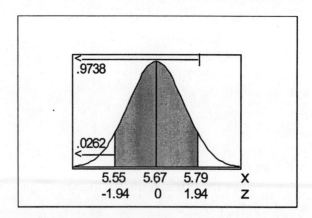

If 0.9476 of the quarters are accepted, then $1 - 0.9476 = 0.0524$ of the quarters are rejected. For 280 quarters, we expect $(0.0524)(280) = 14.7$ of them to be rejected.

b. normal distribution,
   since the original distribution is so

$\mu_{\bar{x}} = \mu = 5.670$

$\sigma_{\bar{x}} = \sigma/\sqrt{n} = 0.062/\sqrt{280} = 0.00371$

$P(5.550 < \bar{x} < 5.790)$

   $= P(-32.39 < z < 32.39)$
   $= P(z < 32.39) - P(z < -32.39)$
   $= 0.9999 - 0.0001$
   $= 0.9998$

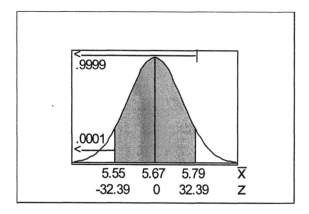

c. Probabilities concerning means do not apply to individuals. It is the information from part (a) that is relevant, since the vending machine deals with quarters one at a time.

21. a. normal distribution,
    since the original distribution is so

$\mu_{\bar{x}} = \mu = 14.4$

$\sigma_{\bar{x}} = \sigma/\sqrt{n} = 1.0/\sqrt{18} = 0.236$

This is a "backwards" normal problem,
   With A = 0.9750 and z = 1.96.

$\bar{x} = \mu_{\bar{x}} + z\sigma_{\bar{x}}$

   $= 14.4 + (1.96)(0.236)$
   $= 14.4 + 0.46$
   $= 14.86$

For 18 men, the total width is (18)(14.86) = 267.5 in.

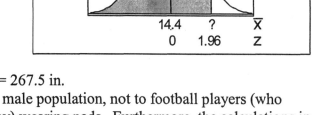

b. The 14.4 inch figure applies to the general male population, not to football players (who would tend to be larger than normal anyway) wearing pads. Furthermore, the calculations in part (a) do not consider any spacing between the players.

23. a. The box below gives the calculations to determine μ=8 and σ=5.385 for the original population of N=6 values.

b. The 15 equally like samples of size n=2 are listed in the first column in the box below.

c. The values of $\bar{x}$ are listed in the second column in the box below.

x	x-μ	(x-μ)^2
2	-6	36
3	-5	25
6	-2	4
8	0	0
11	3	9
18	10	100
48	0	174

$\mu = \Sigma x/N = 48/6 = 8.0$

$\sigma^2 = \Sigma(x-\mu)^2/N = 174/6 = 29$

$\sigma = 5.385$

sample	$\bar{x}$	$\bar{x}-\mu_{\bar{x}}$	$(\bar{x}-\mu_{\bar{x}})^2$
2,3	2.5	-5.5	30.25
2,6	4.0	-4.0	16.00
2,8	5.0	-3.0	9.00
2,11	6.5	-1.5	2.25
2,18	10.0	2.0	4.00
3,6	4.5	-3.5	12.25
3,8	5.5	-2.5	6.25
3,11	7.0	-1.0	1.00
3,18	10.5	2.5	6.25
6,8	7.0	-1.0	1.00
6,11	8.5	0.5	0.25
6,18	12.0	4.0	16.00
8,11	9.5	1.5	2.25
8,18	13.0	5.0	25.00
11,18	14.5	6.5	42.25
	120.0	0.0	174.00

d. Since there are 15 equally likely samples, the desired values can be determined using the box above on the right as follows.

$\mu_{\bar{x}} = \Sigma\bar{x}/15 = 120/15 = 8.0$

$\sigma_{\bar{x}}^2 = \Sigma(\bar{x} - \mu_{\bar{x}})^2/15 = 174/15 = 11.6$

$\sigma_{\bar{x}} = \sqrt{11.6} = 3.406$

e. $\mu = 8 = \mu_{\bar{x}}$

$$\frac{\sigma}{\sqrt{n}}\sqrt{\frac{N-n}{N-1}} = \frac{5.385}{\sqrt{2}}\sqrt{\frac{6-2}{6-1}} = 3.406 = \sigma_{\bar{x}}$$

6-6 Normal as Approximation to Binomial

1. The histogram will be bell-shaped – with the highest frequencies near 0.50, and tapering off evenly toward each end.

3. No. With n=6 and p=0.001, the requirements that np≥5 and nq≥ 5 are not met.

5. The area to the right of 15.5.   In symbols, $P(x>15) = P_C(x>15.5)$.

7. The area to the left of 11.5.   In symbols, $P(x<12) = P_C(x<11.5)$.

9. The area to the left of 4.5.   In symbols $P(x\leq4) = P_C(x<4.5)$.

11. The area between 7.5 and 10.5.   In symbols, $P(8\leq x\leq 10) = P_C(7.5<x<10.5)$.

13. binomial: n=12 and p=0.6
    a. from Table A-1, $P(x=7) = 0.227$
    b. normal approximation not appropriate since
        $nq = 12(0.4) = 4.8 < 5$

15. binomial: n = 11 and p = 0.5
    a. from Table A-1,
        $P(x\geq4) = P(x=4) + P(x=5) +...+ P(x=11)$
                $= 0.161 + 0.226 + 0.226 + 0.161 + 0.081 + 0.027 + 0.005 + 0^+$
                $= 0.887$
    b. normal approximation appropriate since
        $np = 11(0.5) = 5.5 \geq 5$
        $nq = 11(0.5) = 5.5 \geq 5$
        $\mu = np = 11(0.5) = 5.5$
        $\sigma = \sqrt{npq} = \sqrt{11(0.5)(0.5)} = 1.658$
        $P(x\geq4)$
            $= P_C(x>3.5)$
            $= 1 - P(x<3.5)$
            $= 1 - P(z<-1.21)$
            $= 1 - 0.1131$
            $= 0.8869$

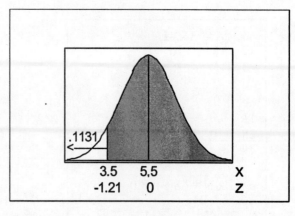

17. Let x = the number of girl births.
    binomial: n=64 and p=0.5
    normal approximation appropriate since
        np = 64(0.5) = 32 ≥5
        nq = 64(0.5) = 32 ≥5
    μ = np = 64(0.5) = 32
    $\sigma = \sqrt{npq} = \sqrt{64(0.5)(0.5)} = 4$
    P(x>36)
        = P_C(x>36.5)
        = 1 – P(x<36.5)
        = 1 – P(z<1.125)
        = 1 – 0.8697 [(0.8686+.8708)/2]
        = 0.1303

No. Since 0.1303 > 0.05, getting more than 36 girls is not an unusual occurrence.

19. Let x = the number that actually voted.
    binomial: n=1002 and p=0.61
    normal approximation appropriate since
        np = 1002(0.61) = 611.22 ≥5
        nq = 1002(0.39) = 390.78 ≥5
    μ = np = 1002(0.61) = 611.22
    $\sigma = \sqrt{npq} = \sqrt{1002(0.61)(0.39)} = 15.439$
    P(x≥701)
        = P_C(x>700.5)
        = 1 – P(x<700.5)
        = 1 – P(z<5.78)
        = 1 – 0.9999
        = 0.0001

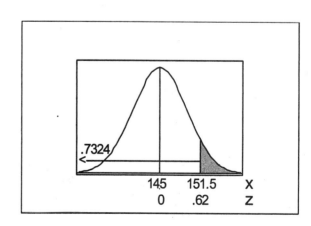

Under the stated conditions, getting 701 persons that actually voted would be a very rare occurrence. It appears that the people in the survey were not telling the truth.

21. Let x = the number with yellow pods.
    binomial: n=580 and p=0.25
    normal approximation appropriate since
        np = 580(0.25) = 145 ≥5
        nq = 580(0.75) = 435 ≥5
    μ = np = 580(0.25) = 145
    $\sigma = \sqrt{npq} = \sqrt{580(0.25)(0.75)} = 10.428$
    P(x≥152)
        = P_C(x>151.5)
        = 1 – P(x<151.5)
        = 1 – P(z<-0.62)
        = 1 – 0.7324
        = 0.2676
No. There is no evidence to suggest that Mendel's 25% rate is incorrect.

23. Let x = the number with such cancer.
binomial: n=420,095 and p=0.00034
normal approximation appropriate since
    np = 420,095(0.00034) = 142.8 ≥5
    nq = 420,095(0.99966) = 419857.2 ≥5
μ = np = 420,095(0.00034) = 142.8
$\sigma = \sqrt{npq} = \sqrt{420095(0.00034)(0.99966)}$
    = 11.949
P(x≤135)
    = P_C(x<135.5)
    = P(z<-0.61)
    = 0.2709

Finding "135 or fewer cancer cases" is not an unusual event, and so the cancer rate among cell phone users appears not to differ significantly from that of the general population.
NOTE: To test the media claim that cell phone usage is associated with a higher rate, one should calculate the probability of finding "135 or more cancer cases."

25. Let x = the number females hired.
binomial: n=62 and p=0.5
normal approximation appropriate since
    np = 62(0.5) = 31 ≥5
    nq = 62(0.5) = 31 ≥5
μ = np = 62(0.5) = 31
$\sigma = \sqrt{npq} = \sqrt{62(0.5)(0.5)}$ = 3.937
P(x≤21)
    = P_C(x<21.5)
    = P(z<-2.41)
    = 0.0080

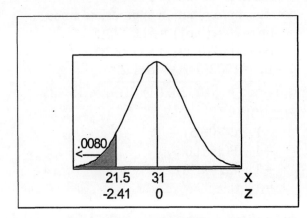

Hiring 21 or fewer females by chance alone would be an unusual event. It appears that factors other than chance are at work, including the possibility of gender discrimination.

27. Let x = the number with Group O blood.
binomial: n=400 and p=0.45
normal approximation appropriate since
    np = 400(0.45) = 180 ≥5
    nq = 400(0.55) = 220 ≥5
μ = np = 400(0.45) = 180
$\sigma = \sqrt{npq} = \sqrt{400(0.45)(0.55)}$ = 9.950
P(x≥177)
    = P_C(x>176.5)
    = 1 – P(x<176.5)
    = 1 – P(z<-0.35)
    = 1 – 0.3632
    = 0.6368

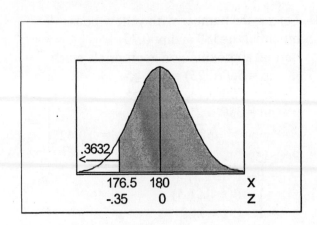

Yes. A pool of 400 is likely to be sufficient, but it might be prudent to try for a larger pool.

29. Let x = the number opposed to cloning.
    binomial: n=1012 and p=0.50
    normal approximation appropriate since
    np = 1012(0.50) = 506 ≥5
    nq = 1012(0.5) = 506 ≥5
    μ = np = 1012(0.50) = 506
    $\sigma = \sqrt{npq} = \sqrt{1012(0.50)(0.50)} = 15.906$
    P(x≥901)
        = P_C(x>900.5)
        = 1 − P(x<900.5)
        = 1 − P(z<24.80)
        = 1 − 0.9999
        = 0.0001

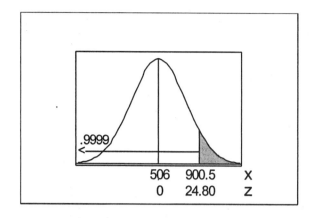

Yes. If only 50% of the people oppose cloning, a very rare event has occurred. The result is strong evidence in favor of the claim that more than 50% of the people are opposed to cloning.

31. Let x = the number of charges over $100.
    binomial: n=30 and p=0.358
    normal approximation appropriate since
    np = 30(0.358) = 10.74 ≥5
    nq = 30(0.642) = 19.26 ≥5
    μ = np = 30(0.358) = 10.74
    $\sigma = \sqrt{npq} = \sqrt{30(0.358)(0.642)} = 2.626$
    P(x≥18)
        = P_C(x>17.5)
        = 1 − P(x<17.5)
        = 1 − P(z<2.57)
        = 1 − 0.9949
        = 0.0051

Yes. If the usual usage pattern still applies, then there has been an unusual occurrence. Verification of correct usage of the card is warranted.

33. Let x = the number of times Marc wins $35.
    binomial: n=200 and p=1/38
    normal approximation appropriate since
    np = 200(1/38) = 5.26 ≥5
    nq = 200(37/38) = 194.76 ≥5
    μ = np = 784(0.079) = 61.936
    $\sigma = \sqrt{npq} = \sqrt{200(1/38)(37/38)} = 2.264$
    Marc needs at least 6 $35 wins for a profit.
    P(x≥6)
        = P_C(x>5.5)
        = 1 − P(x<5.5)
        = 1 − P(z<0.10)
        = 1 − 0.5398
        = 0.4602

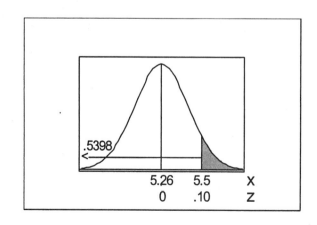

35. a. binomial: n=4 and p=0.350

$P(x \geq 1) = 1 - P(x=0)$
$$= 1 - [4!/(4!0!)](0.350)^0(0.650)^4$$
$$= 1 - 0.1785$$
$$= 0.8215$$

   b. binomial: n=56(4)=224 and p=0.350

normal approximation appropriate since

$np = 224(0.350) = 78.4 \geq 5$
$nq = 224(0.650) = 145.6 \geq 5$
$\mu = np = 224(0.350) = 78.4$
$\sigma = \sqrt{npq} = \sqrt{224(0.350)(0.650)} = 7.139$

$P(x \geq 56)$
$$= P_C(x>55.5)$$
$$= 1 - P(x<55.5)$$
$$= 1 - P(z<-3.20)$$
$$= 1 - 0.0007$$
$$= 0.9993$$

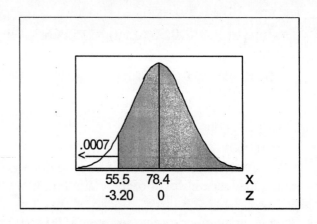

   c. Let H = getting at least one hit in 4 at bats.

$P(H) = 0.8215$ [from part (a)]

For 56 consecutive games, $[P(H)]^{56} = [0.8125]^{56} = 0.0000165$

   d. The solution below employs the techniques and notation of parts (a) and (c).

for $[P(H)]^{56} > 0.10$
$$P(H) > (0.10)^{1/56}$$
$$P(H) > 0.9597$$

for $P(H) = P(x \geq 1) > 0.9597$
$$1 - P(x=0) > 0.9597$$
$$0.0403 > P(x=0)$$
$$0.0403 > [4!/(4!0!)]p^0(1-p)^4$$
$$0.0403 > (1-p)^4$$
$$(0.0403)^{1/4} > 1-p$$
$$p > 1 - (0.0403)^{1/4}$$
$$p > 1 - 0.448$$
$$p > 0.552$$

37. In Excel, BINOMDIST(825,2000,.5,false) = 0. But using the normal approximation, P(824.5<x<825.5) = P(-7.849<z<-7.804) = NORMSDIST(-7.804) – NORMSDIST(-7.849) gives an answer.

6-7 Assessing Normality

1. A normal quantile plot can be used to determine whether sample data come form a normal distribution. In theory, it compares the z scores for the sample data with the z scores for normally distributed data with the same cumulative relative frequencies as the sample data. In practice, it uses the sample data directly – since converting to z scores is a linear transformation that re-labels the scores but does not change their distribution.

3. The 100 points would approximate a straight line and show no systematic pattern.

5. Not normal. There is a systematic pattern.

7. Normal.

NOTE: This table gives the information for working exercises #9-#12 and #13-#16 by hand. While these exercises can be answered using computer software, seeing the actual procedure and calculations involved promotes a better understanding of the concepts and processes of this section. For each problem,

    x = the original scores arranged in order

    cp = the cumulative probability values 1/2n, 3/2n, 5/2n,…, (2n-1)/2n

    z = the normal z scores that have the cp to their left.

#	9. BMI x	cp	z	10. pennies x	cp	z	11. precipitation x	cp	z	12. temperature x	cp	z
1	19.6	0.013	-2.24	3.0050	.104	-2.19	0	0.014	-2.19	26	0.05	-1.64
2	19.9	0.038	-1.78	3.0054	.043	-1.72	0	0.043	-1.72	33	0.15	-1.04
3	20.7	0.063	-1.53	3.0064	.071	-1.47	0	0.071	-1.47	34	0.25	-0.67
4	21.5	0.088	-1.36	3.0131	.100	-1.28	0	0.100	-1.28	39	0.35	-0.39
5.	21.6	0.113	-1.21	3.0185	.129	-1.13	0	0.129	-1.13	48	0.45	-0.13
6	22.7	0.138	-1.09	3.0290	.157	-1.01	0	0.157	-1.01	58	0.55	0.13
7	23.2	0.163	-0.98	3.0357	.186	-0.89	0	0.186	-0.89	66	0.65	0.39
8	23.3	0.188	-0.89	3.0377	.214	-0.79	0	0.214	-0.79	67	0.75	0.67
9	23.4	0.213	-0.80	3.0408	.243	-0.70	0	0.243	-0.70	71	0.85	1.04
10	23.5	0.238	-0.71	3.0476	.271	-0.61	0	0.271	-0.61	72	0.95	1.64
11	23.8	0.263	-0.64	3.0480	.300	-0.52	0	0.300	-0.52			
12	23.8	0.288	-0.56	3.0561	.329	-0.44	0	0.329	-0.44			
13	24.2	0.313	-0.49	3.0570	.357	-0.37	0	0.357	-0.37			
14	24.5	0.338	-0.42	3.0755	.386	-0.29	0	0.386	-0.29			
15	24.6	0.363	-0.35	3.0765	.414	-0.22	0	0.414	-0.22			
16	24.6	0.388	-0.29	3.0786	.443	-0.14	0	0.443	-0.14			
17	25.2	0.413	-0.22	3.0816	.471	-0.07	0	0.471	-0.07			
18	25.5	0.438	-0.16	3.0862	.500	0	0	0.500	0			
19	25.6	0.463	-0.09	3.0936	.529	0.07	0	0.529	0.07			
20	26.2	0.488	-0.03	3.0965	.557	0.14	0.01	0.557	0.14			
21	26.2	0.513	0.03	3.0976	.586	0.22	0.01	0.586	0.22			
22	26.3	0.539	0.09	3.0994	.614	0.29	0.01	0.614	0.29			
23	26.4	0.563	0.16	3.1029	.643	0.37	0.01	0.643	0.37			
24	26.4	0.588	0.22	3.1031	.671	0.44	0.01	0.671	0.44			
25	26.6	0.613	0.29	3.1038	.700	0.52	0.01	0.700	0.52			
26	26.7	0.638	0.35	3.1083	.729	0.61	0.02	0.729	0.61			
27	26.9	0.663	0.42	3.1114	.757	0.70	0.05	0.757	0.70			
28	27.0	0.688	0.49	3.1141	.786	0.79	0.06	0.786	0.79			
29	27.1	0.713	0.56	3.1267	.814	0.89	0.07	0.814	0.89			
30	27.4	0.738	0.64	3.1366	.843	1.01	0.21	0.843	1.01			
31	27.8	0.763	0.71	3.1461	.871	1.13	0.47	0.871	1.13			
32	28.1	0.788	0.80	3.1524	.900	1.28	0.67	0.900	1.28			
33	28.3	0.813	0.89	3.1535	.929	1.47	1.59	0.929	1.47			
34	28.7	0.838	0.98	3.1692	.957	1.72	1.99	0.957	1.72			
35	30.9	0.863	1.09	3.1934	.986	2.19	2.85	0.986	2.19			
36	31.4	0.888	1.21									
37	31.9	0.913	1.35									
38	32.1	0.938	1.53									
39	33.1	0.963	1.78									
40	33.2	0.988	2.24									

9. Yes. The frequency distribution and
   histogram for the male BMI data indicate
   an approximately normal distribution. The
   frequencies taper off from the modal class
   relatively symmetrically in both directions.

BMI value	frequency
18.75 – 21.25	3
21.25 – 23.75	7
23.75 – 26.25	11
26.25 – 28.75	13
28.75 – 31.25	1
31.25 – 33.75	5
	40

11. No. The frequency distribution and
    histogram for the precipitation data do not
    indicate an approximately normal distribution.
    The scores bunch up at the lower end.

precipitation	frequency
-0.375 – 0.375	30
0.375 – 1.125	2
1.125 – 1.875	1
1.875 – 2.625	1
2.625 – 3.375	1
	35

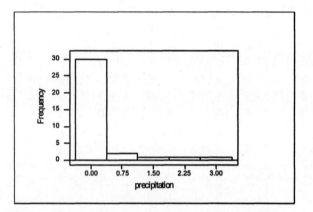

NOTE: The normal quantile plots for exercises #13-#16 may be constructed using the appropriate
columns from the table on the previous page.. *Exercise #16 uses the exercise #12 columns and
most clearly illustrates the process. Each of the 10 scores is 10% of the data set. The first score
covers 0.00 to 0.10, the midpoint of which is 0.05 (the cp value).* Plot the x values on the
horizontal axis and the corresponding normal z values (determined by the cp value) on the vertical
axis. Plotting the z scores "stretches out" the evenly spaced cp values to agree with the cumulative
probabilities of the normal distribution. As with most exercises in this section, the judgment as to
whether the points approximate a straight line is a subjective one.

13. Yes. Since the points lie close to a straight
    line and there is no other obvious pattern,
    conclude that the population distribution is
    approximately normal.

15. No. Since the points do not lie close to a
    straight line, conclude that the population
    distribution is not approximately normal.

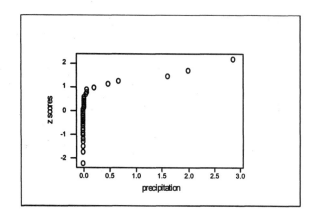

17. The two histograms are shown below. The heights (on the left) appear to be approximately
    normally distributed, while the cholesterol levels (on the right) appear to be positively skewed.
    Many natural phenomena are normally distributed. Height is a natural phenomenon unaffected
    by human input, but cholesterol levels are humanly influenced (by diet, exercise, medication,
    etc.) in ways that might alter any naturally occurring distribution.

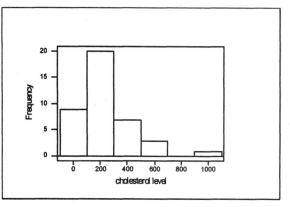

19. The corresponding z scores in the table below were determined as follows.
    (1) Arrange the n scores in order and place them in the x column.
    (2) For each x_i, calculate the cumulative probability using $cp_i = (2i-1)/2n$ for i = 1,2,...,n.
    (3) For each cp_i, find the z_i for which
        $P(z<z_i) = cp_i$ for i = 1,2,...,n.
    The resulting normal quantile plot at the
    right indicates that the data appear to
    come from a population with a normal
    distribution.

i	x	cp	z
1	73	0.10	-1.28
2	78	0.30	-0.52
3	79	0.50	0
4	82	0.70	0.52
5	85	0.90	1.28

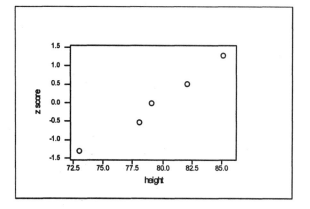

21. No. The z scores from the cumulative probabilities must be used and not the z scores from the
    raw data. The z formula is a linear transformation, $z = (x-\mu)/\sigma = (1/\sigma)x - (\mu/\sigma) = ax + b$. The
    (x,z) pairs all lie exactly on the straight line $z = (1/\sigma)x - (\mu/\sigma)$ regardless of the distribution.

Statistical Literacy and Critical Thinking

1. A normal distribution is one that is symmetric and bell-shaped – more technically, it is one that can be described by the following formula, where μ and σ are specified values,

$$f(x) = \frac{e^{-\frac{1}{2\sigma^2}(x-\mu)^2}}{\sigma\sqrt{2\pi}}.$$

A standard normal distribution is one that has a mean of 0 and a standard deviation of 1 – more technically, it is the distribution that results when $\mu = 0$ and $\sigma = 1$ in the above formula.

2. Regardless of the distribution of grip strengths in the original population, the sample means will be normally distributed around the mean of the grip strengths in the original population.

3. No. The normal distribution formed by his large sample is not necessarily the same normal distribution formed by the population. The normal distribution formed from his friends and relatives could, for example, have a different mean – i.e., it could have the same shape as the population distribution but be centered around a higher or lower value.

4. The Central Limit Theorem tells us that as we take many large independent random samples of size n from any population (i.e., regardless of its shape) with mean μ and standard deviation σ, the resulting sample means will be normally distributed with mean μ and standard deviation $\sigma/\sqrt{n}$.

Review Exercises

1. a. normal distribution
   $\mu = 0$
   $\sigma = 1$
   $P(-0.5 < x < 0.5)$
   $\quad = P(-0.50 < z < 0.50)$
   $\quad = P(x < 0.50) - P(z < -0.50)$
   $\quad = 0.6915 - 0.3085$
   $\quad = 0.3830$

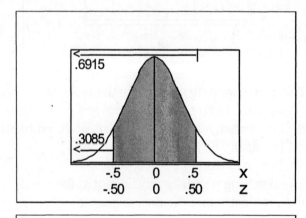

b. normal distribution,
   since the original distribution is so
   $\mu_{\bar{x}} = \mu = 0$
   $\sigma_{\bar{x}} = \sigma/\sqrt{n} = 1/\sqrt{16} = 0.250$
   $P(-0.5 < \bar{x} < 0.5)$
   $\quad = P(-2.00 < z < 2.00)$
   $\quad = P(z < 2.00) - P(z < -2.00)$
   $\quad = 0.9772 - 0.0228$
   $\quad = 0.9544$

c. For P_{90}, A = 0.9000 [0.8997] and z = 1.28
$$x = \mu + z\sigma$$
$$= 0 + (1.28)(1)$$
$$= 0 + 1.28$$
$$= 1.28$$

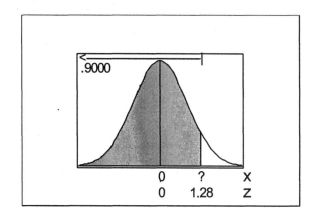

2. a. normal distribution
$$\mu = 69.0$$
$$\sigma = 2.8$$
$$P(x>74.0)$$
$$= 1 - P(x<74.0)$$
$$= 1 - P(z<1.79)$$
$$= 1 - 0.9633$$
$$= 0.0367 \text{ or } 3.67\%$$

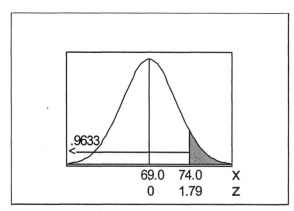

b. normal distribution,
   since the original distribution is so
$$\mu_{\bar{x}} = \mu = 69.0$$
$$\sigma_{\bar{x}} = \sigma/\sqrt{n} = 2.8/\sqrt{4} = 1.4$$
$$P(\bar{x}>74.0)$$
$$= 1 - P(\bar{x}<74.0)$$
$$= 1 - P(z<3.57)$$
$$= 1 - 0.9999$$
$$= 0.0001$$

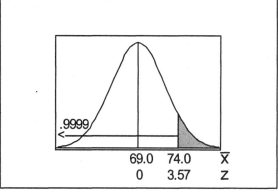

c. For the tallest 10%,
   A = 0.9000 [0.8997] and z = 1.28
$$x = \mu + z\sigma$$
$$= 69.0 + (1.28)(2.8)$$
$$= 69.0 + 3.6$$
$$= 72.6 \text{ in}$$

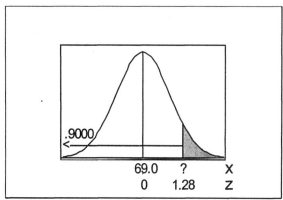

3. a. normal distribution

$\mu = 178.1$

$\sigma = 40.7$

$P(x>260)$

$\quad = 1 - P(x<260)$

$\quad = 1 - P(z<2.01)$

$\quad = 1 - 0.9778$

$\quad = 0.0222$

b. normal distribution

$\mu = 178.1$

$\sigma = 40.7$

$P(170<x<200)$

$\quad = P(-0.20<z<0.54)$

$\quad = P(z<0.54) - P(z<-0.20)$

$\quad = 0.7054 - 0.4207$

$\quad = 0.2847$

c. normal distribution,

since the original distribution is so

$\mu_{\bar{x}} = \mu = 178.1$

$\sigma_{\bar{x}} = \sigma/\sqrt{n} = 40.7/\sqrt{9} = 13.567$

$P(170<\bar{x}<200)$

$\quad = P(-0.60<z<1.61)$

$\quad = P(z<1.61) - P(z<-0.60)$

$\quad = 0.9463 - 0.2743$

$\quad = 0.6720$

d. For the top 3%,

$\quad$ A = 0.9700 [0.9699] and z = 1.88

$x = \mu + z\sigma$

$\quad = 178.1 + (1.88)(40.7)$

$\quad = 178.1 + 76.5$

$\quad = 254.6$

4. Let x = the number incorrect responders who are female.
   binomial problem: n=20 and p=0.5, use the binomial formula $P(x) = \{n!/[(n-x)!x!]\}p^x q^{n-x}$
   $P(x \geq 18) = P(x=18) + P(x=19) + P(x=20)$
   $= [20!/(2!18!)](0.5)^{18}(0.5)^2 + [20!/(1!19!)](0.5)^{19}(0.5)^1 + [20!/(0!20!)](0.5)^{20}(0.5)^0$
   $= [190](0.5)^{18}(0.25) + [20](0.5)^{19}(0.5) + [1](0.5)^{20}(1)$
   $= 0.000181 + 0.000019 + 0.000001 = 0.000201$

   normal approximation appropriate since

   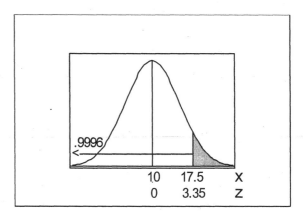

   $np = 20(0.5) = 10 \geq 5$
   $nq = 20(0.5) = 10 \geq 5$
   $\mu = np = 20(0.5) = 10.0$
   $\sigma = \sqrt{np(1-p)} = \sqrt{20(.5)(.5)} = 2.236$
   $P(x \geq 18)$
   $= P_C(x>17.5)$
   $= 1 - P(x<17.5)$
   $= 1 - P(z<3.35)$
   $= 1 - 0.9996$
   $= 0.0004$

Yes. By either the exact binomial or the normal approximation, having 18 or more females among the 20 incorrect responders would be a very unusual event. This is evidence that the question is more often answered correctly by men than by women, and in that sense the question favors men.

5. Let x = the number of females hired.
   binomial problem: n=20 and p=0.3, use the binomial formula $P(x) = \{n!/[(n-x)!x!]\}p^x q^{n-x}$
   $P(x \leq 3) = P(x=0) + P(x=1) + P(x=2) + P(x=3)$
   $= [20!/(20!0!)](0.3)^0(0.7)^{20} + [20!/(19!1!)](0.3)^1(0.7)^{19} + [20!/(18!2!)](0.3)^2(0.7)^{18} + [20!/(17!3!)](0.3)^3(0.7)^{17}$
   $= [1](1)(0.7)^{20} + [20](0.3)(0.7)^{19} + [190](0.09)(0.7)^{18} + [1140](0.027)(0.7)^{17}$
   $= 0.000798 + 0.006839 + 0.027846 + 0.071604 = 0.107087$

   normal approximation appropriate since

   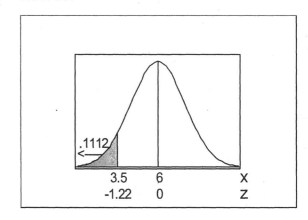

   $np = 20(0.3) = 6 \geq 5$
   $nq = 20(0.7) = 14 \geq 5$
   $\mu = np = 20(0.5) = 10.0$
   $\sigma = \sqrt{np(1-p)} = \sqrt{20(.3)(.7)} = 2.049$
   $P(x \leq 3)$
   $= P_C(x<3.5)$
   $= P(z<-1.22)$
   $= 0.1112$

No. By either the exact binomial or the normal approximation, having 3 or fewer females among the 20 persons hired would not be an unusual event. There is not evidence that the company is discriminating based on gender.

6. Arranged in order the 19 scores are:
   1 4 4 4 10 13 18 19 34 44 45 48 64 67 68 103 124 125 125.
   Two reasonable frequency distributions for the data are given below.

days	frequency		days	frequency
1 – 24	8		1 – 29	8
25 – 49	4		30 – 59	4
50 – 74	3		60 – 89	3
75 – 99	0		90 –129	4
100 –124	2			19
125 –149	2			
	19			

No. The lengths of time do not appear to come from a population with a normal distribution.
It appears that the most frequently occurring values are at the lower end, and not near the
middle of the distribution.

Cumulative Review Exercises

1. The n=8 ordered scores are: 0 1 3 7 10 12 24 27.
   summary statistics: $\Sigma x = 84$ $\Sigma x^2 = 1608$

   a. $\overline{x} = \Sigma x/n = 84/8 = 10.5$ g

   b. $\tilde{x} = (7+10)/2 = 8.5$ g

   c. $s^2 = [n(\Sigma x^2) – (\Sigma x)^2]/[n(n-1)]$
      $= [8(1608) – (84)^2]/[8(7)]$
      $= 5808/56 = 103.714$
      $s = 10.184$, rounded to 10.2 g

   d. $s^2 = 103.71$ g^2 [from part (c)]

   e. $z = (x – \overline{x})/s$
      $z_3 = (3 – 10.5)/10.184 = -0.74$

   f. The actual percentage of scores that exceeds 3 is 5/8 = 62.5%.

   g. Assuming a normal distribution and using
      the sample values for $\overline{x}$ and s to estimate μ
      and σ, the percentage of the population that
      exceeds 3 is
      $P(x>3) = P(z>-0.74)$
      $\quad\quad = 1 – P(z<-0.74)$
      $\quad\quad = 1 – 0.2296$
      $\quad\quad = 0.7704$

   h. Ratio, since differences are meaningful and
      there is a natural zero.

   i. Continuous, since the exact amounts can be
      any value on a continuum.

   j. No. The items were selected for analysis because they represent an interesting variety, not
      because they are proportionately representative of human consumption.

2. a. Let L = a person is left-handed.
   P(L) = 0.10, for each random selection
   $P(L_1 \text{ and } L_2 \text{ and } L_3) = P(L_1) \cdot P(L_2) \cdot P(L_3)$
   $= (0.10)(0.10)(0.10)$
   $= 0.001$

   b. Let N = a person is not left-handed.
   P(N) = 0.90, for each random selection
   P(at least one left-hander) = 1 – P(no left-handers)
   $= 1 - P(N_1 \text{ and } N_2 \text{ and } N_3)$
   $= 1 - P(N_1) \cdot P(N_2) \cdot P(N_3)$
   $= 1 - (0.90)(0.90)(0.90)$
   $= 1 - 0.729$
   $= 0.271$

   c. binomial: n=3 and p=0.10
   normal approximation not appropriate since
   $np = 3(0.10) = 3 < 5$

   d. binomial: n=50 and p=0.10
   $\mu = np = 50(0.10) = 5$

   e. binomial problem: n=50 and p=0.10
   $\sigma = \sqrt{npq} = \sqrt{50(0.10)(0.90)} = 2.121$

   f. There are two previous approaches that may be used to answer this question.
   (1) An unusual score is one that is more than two standard deviations from the mean. Use
   the values for μ and σ from parts (d) and (e).
   $z = (x - \mu)/\sigma$
   $z_8 = (8 - 5)/2.121$
   $= 1.41$
   Since 8 is 1.41<2 standard deviations from the mean, it would not be an unusual result.
   (2) A score is unusual if the probability of getting that result or a more extreme result is less
   than or equal to 0.05.
   binomial: n = 50 and p = 0.1
   normal approximation appropriate since
   $np = 50(0.1) = 5 \geq 5$
   $nq = 50(0.9) = 45 \geq 5$
   Use the values for μ and σ
   from parts (d) and (e).
   $P(x \geq 8) = P_C(x > 7.5)$
   $= 1 - P(x < 7.5)$
   $= 1 - P(z < 1.18)$
   $= 1 - 0.8810$
   $= 0.1190$

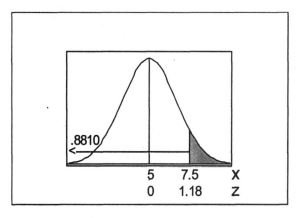

Since 0.1190>0.05, getting 8 left-handers in a group of 50 is not an unusual event.

Chapter 7

Estimates and Sample Sizes

7-2 Estimating a Population Proportion

1. A critical value for a normal distribution is the z score that separates, at some prescribed level, the more extreme (i.e., less likely) high values from the more common (i.e., more likely) ones. The level is indicated by a subscript, and so if $\alpha = 0.10$ and $\alpha/2 = 0.05$, then $z_{\alpha/2} = 1.645$ is the critical value that separates the highest $0.05 = 5\%$ of the scores from the others. Since the standard normal distribution is symmetric around 0, the critical value for the lower tail of the distribution is the negative of the critical value for the upper tail – and so if $\alpha = 0.10$ and $\alpha/2 = 0.05$, then $-z_{\alpha/2} = -1.645$ is the critical value that separates the lowest $0.05 = 5\%$ of the scores from the others.

NOTE: It is also acceptable to use the actual numerical value as the subscript – and so if $\alpha = 0.10$ and $\alpha/2 = 0.05$, then one may indicate that $z_{\alpha/2} = z_{0.05} = 1.645$.

3. By including a statement of the maximum likely error, a confidence interval provides information about the accuracy of an estimate.

5. For 99% confidence, $\alpha = 1 - 0.99 = 0.01$ and $\alpha/2 = 0.01/2 = 0.005$.
   For the upper 0.005, A = 0.9950 and z = 2.575.
   $z_{\alpha/2} = 2.575$

7. For 98% confidence, $\alpha = 1 - 0.98 = 0.02$ and $\alpha/2 = 0.02/2 = 0.01$.
   For the upper 0.01, A = 0.9900 [0.9901] and z = 2.33.
   $z_{\alpha/2} = 2.33$

9. Let L = the lower confidence limit; U = the upper confidence limit.
   $\hat{p} = (L+U)/2 = (0.222+0.444)/2 = 0.666/2 = 0.333$
   $E = (U-L)/2 = (0.444-0.222)/2 = 0.222/2 = 0.111$
   The interval can be expressed as 0.333 ± 0.111.

11. Let L = the lower confidence limit; U = the upper confidence limit.
    $\hat{p} = (L+U)/2 = (0.206+0.286)/2 = 0.492/2 = 0.246$
    $E = (U-L)/2 = (0.286-0.206)/2 = 0.080/2 = 0.040$
    The interval can be expressed as 0.246 ± 0.040.

13. Let L = the lower confidence limit; U = the upper confidence limit.
    $\hat{p} = (L+U)/2 = (0.868+0.890)/2 = 1.758/2 = 0.879$
    $E = (U-L)/2 = (0.890-0.868)/2 = 0.022/2 = 0.011$

15. Let L = the lower confidence limit; U = the upper confidence limit.
    $\hat{p} = (L+U)/2 = (0.607+0.713)/2 = 1.320/2 = 0.660$
    $E = (U-L)/2 = (0.713-0.607)/2 = 0.106/2 = 0.053$

IMPORTANT NOTE: When calculating $E = z_{\alpha/2}\sqrt{\hat{p}\hat{q}/n}$ do not round off in the middle of the problem. This, and the subsequent calculations of $U = \hat{p} + E$ and $L = \hat{p} - E$ may accomplished conveniently on most calculators having a memory as follows.

   (1) Calculate $\hat{p} = x/n$ and STORE the value.

   (2) Calculate E as $1 - $ RECALL $= *$ RECALL $= \div\, n = \sqrt{} * z_{\alpha/2} =$

   (3) With the value for E showing on the display, the upper confidence limit U can be calculated by using $+$ RECALL $=$.

   (4) With the value for U showing on the display, the lower confidence limit L can be calculated by using $-$ RECALL $\pm$ $+$ RECALL.

THE MANUAL USES THIS PROCEDURE, AND ROUNDS THE FINAL ANSWER TO 3 SIGNIFICANT DIGITS, EVEN THOUGH IT REPORTS INTERMEDIATE STEPS WITH A FINITE NUMBER OF DECIMAL PLACES. If the above procedure does not work on your calculator, or to find out if some other procedure would be more efficient on your calculator, ask your instructor for assistance. You must become familiar with your own calculator – and be sure to do your homework on the same calculator you will use for the exams.

17. $\alpha = 0.05$ and $\hat{p} = x/n = 200/500 = 0.40$

   $E = z_{\alpha/2}\sqrt{\hat{p}\hat{q}/n} = 1.96\sqrt{(0.40)(0.60)/500} = 0.0429$

19. $\alpha = 0.02$ and $\hat{p} = x/n = [267]/1068 = 0.25$

   $E = z_{\alpha/2}\sqrt{\hat{p}\hat{q}/n} = 2.33\sqrt{(0.25)(0.75)/1068} = 0.0309$

   NOTE: The value x=[267] was not given. In truth, any $262 \leq x \leq 272$ rounds to the given $\hat{p} = x/1068 = 25\%$. For want of a more precise value, $\hat{p} = 0.25$ is used in the calculation of E.

21. $\alpha = 0.05$ and $\hat{p} = x/n = 200/500 = 0.4000$

      $\hat{p} \pm z_{\alpha/2}\sqrt{\hat{p}\hat{q}/n}$

   $0.4000 \pm 1.96\sqrt{(0.4000)(0.6000)/500}$

   0.4000 ± 0.0429

   $0.357 < p < 0.443$

23. $\alpha = 0.02$ and $\hat{p} = x/n = 267/1068 = 0.2500$

      $\hat{p} \pm z_{\alpha/2}\sqrt{\hat{p}\hat{q}/n}$

   $0.2500 \pm 2.33\sqrt{(0.2500)(0.7500)/1068}$

   0.2500 ± 0.0309

   $0.219 < p < 0.281$

25. $\alpha = 0.05$ and $E = 0.020$; $\hat{p}$ unknown, use $\hat{p} = 0.5$

   $n = [(z_{\alpha/2})^2\hat{p}\hat{q}]/E^2$

     $= [(1.96)^2(0.5)(0.5)]/(0.020)^2 = 2401$

27. $\alpha = 0.05$ and $E = 0.03$; $\hat{p}$ estimated to be 0.27

   $n = [(z_{\alpha/2})^2\hat{p}\hat{q}]/E^2$

     $= [(1.96)^2(0.27)(0.73)]/(0.03)^2 = 841.31$, rounded up to 842

29. Let x = the number of girls born using the method.
    $\alpha = 0.01$ and $\hat{p} = x/n = 295/325 = 0.9077$

$$\hat{p} \pm z_{\alpha/2}\sqrt{\hat{p}\hat{q}/n}$$
$$0.9077 \pm 2.575\sqrt{(0.9077)(0.0922)/325}$$

0.9077 ± 0.0413
0.866 < p < 0.949
Yes. Since 0.5 is not within the confidence interval, and below the interval, we can be 99% certain that the method is effective.

31. Let x = the number of deaths in the week before Thanksgiving.
    $\alpha = 0.05$ and $\hat{p} = x/n = 6062/12000 = 0.5052$

$$\hat{p} \pm z_{\alpha/2}\sqrt{\hat{p}\hat{q}/n}$$
$$0.5052 \pm 1.96\sqrt{(0.5052)(0.4948)/12000}$$

0.5052 ± 0.0089
0.496 < p < 0.514
No. Since 0.5 is within the confidence interval, it is a reasonable possibility for the true population value.

33. Let x = the number with yellow pods. There are 428 + 152 = 580 total trials.
    a. $\alpha = 0.05$ and $\hat{p} = x/n = 152/580 = 0.2621$

$$\hat{p} \pm z_{\alpha/2}\sqrt{\hat{p}\hat{q}/n}$$
$$0.2621 \pm 1.96\sqrt{(0.2621)(0.7379)/580}$$

0.2621 ± 0.0358
0.226 < p < 0.298
    b. No. Since 0.25 is within the confidence interval, it is a reasonable possibility for the true population value. The results do not contradict the theory.

35. Let x = the number that develop those types of cancer.
    a. $\alpha = 0.05$ and $\hat{p} = x/n = 135/420095 = 0.0003214$

$$\hat{p} \pm z_{\alpha/2}\sqrt{\hat{p}\hat{q}/n}$$
$$0.0003214 \pm 1.96\sqrt{(0.0003214)(0.9996786)/420095}$$

0.0003214 ± 0.0000542
0.000267 < p < 0.000376
    b. No. Since 0.0340% = 0.000340 is within the confidence interval, it is a reasonable possibility for the true population value. The results do not provide evidence that cell phone users have a different cancer rate than the general population.

37. NOTE: Since the actual value of x is not given, this problem is limited to the two decimal accuracy reported in 39%. Any $335 \le x \le 343$ rounds to $\hat{p} = x/n = x/870 = 0.39$. All that can be said is that $0.38506 \le \hat{p} \le 0.39425$, and 3 decimal confidence interval accuracy is not possible.
    Let x = the number of Mexican-Americans selected for grand jury duty.
    $\alpha = 0.01$ and $\hat{p} = x/n = x/870 = 0.39$

$\hat{p} \pm z_{\alpha/2}\sqrt{\hat{p}\hat{q}/n}$

$0.39 \pm 2.575\sqrt{(0.39)(0.61)/870}$

0.39 ± 0.04

$0.35 < p < 0.43$

NOTE: using x = 339, $\hat{p}$ = 339/870 = 0.3897

$0.3897 \pm 2.575\sqrt{(0.3897)(0.6103)/870}$

0.3897 ± 0.0426

$0.347 < p < 0.432$

Yes. Since 0.791 is not in the confidence interval, we can be 99% certain that the selection process does not reflect the general population.

39. NOTE: Since the actual value of x is not given, this problem is limited to the two decimal accuracy reported in 94%. Any $3999 \leq x \leq 4040$ rounds to $\hat{p} = x/n = x/4276 = 0.94$. All that can be said is that $0.93522 \leq \hat{p} \leq 0.94480$, and 3 decimal confidence interval accuracy is not possible.

    Let x = the number of households with at least one telephone.

    $\alpha = 0.01$ and $\hat{p} = x/n = x/4276 = 0.94$

    $\hat{p} \pm z_{\alpha/2}\sqrt{\hat{p}\hat{q}/n}$

    $0.94 \pm 2.575\sqrt{(0.94)(0.06)/4276}$

    0.94 ± 0.01

    $0.93 < p < 0.95$

    NOTE: using x = 4019, $\hat{p}$ = 4019/4276 = 0.93990

    $0.93990 \pm 2.575\sqrt{(0.93990)(0.06010)/4276}$

    0.93990 ± 0.00936

    $0.931 < p < 0.949$

    Yes. Since 0.35 is not within the confidence interval, and is below it, we can be 99% certain that the percentage of households with telephones is now greater than the 35% rate of 1920.

41. $\alpha = 0.01$ and E = 0.02; $\hat{p}$ unknown, use $\hat{p} = 0.5$

    $n = [(z_{\alpha/2})^2 \hat{p}\hat{q}]/E^2$

    $= [(2.575)^2 (0.5)(0.5)]/(0.02)^2 = 4144.14$, rounded up to 4145

43. $\alpha = 0.02$ and E = 0.03; $\hat{p}$ unknown, use $\hat{p} = 0.5$

    $n = [(z_{\alpha/2})^2 \hat{p}\hat{q}]/E^2$

    $= [(2.33)^2 (0.5)(0.5)]/(0.03)^2 = 1508.03$, rounded up to 1509

45. $\alpha = 0.05$ and $\hat{p} = x/n = 27/100 = 0.2700$

    $\hat{p} \pm z_{\alpha/2}\sqrt{\hat{p}\hat{q}/n}$

    $0.2700 \pm 1.96\sqrt{(0.2700)(0.7300)/100}$

    0.2700 ± 0.0870

    $0.183 < p < 0.357$

    Yes. Since 0.24 is within the confidence interval, this result is consistent with the reported value.

47. a. There was precipitation on 16 of the 53 Wednesdays.

    $\alpha = 0.05$ and $\hat{p} = x/n = 16/53 = 0.3019$

    $\hat{p} \pm z_{\alpha/2}\sqrt{\hat{p}\hat{q}/n}$

    $0.3019 \pm 1.96\sqrt{(0.3019)(0.6981)/53}$

    0.3019 ± 0.1236

    $0.178 < p < 0.425$

b. There was precipitation on 15 of the 52 Sundays.

$\alpha = 0.05$ and $\hat{p} = x/n = 15/52 = 0.2885$

$$\hat{p} \pm z_{\alpha/2}\sqrt{\hat{p}\hat{q}/n}$$

$0.2885 \pm 1.96\sqrt{(0.2885)(0.7115)/52}$

0.2885 ± 0.1231

$0.165 < p < 0.412$

No. The confidence intervals are similar. Precipitation does not appear to occur more often on either day.

49. $\alpha = 0.05$ and $E = 0.04$; $\hat{p}$ unknown, use $\hat{p} = 0.5$

$$n = \frac{N\hat{p}\hat{q}[z_{\alpha/2}]^2}{\hat{p}\hat{q}[z_{\alpha/2}]^2 + (N-1)E^2}$$

$$= \frac{(1250)(0.5)(0.5)[1.96]^2}{(0.5)(0.5)[1.96]^2 + (1249)(0.04)^2} = \frac{1200.5}{2.9588} = 405.74, \text{ rounded up to } 406$$

51. $\alpha = 0.05$ and $\hat{p} = x/n = 3/8 = 0.3750$

$$\hat{p} \pm z_{\alpha/2}\sqrt{\hat{p}\hat{q}/n}$$

$0.3750 \pm 1.96\sqrt{(0.3750)(0.6250)/8}$

0.3750 ± 0.3355

$0.0395 < p < 0.710$

Yes. The results are "reasonably close" – being shifted down 4.5% from the correct interval $0.085 < p < 0.755$. But depending on the context, such an error could be serious.

53. a. If $\hat{p} = x/n = 0/n = 0$, then

(1) $np \approx 0 < 5$, and the normal approximation to the binomial does not apply.

(2) $E = z_{\alpha/2}\sqrt{\hat{p}\hat{q}/n} = 0$, and there is no meaningful interval.

b. Since $\hat{p} = x/n = 0/20 = 0$, use the upper limit $3/n = 3/20 = 0.15$ to produce the interval $0 \leq p < 0.15$. NOTE: Do not use $0 < p < 0.15$, because the failure to observe any successes in the sample does not rule out $p = 0$ as the true population proportion.

7-3 Estimating a Population Mean: σ Known

1. In repeated sampling, we expect the procedure we used to produce an interval that includes the value of the population mean 95% of the time – in particular, we have 95% confidence that the interval from 2.5 to 6.0 includes the value of the population mean.

3. No. The students at the science fair are a convenience sample that is not necessarily representative of the entire population. From a practical point of view, however, such a sample may be a reasonable staring point for such an investigation for the following reasons:

(1) The students are present for a science fair, and taking such measurements would be a natural part of the day's agenda. Trying to take measurements from a representative random sample of general riders would be awkward and inconvenient.

(2) Since there is concern about pressure on leg restraints, the ride is likely to be an active one not appropriate for young children or older patrons – i.e., the teenagers might be reasonably representative of the typical patrons of that particular ride.

(3) Since the objective is probably to determine the maximum likely pressure on the leg restraints, it would make sense to take measurements from that part of the general population of riders that is strong and healthy and could reasonable be expected to exert larger amounts of pressure on the restraints – and teenagers meet that description.

But while taking measurements from teenagers makes a certain amount of sense, the students attending a science fair may not be representative of the size, strength and general athleticism of the teenage population that patronizes the ride.

5. For 95% confidence, $\alpha = 1-0.95 = 0.05$ and $\alpha/2 = 0.05/2 = 0.025$.
   For the upper 0.025, A = 0.0250 and z = 1.96.
   $z_{\alpha/2} = 1.96$

7. For 92% confidence, $\alpha = 1-0.92 = 0.08$ and $\alpha/2 = 0.08/2 = 0.04$.
   For the upper 0.04, A = 0.9600 [0.9599] and z = 1.75.
   $z_{\alpha/2} = 1.75$

9. $\alpha = 0.05$
   $E = z_{\alpha/2}\sigma/\sqrt{n} = 1.96(15)/\sqrt{100} = 2.94$

11. $\alpha = 0.01$
    $E = z_{\alpha/2}\sigma/\sqrt{n} = 2.575(15)/\sqrt{9} = 12.875$

13. σ is known and n>30. Assuming that the data come from a random sample,
    $\alpha = 0.05$
    $$\bar{x} \pm z_{\alpha/2}\sigma/\sqrt{n}$$
    $67200 \pm 1.96(18277)/\sqrt{41}$
    67200 ± 5595
    $61605 < \mu < 72795$

15. σ is known and n>30. Assuming that the data come from a random sample,
    $\alpha = 0.01$
    $$\bar{x} \pm z_{\alpha/2}\cdot\sigma/\sqrt{n}$$
    $688 \pm 2.575(68)/\sqrt{70}$
    688 ± 21
    $667 < \mu < 709$

17. $\alpha = 0.05$
    $n = [z_{\alpha/2}\cdot\sigma/E]^2$
    $= [(1.96)(2.5)/(0.5)]^2 = 96.04$, rounded up to 97

19. $\alpha = 0.10$
    $n = [z_{\alpha/2}\cdot\sigma/E]^2$
    $= [(1.645)(12)/(1)]^2 = 389.67$, rounded up to 390

21. $\bar{x} = 67.3849$ mph

23. First find the value of E as follows: width = $68.1144 - 66.6554 = 2\cdot E$
    $1.4590 = 2\cdot E$
    $0.7295 = E$
    Then use $\bar{x}$ and E to give the interval as: 67.3849 ± 0.7295 (mph)

25. $\alpha = 0.05$

$\bar{x} \pm z_{\alpha/2} \cdot \sigma / \sqrt{n}$

$4.8 \pm 1.96(2.2)/\sqrt{80}$

4.8 ± 0.5

$4.3 \text{ yrs} < \mu < 5.3 \text{ yrs}$

No. The proportion of the population falling below a certain value is determined by the shape of the distribution. The confidence interval makes a statement about the mean of the distribution, not about its shape. The most we can say is that since 4 falls below the confidence interval for μ, we expect less than half of the values to fall below 4 – and the 39% figure is consistent with that.

27. $\alpha = 0.05$

$\bar{x} \pm z_{\alpha/2} \cdot \sigma / \sqrt{n}$

$58.3 \pm 1.96(9.5)/\sqrt{40}$

58.3 ± 2.9

$55.4 \text{ sec} < \mu < 61.2 \text{ sec}$

Yes. Since the confidence interval includes 60, and since it does not extend very far below and above 60, it is likely that the mean value for the population of all statistics students is reasonably close to 60 seconds.

29. summary statistics: n = 14 $\Sigma x = 1875$ $\bar{x} = 133.93$ $\alpha = 0.05$

$\bar{x} \pm z_{\alpha/2} \cdot \sigma / \sqrt{n}$

$133.93 \pm 1.96(10)/\sqrt{14}$

133.93 ± 5.24

$128.7 \text{ mm} < \mu < 139.2 \text{ mm}$

Ideally, there is a sense in which all the measurements should be the same – and in that case there would be no need for a confidence interval. It is unclear what the given $\sigma = 10$ represents in this situation. Is it the true standard deviation in the values of all people in the population (in which case it would not be appropriate in this context where only a single person is involved)? Is it the true standard deviation in momentary readings on a single person (due to constant biological fluctuations)? Is it the true standard deviation in readings from evaluator to evaluator (when they are supposedly evaluating the same thing)? Using the methods of this section and assuming $\sigma = 10$, the confidence interval would be $128.7 < \mu < 139.2$ as given above even if all the readings were the same.

31. summary statistics: n = 40 $\Sigma x = 225.5719$ $\bar{x} = 5.6392975$ $\alpha = 0.01$

$\bar{x} \pm z_{\alpha/2} \cdot \sigma / \sqrt{n}$

$5.6392975 \pm 2.575(0.068)/\sqrt{40}$

5.6392975 ± 0.0276857

$5.61161 \text{ g} < \mu < 5.66698 \text{ g}$

The proportion of the population falling between certain values is determined by the shape of the distribution. The confidence interval makes a statement about the mean of the distribution, not about its shape. Even though the mean weight is apparently well within the prescribed limits, there is no guarantee that this is true for individual quarters.

33. $\alpha = 0.05$

$n = [z_{\alpha/2} \cdot \sigma / E]^2$

$= [(1.96)(15)/(2)]^2 = 216.09$, rounded up to 217

35. $\alpha = 0.05$

$n = [z_{\alpha/2} \cdot \sigma/E]^2$

$= [(1.96)(10.6)/(0.25)]^2 = 6906.27$, rounded up to 6907

37. Using the range rule of thumb, $\sigma \approx (range)/4 = (70,000-12,000)/4 = 58,000/4 = 14,500$.

$\alpha = 0.05$

$n = [z_{\alpha/2} \cdot \sigma/E]^2$

$= [(1.96)(14500)/(100)]^2 = 80,769.64$, rounded up to 80,770

No; this sample size is not practical. There are only 3 terms in the formula for n: $z_{\alpha/2}$ and E are set by the expectations of the researcher, while σ is determined by the variable under consideration. Lower the confidence level (to make $z_{\alpha/2}$ less than 1.96, thus decreasing n) and/or increase the margin of error (to make E larger than 100, thus decreasing n).

39. Since $n/N = 35/110 = 0.318 > 0.05$, use the finite population correction factor.

$\alpha = 0.05$

$$\bar{x} \pm [z_{\alpha/2}\sigma/\sqrt{n}] \cdot \sqrt{(N-n)/(N-1)}$$

$110 \pm [1.96(15)/\sqrt{35}] \cdot \sqrt{(250-35)/(250-1)}$

$110 \pm [4.9695] \cdot [0.9292]$

110 ± 4.617

$105 < \mu < 115$

7-4 Estimating a Population Mean: σ Not Known

1. According to the point estimate ("average"), the parameter of interest is a population mean. But according to the margin of error ("percentage points"), the parameter of interest is a population proportion. It is possible that the margin of error the paper intended to communicate was 1% of $483 (or $4.83, which in a 95% confidence interval would correspond to a sample standard deviation of $226.57) – but the proper units for the margin of error in a situation like this are "dollars" and not "percentage points."

3. We are 99% confident that the interval from 114.4 mmHg to 123.4 mmHg includes the true mean systolic blood pressure for the population of males.

IMPORTANT NOTE: This manual uses the following conventions.
(1) The designation "df" stands for "degrees of freedom."
(2) Since the t value depends on the degrees of freedom, a subscript may be used to clarify which t distribution is being used. For df =15 and $\alpha/2$ =0.025, for example, one may indicate $t_{15,\alpha/2} = 2.132$. As with the z distribution, it is also acceptable to use the actual numerical value within the subscript and indicate $t_{15,.025} = 2.132$.
(3) Always use the closest entry in Table A-3. When the desired df is exactly halfway between the two nearest tabled values, be conservative and choose the one with the lower df.
(4) As the degrees of freedom increase, the t distribution approaches the standard normal distribution – and the "large" row of the t table actually gives z values. Consequently the z score for certain "popular" α and $\alpha/2$ values may be found by reading Table A-3 "frontwards" instead of Table A-2 "backwards." This is not only easier but also more accurate – since Table A-3 includes one more decimal place. Note the following examples.

For "large" df and $\alpha/2 = 0.05$, $t_{\alpha/2} = 1.645 = z_{\alpha/2}$ (as found in the z table).

For "large" df and $\alpha/2 = 0.01$, $t_{\alpha/2} = 2.326 = z_{\alpha/2}$ (more accurate than the 2.33 in the z table).

This manual uses this technique from this point on. [For df = "large" and $\alpha/2 = 0.005$,

$t_{\alpha/2} = 2.576 \neq 2.575 = z_{\alpha/2}$ (as found in the z table). This is a discrepancy caused by using different mathematical approximation techniques to construct the tables, and not a true difference. This manual will continue to use 2.575.]

5. σ unknown and population approximately normal
   If the sample is a simple random sample, use t.
   $\alpha = 0.05$ and df = 11
   $t_{\alpha/2} = 2.201$

7. σ known and population not approximately normal
   Since n<30, neither the z distribution nor the t distribution applies.

9. σ unknown and population approximately normal

   If the sample is a simple random sample, use t.
   $\alpha = 0.10$ and df = 199 [200]
   $t_{\alpha/2} = 1.653$

11. σ known and population approximately normal
    If the sample is a simple random sample, use z.
    $\alpha = 0.02$
    $z_{\alpha/2} = 2.326$

13. $\alpha = 0.05$ and df = 39 [38]
    a. $E = t_{\alpha/2} \cdot s / \sqrt{n}$
       $= 2.024(4.9) / \sqrt{40}$
       $= 1.6$ kg
    b. $\bar{x} \pm E$
       3.0 ± 1.6
       1.4 kg $< \mu < 4.6$ kg

15. $113.583 < \mu < 122.417$
    We have 95% confidence that the interval from 113.583 to 122.417 contains the true mean IQ of the population of statistics students.

17. σ unknown and n>30, use t
    $\alpha = 0.05$ and df = 185
       $\bar{x} \pm t_{\alpha/2} \cdot s / \sqrt{n}$
    $3103 \pm 1.972(696) / \sqrt{186}$
    3103 ± 101
    $3002 < \mu < 3204$ (grams)
    The confidence interval for the mean birth weight of babies born to mothers who used cocaine is $2608 < \mu < 2792$ (grams) – which is significantly lower than the interval above. Yes; cocaine use appears to be associated with lower birth weights – although the relationship might not be cause and effect due to the cocaine alone, as those who use cocaine may also have other dietary and healthcare issues.

19. a. σ unknown and n>30, use t
       $\alpha = 0.01$ and df = 34
          $\bar{x} \pm t_{\alpha/2} \cdot s / \sqrt{n}$
       $-1.3 \pm 2.728(4.7) / \sqrt{35}$
       -1.3 ± 2.2
       $-3.5 < \mu < 0.9$ (°F)
    b. Yes, the confidence interval includes 0°F. Since the confidence interval includes zero, it is possible that the forecasts are correct on the average. While the meteorologist may want to

do further investigation to see if the pattern of slightly high predictions prevails, the above interval does not provide evidence to support his claim.

21. preliminary values: $n = 6$, $\Sigma x = 9.23$, $\Sigma x^2 = 32.5197$

$\bar{x} = (\Sigma x)/n = (9.23)/6 = 1.538$

$s^2 = [n(\Sigma x^2) - (\Sigma x)^2]/[n(n-1)]$

$ = [6(32.5197) - (9.23)^2]/[6(5)] = 3.664$

$s = 1.914$

σ unknown (and assuming the distribution is approximately normal), use t

$\alpha = 0.05$ and df = 5

$\bar{x} \pm t_{\alpha/2} \cdot s/\sqrt{n}$

$1.538 \pm 2.571(1.914)/\sqrt{6}$

1.538 ± 2.009

$-0.471 < \mu < 3.547$ [which should be adjusted, since negative values are not possible]

$0 < \mu < 3.547$ (micrograms/cubic meter)

Yes. The fact that 5 of the 6 sample values are below $\bar{x}$ raises a question about whether the data meet the requirement that the underlying distribution is normal.

23. preliminary values: $n = 16$, $\Sigma x = 10565$, $\Sigma x^2 = 7114147$

$\bar{x} = (\Sigma x)/n = (10565)/16 = 660.3$

$s^2 = [n(\Sigma x^2) - (\Sigma x)^2]/[n(n-1)]$

$ = [16(7114147) - (10565)^2]/[16(15)] = 9196.3625$

$s = 95.898$

σ unknown (and assuming the distribution is approximately normal), use t

$\alpha = 0.01$ and df = 15

$\bar{x} \pm t_{\alpha/2} \cdot s/\sqrt{n}$

$660.3 \pm 2.947(95.898)/\sqrt{16}$

660.3 ± 70.7

$589.7 < \mu < 731.0$ (FICO units)

No. The proportion of the population falling above a certain value is determined by the shape of the distribution. The confidence interval makes a statement about the mean of the distribution, not about its shape. If the distribution is approximately normal with $\mu = 660.3$ and $\sigma = 95.9$, then $P(x>620) \approx P(z>-0.42) = 1 - 0.3372 = 0.6628$, which is certainly not "almost all."

25. preliminary values: $n = 7$, $\Sigma x = 0.85$, $\Sigma x^2 = 0.1123$

$\bar{x} = (\Sigma x)/n = (0.85)/7 = 0.121$

$s^2 = [n(\Sigma x^2) - (\Sigma x)^2]/[n(n-1)]$

$ = [7(0.1123) - (0.85)^2]/[7(6)] = 0.001514$

$s = 0.0389$

σ unknown (and assuming the distribution is approximately normal), use t

$\alpha = 0.02$ and df = 6

$\bar{x} \pm t_{\alpha/2} \cdot s/\sqrt{n}$

$0.121 \pm 3.143(0.0389)/\sqrt{7}$

$0.121 \pm .046$

$0.075 < \mu < 0.168$ (grams/mile)

No. Since the confidence interval includes values greater than 0.165, there is a reasonable possibility that the true mean emission amount is greater than that.

27. a. preliminary values: n = 40, $\Sigma x = 2776$, $\Sigma x^2 = 197632$

$\bar{x} = (\Sigma x)/n = (2776)/40 = 69.4$

$s^2 = [n(\Sigma x^2) - (\Sigma x)^2]/[n(n-1)] = [40(197632) - (2776)^2]/[40(39)] = 127.631$

$s = 11.297$

σ unknown and n>30, use t

$\alpha = 0.05$ and df = 39

$\bar{x} \pm t_{\alpha/2} \cdot s/\sqrt{n}$

$69.4 \pm 2.024(11.297)/\sqrt{40}$

69.4 ± 3.6

$65.8 < \mu < 73.0$ (beats/min)

b. preliminary values: n = 40, $\Sigma x = 3052$, $\Sigma x^2 = 238960$

$\bar{x} = (\Sigma x)/n = (3052)/40 = 76.3$

$s^2 = [n(\Sigma x^2) - (\Sigma x)^2]/[n(n-1)] = [40(238960) - (3052)^2]/[40(39)] = 156.215$

$s = 12.499$

σ unknown and n>30, use t

$\alpha = 0.05$ and df = 39

$\bar{x} \pm t_{\alpha/2} \cdot s/\sqrt{n}$

$76.3 \pm 2.024(12.499)/\sqrt{40}$

76.3 ± 4.0

$72.3 < \mu < 80.3$ (beats/min)

c. Since the two confidence intervals overlap, we cannot conclude that the two population means are different.

29. preliminary values: n = 29, $\Sigma x = 1777$, $\Sigma x^2 = 346863$

$\bar{x} = (\Sigma x)/n = (1777)/29 = 61.3$

$s^2 = [n(\Sigma x^2) - (\Sigma x)^2]/[n(n-1)] = [29(346863) - (1777)^2]/[29(28)] = 8499.135$

$s = 92.191$

σ unknown (and assuming the distribution is approximately normal), use t

$\alpha = 0.05$ and df = 28

$\bar{x} \pm t_{\alpha/2} \cdot s/\sqrt{n}$

$61.3 \pm 2.048(92.191)/\sqrt{29}$

61.3 ± 35.1

$26.2 < \mu < 96.3$ (years)

This is a considerable change from the $42.6 < \mu < 46.4$ (years) of exercise #22. Confidence intervals can be quite sensitive to extreme outliers. If an outlier is known to be an error and the correct value cannot be determined, it should be discarded. If an outlier may be correct, one could construct confidence intervals with and without the outlier and then make an assessment.

31. <u>assuming a large population</u>

$\alpha = 0.05$ and df = 99

$E = t_{\alpha/2} \cdot s/\sqrt{n}$

$= 1.984(0.0518)/\sqrt{100}$

$= 0.0103$ g

$\bar{x} \pm E$

0.8565 ± 0.0103

$0.8462 < \mu < 0.8668$ (grams)

<u>using the finite population N = 465</u>

$\alpha = 0.05$ and df = 99

$E = [t_{\alpha/2} \cdot s/\sqrt{n}] \times \sqrt{(N-n)/(N-1)}$

$= [1.984(0.0518)/\sqrt{100}] \times \sqrt{365/464}$

$= 0.0091$ g

$\bar{x} \pm E$

0.8565 ± 0.0091

$0.8474 < \mu < 0.8656$ (grams)

The second confidence interval is narrower, reflecting the fact that there are more restrictions and less variability (and more certainty) in the finite population situation when n>.05N.

33. a. In general, one sample value gives no information about the variation of the variable. It is possible, however, that one value plus other considerations can give some insight. If one knows that 0 is a possible value, for example, then one large sample value would indicate a large variance. [For example: If you take a sample of n=1 of the daily snowfall in a US city and find that 10.0 feet of snow fell that day, you would assume that there are days with no snow and that there must be a large variability in the amounts of daily snowfall.]

b. The formula for E requires a value for s and a t score with n-1 degrees of freedom. When n=1, the formula for s fails to produce a value [because there is an (n-1) in the denominator] and there is no df=0 row for the t statistic. No confidence interval can be constructed.

c. $\bar{x} \pm 9.68|x|$

$12.0 \pm 9.68|12.0|$

12.0 ± 116.2

$-104.2 < \mu < 128.2$ [which should be adjusted, since negative heights are not possible]

$0 < \mu < 128.2$ (feet)

Is it likely that some other randomly selected Martian may be 50 feet tall?

No, if "likely" is understood to be "highly probable." The range for individual heights would be even larger than the $0 - 128$ given for the mean. With so many possibilities over such a wide range, 50 (or any other individual value) is not highly probable.

Yes, if "likely" is understood to be "reasonable." Since the confidence interval includes the value 50, it is a reasonable possibility for the mean height of all Martians – and any possible mean height would be a possible individual height.

7-5 Estimating a Population Variance

1. We have 95% confidence that the interval from 2.25 to 3.52 (inches) contains the true value for the standard deviation of the heights of all women.

3. No. The researcher is dealing with the standard deviation of the $\bar{x}$'s, not with the standard deviation of the individual x's. The Central Limit Theorem states that $\sigma_{\bar{x}} = \sigma/\sqrt{n}$, and the standard deviation on which the researcher is placing a confidence interval would be too small by a factor of $\sqrt{n}$.

NOTE: Since the χ^2 value depends on the degrees of freedom, a subscript may be used to clarify which χ^2 distribution is being used. For df =26 and $\alpha/2$ =0.025, for example, one may indicate $\chi^2_{26,R} = 41.923$. Note that $\chi^2_R = \chi^2_{\alpha/2}$ and $\chi^2_L = \chi^2_{1-\alpha/2}$. As with the z and t distributions, one may also use clarifying numerical values in the subscripts. For df =26 and α =0.05, for example, one may indicate $\chi^2_R = \chi^2_{26,.025} = 41.923$ and $\chi^2_L = \chi^2_{26,.975} = 13.844$.

5. $\alpha = 0.05$ and df $= 26$ $\chi^2_L = 13.844$ $\chi^2_R = 41.923$

7. $\alpha = 0.01$ and df $= 40$ $\chi^2_L = 20.707$ $\chi^2_R = 66.766$

9. $\alpha = 0.05$ and df $= 40$

$$(n-1)s^2/\chi^2_R < \sigma^2 < (n-1)s^2/\chi^2_L$$

$$(40)(18277)^2/59.342 < \sigma^2 < (40)(18277)^2/24.433$$

$$225168501 < \sigma^2 < 546881233$$

$$15006 < \sigma < 23385 \ (\$)$$

11. $\alpha = 0.01$ and df $= 69$

$$(n-1)s^2/\chi_R^2 < \sigma^2 < (n-1)s^2/\chi_L^2$$
$$(69)(68)^2/104.215 < \sigma^2 < (69)(68)^2/43.275$$
$$3061.52 < \sigma^2 < 7372.76$$
$$55 < \sigma < 86 \ \text{(FICO units)}$$

13. From the upper right section of Table 7-2, n = 767.

15. From the lower left section of Table 7-2, n = 1401.
   Whether or not this is a practical sample size depends on the nature of the problem – i.e., on how hard it is to select 1401 individuals from the population, and on how difficult and/or expensive it is to get the desired information/measurement from each individual.

17. $\alpha = 0.05$ and df $= 189$

$$(n-1)s^2/\chi_R^2 < \sigma^2 < (n-1)s^2/\chi_L^2$$
$$(189)(645)^2/228.9638 < \sigma^2 < (189)(645)^2/152.8222$$
$$343411 < \sigma^2 < 514511$$
$$586 < \sigma < 717 \ \text{(grams)}$$

No. Since the confidence interval includes 696, it is a reasonable possibility for σ.

19. $\alpha = 0.01$ and df $= 105$

$$(n-1)s^2/\chi_R^2 < \sigma^2 < (n-1)s^2/\chi_L^2$$
$$(105)(0.62)^2/140.169 < \sigma^2 < (105)(0.62)^2/67.328$$
$$0.2880 < \sigma^2 < 0.5995$$
$$0.54 < \sigma < 0.77 \ (\text{°F})$$

Yes. Since the confidence interval does not include 2.10, and is so far below 2.10, we can safely conclude that the population standard deviation is less than 2.10°F.

21. preliminary values: n = 16, $\Sigma x = 10565$, $\Sigma x^2 = 7114147$
   $\bar{x} = (\Sigma x)/n = (19565)/16 = 660.3$
   $s^2 = [n(\Sigma x^2) - (\Sigma x)^2]/[n(n-1)] = [16(7114147) - (10565)^2]/[16(15)] = 9196.3625$
   $s = 95.898$
   $\alpha = 0.01$ and df $= 15$

$$(n-1)s^2/\chi_R^2 < \sigma^2 < (n-1)s^2/\chi_L^2$$
$$(15)(9196.3625)/32.801 < \sigma^2 < (15)(9196.3625)/4.601$$
$$4205.52 < \sigma^2 < 29981.62$$
$$64.9 < \sigma < 173.2 \ \text{(FICO units)}$$

23. preliminary values: n = 6, $\Sigma x = 9.23$, $\Sigma x^2 = 32.5197$
   $\bar{x} = (\Sigma x)/n = (9.23)/6 = 1.538$
   $s^2 = [n(\Sigma x^2) - (\Sigma x)^2]/[n(n-1)] = [6(32.5197) - (9.213)2]/[6(5)] = 3.664$
   $s = 1.914$
   $\alpha = 0.05$ and df $= 5$

$$(n-1)s^2/\chi_R^2 < \sigma^2 < (n-1)s^2/\chi_L^2$$
$$(5)(3.664)/12.833 < \sigma^2 < (5)(3.664)/0.831$$
$$1.4276 < \sigma^2 < 22.0468$$
$$1.195 < \sigma < 4.695 \ \text{(micrograms per cubic meter)}$$

Yes. One of the requirements to use the methods of this section is that the original distribution be approximately normal, and the fact that 5 of the 6 sample values are less than the mean suggests that the original distribution is not normal.

25. a. summary statistics: $n = 40$, $\bar{x} = 25.997$, $s = 3.431$

    $\alpha = 0.01$ and $df = 39$

$$(n-1)s^2/\chi_R^2 < \sigma^2 < (n-1)s^2/\chi_L^2$$
$$(39)(3.431)^2/66.766 < \sigma^2 < (39)(3.431)^2/20.707$$
$$6.876 < \sigma^2 < 22.171$$
$$2.62 < \sigma < 4.71$$

  b. summary statistics: $n = 40$, $\bar{x} = 25.740$, $s = 6.166$

    $\alpha = 0.01$ and $df = 39$

$$(n-1)s^2/\chi_R^2 < \sigma^2 < (n-1)s^2/\chi_L^2$$
$$(39)(6.166)^2/66.766 < \sigma^2 < (39)(6.166)^2/20.707$$
$$22.208 < \sigma^2 < 71.607$$
$$4.71 < \sigma < 8.46$$

  c. The intervals appear not to overlap. It appears that the female BMI values are more variable than the male BMI values.

    NOTE: In some statistical packages these intervals overlap slightly at the 99% level, indicating the standard deviations could be the same and are not significantly different.

27. Applying the given formula yields the following χ_L^2 and χ_R^2 values.

$$\chi^2 = (1/2)[\pm z_{\alpha/2} + \sqrt{2(df) - 1}]^2$$
$$= (1/2)[\pm 1.96 + \sqrt{2(189) - 1}]^2$$
$$= (1/2)[\pm 1.96 + 19.416]^2$$
$$= (1/2)[17.456]^2 \quad \text{and} \quad (1/2)[21.376]^2$$
$$= 152.3645 \quad\quad \text{and} \quad 228.4771$$

These are close to the 152.8222 and 228.9638 given in exercise #17.

Statistical Literacy and Critical Thinking

1. The critical value $z_{\alpha/2}=1.96$ is the z score with $\alpha/2 = 0.025$ of the distribution above it. Since the standard normal distribution of z scores is symmetric around 0, there is also 0.025 of the distribution below a z score of -1.96. And so the portion of the distribution between z scores of -1.96 and +1.96 is $1 - (0.025+0.025) = 1 - 0.05 = 0.95 = 95\%$.

2. The confidence interval gives an indication of the accuracy of the estimate. A wide confidence interval indicates that there is considerable variability in the problem and the point estimate obtained from the sample may not be close to the true value of the parameter in the population. A narrow confidence interval indicates that the point estimate obtained from the sample is likely close to the true value of the parameter in the population.

3. 0.65 ± 0.03 or $0.62 < p < 0.68$

4. We are 95% confident that the interval from 0.62 to 0.68 includes the true population value for the parameter p.

Review Exercises

1. NOTE. Since the actual value of x is not given, this problem is limited to the two decimal accuracy reported in 93%. Any $929 \leq x \leq 938$ rounds to $\hat{p} = x/n = x/1004 = 0.93$. All that can be said is that $0.92530 \leq \hat{p} \leq 0.93426$, and 3 decimal confidence interval accuracy is not possible. Let x = the number of adults who think that service should be refused as indicated.
   $\alpha = 0.05$ and $\hat{p} = x/n = x/1004 = 0.93$

   $\hat{p} \pm z_{\alpha/2} \sqrt{\hat{p}\hat{q}/n}$ NOTE: using x = 934, $\hat{p} = 934/1004 = 0.9303$

   $0.93 \pm 1.96 \sqrt{(0.93)(0.07)/1004}$ $0.9303 \pm 1.96 \sqrt{(0.9303)(0.0697)/1004}$

   $0.93 \pm .02$ 0.9303 ± 0.0158

   $0.91 < p < 0.95$ $0.915 < p < 0.946$

   We are 95% confident that the interval from 0.91 to 0.95 includes the true proportion of all adults who think that service should be refused as indicated.

2. $\alpha = 0.05$ and E = 0.04; $\hat{p}$ estimated to be 0.93
   $n = [(z_{\alpha/2})^2 \hat{p}\hat{q}]/E^2$
      $= [(1.96)^2(0.93)(0.07)]/(0.04)^2 = 156.31$, rounded up to 157

3. summary statistics: n = 16, $\overline{x} = 10.89$, s = 1.56
   σ unknown (and assuming the distribution is approximately normal), use t
   $\alpha = 0.05$ and df = 15

   $\overline{x} \pm t_{\alpha/2} \cdot s/\sqrt{n}$

   $10.89 \pm 2.131(1.56)/\sqrt{16}$

   10.89 ± 0.83

   $10.06 < \mu < 11.72$ (oz.)

   No. Since the confidence interval does not include 12, we can be confident that the population mean is not as required and that the machines are not working properly. In addition, there is far too much variation in the individual amounts. It appears that a customer could easily (since these values appeared in a small sample of size n=16) get as little as 6.8 oz. or as much as 13.2 oz. (which is almost twice as much as 6.8!) – and that amount of variation is unacceptable.

4. summary statistics: n = 16, $\overline{x} = 10.89$, s = 1.56
   $\alpha = 0.05$ and df = 15

   $(n-1)s^2/\chi_R^2 < \sigma^2 < (n-1)s^2/\chi_L^2$

   $(15)(1.56)^2/27.488 < \sigma^2 < (15)(1.56)^2/6.262$

           $1.3280 < \sigma^2 < 5.8294$

              $1.15 < \sigma < 2.41$ (oz.)

   No. Since the confidence interval does not include 0.25, it is not a reasonable possibility for the true population standard deviation. Modifications are required.

5. a. $\alpha = 0.05$ and E = 0.02; $\hat{p}$ unknown, use $\hat{p} = 0.5$
   $n = [(z_{\alpha/2})^2 \hat{p}\hat{q}]/E^2$
      $= [(1.96)^2(0.5)(0.5)]/(0.02)^2 = 2401$

   b. $\alpha = 0.05$ and E = 0.5; assume σ = 8.7
   $n = [z_{\alpha/2} \cdot \sigma/E]^2$
      $= [(1.96)(8.7)/(0.5)]^2 = 1163.08$, rounded up to 1164

c. To meet the requirements in both (a) and (b) simultaneously, choose the larger of the two required sample sizes – i.e., use n = 2401.

6. summary statistics: n = 25, $\bar{x}$ = 7.01, s = 3.74
   a. σ unknown (and assuming the distribution is approximately normal), use t
      α = 0.05 and df = 24

   $$\bar{x} \pm t_{\alpha/2} \cdot s/\sqrt{n}$$
   $$7.01 \pm 2.064(3.74)/\sqrt{25}$$
   $$7.01 \pm 1.54$$
   $$5.47 < \mu < 8.55 \ \text{(years)}$$

   b. α = 0.05 and df = 24

   $$(n-1)s^2/\chi_R^2 < \sigma^2 < (n-1)s^2/\chi_L^2$$
   $$(24)(3.74)^2/39.364 < \sigma^2 < (24)(3.74)^2/12.401$$
   $$8.5282 < \sigma^2 < 27.0706$$
   $$2.92 < \sigma < 5.20 \ \text{(years)}$$

   c. α = 0.01

   $$n = [z_{\alpha/2} \cdot \sigma/E]^2$$
   $$= [(2.575)(3.74)/(0.25)]^2 = 1483.94, \text{ rounded up to } 1484$$

   d. No. Since persons who purchase GM cars do not necessarily have the same expectations and experience as those who purchase other cars, the sample would not be representative of all car owners.

7. a. Let x = the number of college graduates who are smokers.
      α = 0.02 and $\hat{p} = x/n = 144/785 = 0.183$

   $$\hat{p} \pm z_{\alpha/2} \sqrt{\hat{p}\hat{q}/n}$$
   $$0.183 \pm 2.326 \sqrt{(0.183)(0.817)/785}$$
   $$0.183 \pm 0.032$$
   $$0.151 < p < 0.216$$

   b. Yes. Since the confidence interval does not include 0.27, the smoking rate for those with four years of college appears to be different than the rate for the general population.

8. a. summary statistics: n = 20, $\bar{x}$ = 9004, s = 5629
      σ unknown and the distribution is approximately normal, use t
      α = 0.01 and df = 19

   $$\bar{x} \pm t_{\alpha/2} \cdot s/\sqrt{n}$$
   $$9004 \pm 2.861(5629)/\sqrt{20}$$
   $$9004 \pm 3601$$
   $$5403 < \mu < 12605 \ (\$)$$

   b. A conservative (i.e., worst case scenario) estimate for the mean hospital costs for an accident victim wearing seat belts is $12,605 – but the worst case scenario estimate for an individual would be considerably higher.

Cumulative Review Exercises

1. Begin by making a stem-and-leaf plot and calculating summary statistics.

    10|5
    11| $n = 9$
    11|599 $\Sigma x = 1089$
    12|3 $\Sigma x^2 = 132223$
    12|5788

   a. $\bar{x} = (\Sigma x)/n = 1089/9 = 121.0$ lbs

   b. $\tilde{x} = 123$ lbs

   c. $M = 119$ lbs, 128 lbs (bi-modal)

   d. m.r. $= (105 + 128)/2 = 116.5$ lbs

   e. $R = 128 - 105 = 23$ lbs

   f. $s^2 = [n(\Sigma x^2) - (\Sigma x)^2]/[n(n-1)]$
       $= [9(132223) - (1089)^2]/[9(8)]$
       $= 4086/72 = 56.8$ lbs^2

   g. $s = 7.5$ lbs

   h. for $Q_1 = P_{25}$, $L = (25/100)(9) = 2.25$, round up to 3
       $Q_1 = x_3 = 119$ lbs

   i. for $Q_2 = P_{50}$, $L = (50/100)(9) = 4.50$, round up to 5
       $Q_2 = x_5 = 123$ lbs

   j. for $Q_3 = P_{75}$, $L = (75/100)(9) = 6.75$, round up to 7
       $Q_3 = x_7 = 127$ lbs

   k. Ratio, since differences are consistent and there is a meaningful zero.

   l. The boxplot is given below.

weight (lbs)

NOTE: Both the stem-and-leaf plot and the boxplot indicate that the weights do not appear to come from a normal distribution – but from a distribution that is truncated at some upper limit and skewed to the left. This means that the given sample of size $n=9$ should not be used with the methods of this chapter to construct confidence intervals for the mean and the standard deviation. The intervals constructed in parts (m) and (n) are given as a review of the techniques, and not because those techniques apply.

m. σ unknown (and assuming the distribution is approximately normal), use t

$\alpha = 0.01$ and df = 8

$$\bar{x} \pm t_{\alpha/2} \cdot s/\sqrt{n}$$

$121.0 \pm 3.355(7.5)/\sqrt{9}$

121.0 ± 8.4

$112.6 < \mu < 129.4$ (1bs)

n. $\alpha = 0.01$ and df = 8

$$(n-1)s^2/\chi_R^2 < \sigma^2 < (n-1)s^2/\chi_L^2$$

$(8)(56.75)/21.995 < \sigma^2 < (8)(56.75)/1.344$

$20.68 < \sigma^2 < 337.80$

$4.5 < \sigma < 18.4$ (lbs)

o. $\alpha = 0.01$

$n = [z_{\alpha/2} \cdot \sigma/E]^2$

$= [(2.575)(7.5)/2]^2 = 94.07$, rounded up to 95

p. For the general female population, an usually low weight would be one below 85 lbs (i.e., more than 2σ below μ. Individually, none of the supermodels has an unusually low weight. As a group, however, their weights are well below the general population in both mean and standard deviation. Yes; supermodels tend to weigh less than the general female population – and they are a more homogeneous group than the general female population.

2. a. binomial: n=200 and p=0.25

normal approximation is appropriate since

$np = 200(0.25) = 50 \geq 5$

$nq = 200(0.75) = 150 \geq 5$

use $\mu = np = 200(0.25) = 50$

$$\sigma = \sqrt{npq} = \sqrt{200(0.25)(0.75)}$$

$= 6.214$

$P(x \geq 65)$

$= P_C(x > 64.5)$

$= P(z > 2.37)$

$= 1 - P(z < 2.37)$

$= 1 - 0.9911$

$= 0.0089$

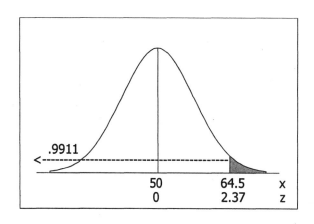

b. $\alpha = 0.05$ and $\hat{p} = x/n = 65/200 = 0.325$

$$\hat{p} \pm z_{\alpha/2}\sqrt{\hat{p}\hat{q}/n}$$

$0.325 \pm 1.96\sqrt{(0.325)(0.675)/200}$

0.325 ± 0.065

$0.260 < p < 0.390$

c. No. The expert's value of 0.25 does not seem correct for two reasons.

From part (a): if the expert is correct, then the probability of getting the sample obtained is very small – less than 1%.

From part (b): there is 95% confidence that the interval from 0.26 to 0.39 includes the true value – i.e., there is 95% confidence that the true value is not 0.25.

3. NOTE: Detailed information about the sampling procedure is not given. It is unclear whether the methods of this text and the given information can answer the specific questions asked in parts (a) and (b). We proceed under the following reasonable scenario. *On one particular day, 1233 persons taking a trip were randomly selected from all persons taking a trip that day. Of those selected, 111 indicated that the particular trip which they were taking involved a visit to a theme park.* A more thorough discussion is given in the EXTENDED NOTE appearing at the conclusion of the exercise.

a. $\hat{p} = x/n = 111/1233 = 0.090$ or 9.0%

b. $\alpha = 0.05$

$$\hat{p} \pm z_{\alpha/2} \sqrt{\hat{p}\hat{q}/n}$$

$0.090 \pm 1.96 \sqrt{(0.090)(0.910)/1233}$

0.090 ± 0.016

$0.074 < p < 0.106$ or $7.4\% < p < 10.6\%$

c. $\alpha = 0.01$ and $E = 0.025$; $\hat{p}$ unknown, use $\hat{p} = 0.5$

$n = [(z_{\alpha/2})^2 \hat{p}\hat{q}]/E^2$

$= [(2.575)^2(0.5)(0.5)]/(0.025)^2 = 2652.25$, rounded up to 2653

EXTENDED NOTE: The goal appears to be to estimate "the percentage of all people who visit a theme park when they take a trip." Even the reasonable scenario suggested above is limited in that the estimate would be valid only for the time of year the people were sampled – and the proportion of trips resulting in visits to theme parks varies according to the season of the year, whether schools are in session, etc.

The wording that "there were 1122 other *respondents*" implies that the data were gathered from people who responded. Since people on trips can't be required to answer surveys, the data may have come only from those who chose to respond – e.g., by people getting pre-stamped cards while they are on a trip and filling them out when they get home. If the sample was not a simple random sample (e.g., a self-selected sample or a convenience sample), then it is not representative of the population and the estimate is not valid.

To see what other difficulties are raised by various interpretations, consider the following population of 5 people and their trip data for the past year.

person	total number of trips	number of trips involving theme parks
A	4	1
B	8	0
C	3	1
D	5	3
E	5	3

One approach is to say that the 5 people took 25 trips, 8 of which involved visits to theme parks – and so $p = 8/25 = 0.32$. But what if D and E are a couple who always travel together? In that case there were really only 20 different trips, 5 of which involved theme parks – and so $p = 5/20 = 0.25$.

In the wording of the exercise for the above population, one could say that: *A survey includes 4 people (A,C,D,E) who took trips that included visits to theme parks, and there was 1 other respondent (B) who took trips without visits to a theme park.* That yields $p = 4/5 = 0.80$.

Were the people surveyed at a particular time of the year and/or with regard to one particular trip, or were they asked about all the trips they had taken last year? This exercise illustrates the importance of clearly defining the goal of the problem and method of sampling.

Chapter 8

Hypothesis Testing

8-2 Basics of Hypothesis Testing

1. In general, a survey that does not examine the entire population is always subject to the possibility of error – i.e., to the possibility that the sample does not provide an exact picture of the population. The only way to definitively "prove" that a value is correct is to examine the entire population. Furthermore, hypothesis testing cannot be used with survey data to conclude that a specific value correct – the proper statement would be that there is not enough evidence to conclude that the 50% figure is wrong.

3. The test statistic is determined by the sample data. The critical value is determined by the set-up of the test of hypothesis – before the sample is even taken. The decision about the null hypothesis is made by comparing the test statistic to the critical value.

5. If the claim were not true, and $p \leq 0.5$, then getting 11 heads in 20 tosses would not be an unusual event. There is not sufficient evidence to support the claim.

7. If the claim were not true, and $\mu \leq 60$, then getting a mean pulse rate of 69.4 in a sample of 400 would be an unusual event. There is sufficient evidence to support the claim..

9. original claim: $p > 0.25$
   H_o: $p = 0.25$
   H_1: $p > 0.25$

11. original claim: $\mu = 121$ lbs
    H_o: $\mu = 121$ lbs
    H_1: $\mu \neq 121$ lbs

13. original claim: $\sigma < 15$
    H_o: $\sigma = 15$
    H_1: $\sigma < 15$

15. original claim: $\mu \geq 0.8535$ g
    H_o: $\mu = 0.8535$ g
    H_1: $\mu < 0.8535$ g

17. Two-tailed test; place $\alpha/2$ in each tail.
    Use $A = 1-\alpha/2 = 1-0.0250 = 0.9750$ and $z = 1.96$.
    critical values: $\pm z_{\alpha/2} = \pm1.96$

19. Right-tailed test; place α in the upper tail.
    Use $A = 1-\alpha = 1-0.0100 = 0.9900$ [closest entry = 0.9901] and $z = 2.33$.
    critical value: $+z_\alpha = +2.33$ [or 2.326 from the "large" row of the t table]

21. Two-tailed test; place $\alpha/2$ in each tail.
    Use $A = 1-\alpha/2 = 1-0.0500 = 0.9500$ and $z = 1.645$.
    critical values: $\pm z_{\alpha/2} = \pm1.645$

23. Left-tailed test; place α in the lower tail.
    Use A = α = 0.0200 [closest entry = 0.0202] and z = -2.05.
    critical value: $-z_\alpha$ = -2.05

25. $\hat{p}$ = x/n = 224/1018 = 0.220

    $z\hat{p} = (\hat{p} - p)/\sqrt{pq/n}$

    $= (0.220 - 0.250)/\sqrt{(0.25)(0.75)/1018}$ = -0.030/0.0136 = -2.21

27. $\hat{p}$ = x/n = 516/580 = 0.890

    $z\hat{p} = (\hat{p} - p)/\sqrt{pq/n}$

    $= (0.890 - 0.750)/\sqrt{(0.75)(0.25)/580}$ = 0.140/0.0180 = 7.77

29. P-value
    = P(z>1.00)
    = 1 − P(z<1.00)
    = 1 − 0.8413
    = 0.1587

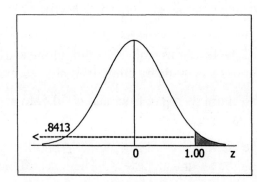

31. P-value
    = 2·P(z>1.96)
    = 2·[1 − P(z<1.96)]
    = 2·[1 − 0.9750]
    = 2·[0.0250]
    = 0.0500

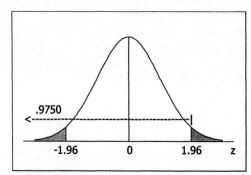

33. P-value
    = P(z>1.50)
    = 1 − P(z<1.50)
    = 1 − 0.9332
    = 0.0668

35. P-value
$$= 2 \cdot P(z < -1.75)$$
$$= 2 \cdot (0.0401)$$
$$= 0.0802$$

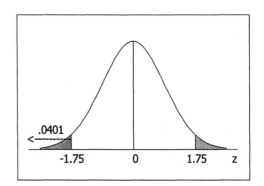

37. NOTE: It is unclear whether "male golfers" is intended to mean "males that are golfers" or "golfers that are male." The final conclusion is stated using the same unclear words given the statement of the original claim.
original claim: $p < 0.5$
competing idea: $p \geq 0.5$
$\quad H_o$: $p = 0.5$
$\quad H_1$: $p < 0.5$
initial conclusion: reject H_o
final conclusion: There is sufficient evidence to support the claim that the proportion of male golfers is less than 0.5.

39. original claim: $p \neq 0.13$
competing idea: $p = 0.13$
$\quad H_o$: $p = 0.13$
$\quad H_1$: $p \neq 0.13$
initial conclusion: fail to reject H_o
final conclusion: There is not sufficient evidence to support the claim that the proportion of M&M's that is red is different from 0.13.

41. type I error: rejecting the hypothesized claim that the proportion of settled malpractice suits is 0.25, when 0.25 is the correct proportion.
type II error: failing to reject the hypothesized claim that the proportion of settled malpractice suits is 0.25, when 0.25 is not the correct proportion.

43. type I error: rejecting the hypothesized claim that the proportion of murders cleared by arrests is 0.62, when 0.62 is the correct proportion.
type II error: failing to reject the hypothesized claim that the proportion of murders cleared by arrests is 0.62, when 0.62 is not the correct proportion.

45. original claim: $p > 0.5$
competing idea: $p \leq 0.5$
$\quad H_o$: $p = 0.5$
$\quad H_1$: $p > 0.5$
$\hat{p} = x/n = x/491 = 0.27$

$z_{\hat{p}} = (\hat{p} - p)/\sqrt{pq/n} = (0.27 - 0.50)/\sqrt{(0.50)(0.50)/491} = -0.23/0.0226 = -10.19$
P-value $= P(z > -10.19) = 1 - P(z < -10.19) = 1 - 0.0001 = 0.9999$
The claim is that $p > .5$. Consider the null and alternative hypotheses given above. Only $\hat{p}$ values so much larger than 0.5 that they are unlikely to occur by chance if $p = 0.5$ is true give statistical support to the claim. A $\hat{p}$ smaller than 0.5 does not give any support to the claim.

47. The test of hypothesis is given below
and illustrated by the figure at the right.
H_o: $p = 0.50$
H_1: $p > 0.50$
$\alpha = 0.05$
C.V. $z = z_\alpha = 1.645$
The c corresponding to $z = 1.645$ is
found by solving $z_c = (c-p)/\sqrt{pq/n}$
for c as follows:

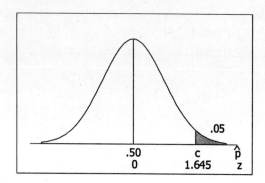

$c = p + z_c \cdot \sqrt{pq/n}$

$\quad = 0.50 + (1.645) \cdot \sqrt{(0.50)(0.50)/64}$

$\quad = 0.50 + (1.645) \cdot (0.0625)$

$\quad = 0.6028$

a. The power calculations are given below
and illustrated by the figure at the right.
power $= P(\text{rejecting } H_o | H_o \text{ is false})$
$\quad = P(\hat{p} > .6028 | p=0.65)$
$\quad = P(z > -0.79)$
$\quad = 1 - P(z < -0.79)$
$\quad = 1 - 0.2148 = 0.7852$
The z corresponding to $c = 0.6028$ is
found as follows:

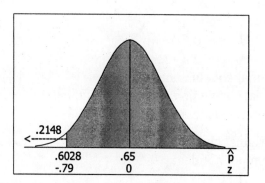

$z_c = (c-p)/\sqrt{pq/n}$

$\quad = (0.6028-0.65)/\sqrt{(0.65)(0.35)/64}$

$\quad = -0.0472/0.0596$

$\quad = -0.79$

b. $\beta = P(\text{type II error}) = P(\text{ not rejecting } H_o | H_o \text{ is false})$
$\qquad\qquad\qquad\quad = 1 - P(\text{rejecting } H_o | H_o \text{ is false})$
$\qquad\qquad\qquad\quad = 1 - 0.7852 = 0.2148$

8-3 Testing a Claim About a Proportion

NOTE: To reinforce the concept that all z scores are standardized rescalings obtained by
subtracting the mean and dividing by the standard deviation, the manual uses the "usual" z formula
written to apply to $\hat{p}$'s

$\quad z\hat{p} = (\hat{p} - \mu\hat{p})/\sigma\hat{p}$.

When the normal approximation to the binomial applies, the $\hat{p}$'s are normally distributed with

$\quad \mu\hat{p} = p$ and $\sigma\hat{p} = \sqrt{pq/n}$.

And so the formula for the z statistic may also be written as $z\hat{p} = (\hat{p} - p)/\sqrt{pq/n}$.

1. The requirements of this section, that $np \geq 5$ and $nq \geq 5$, are precisely the requirements for using
a normal distribution of x's to approximate a binomial distribution of x's. Since dividing each
score by a constant does not change the shape of the distribution, the $\hat{p} = x/n$ values will also
follow a normal distribution whenever those requirements are satisfied.

3. No, for at least two reasons: Internet users are not necessarily representative of the general population, and those who answered constitute a voluntary-response sample.

5. a. $z\hat{p} = (\hat{p} - \mu\hat{p})/\sigma\hat{p} = (\hat{p} - p)/\sqrt{pq/n}$

$$= (0.2494 - 0.25)/\sqrt{(0.25)(0.75)/8023} = -0.0006/0.0048 = -0.12$$

b. $z = \pm z_{\alpha/2} = \pm 1.96$

c. P-value $= 2 \cdot P(z < -0.12) = 2 \cdot (0.4522) = 0.9044$

d. Do not reject H_o; there is not sufficient evidence to conclude that $p \neq 0.25$.

e. No. A hypothesis test will either "reject" or "fail to reject" a claim that a population parameter is equal to a specified value.

7. original claim: $p > 0.29$;

$\hat{p} = x/n = 31{,}969/109{,}857 = 0.2910$

H_o: $p = 0.29$

H_1: $p > 0.29$

$\alpha = 0.05$ [assumed]

C.V. $z = z_\alpha = 1.645$

calculations:

$z\hat{p} = (\hat{p} - \mu\hat{p})/\sigma\hat{p} = 0.73$ [TI-83/84 Plus]

P-value $= 0.2313$ [TI-83/84 Plus]

conclusion:

Do not reject H_o; there is not sufficient evidence to conclude that $p > 0.29$.

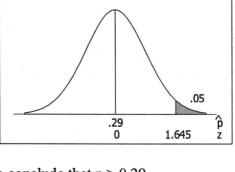

9. original claim: $p > 0.50$

$\hat{p} = x/n = 295/325 = 0.9077$

H_o: $p = 0.50$

H_1: $p > 0.50$

$\alpha = 0.01$

C.V. $z = z_\alpha = 2.326$

calculations:

$z\hat{p} = (\hat{p} - \mu\hat{p})/\sigma\hat{p}$

$$= (0.9077 - 0.50)/\sqrt{(0.50)(0.50)/325}$$

$$= 0.4077/0.0277 = 14.70$$

P-value $= P(z > 14.70) = 1 - P(z < 14.70) = 1 - 0.9999 = 0.0001$

conclusion:

Reject H_o; there is sufficient evidence to conclude that $p > 0.50$.

Yes. The method appears to work.

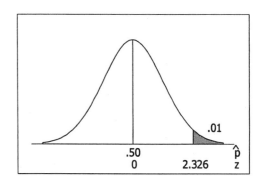

11. original claim: $p > 0.50$

$\hat{p} = x/n = 5720/11000 = 0.5200$

H_o: $p = 0.50$

H_1: $p > 0.50$

$\alpha = 0.01$

C.V. $z = z_\alpha = 2.326$

calculations:

$z\hat{p} = (\hat{p} - \mu\hat{p})/\sigma\hat{p}$

$$= (0.5200 - 0.50)/\sqrt{(0.50)(0.50)/11000}$$

$$= 0.0200/0.0048 = 4.20$$

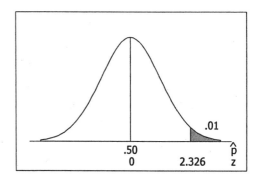

P-value = P(z>4.20) = 1 − P(z<4.20) = 1 − 0.9999 = 0.0001
conclusion:

Reject H_o; there is sufficient evidence to conclude that p > 0.50.

Probably not. The insurance industry is closely monitored, and there would be little room for subjectivity or manipulation in the type of data being analyzed. When those sponsoring a survey have a vested financial interest in the results, however, there is always the possibility that the results might have been presented in a manner as favorable as possible to the sponsor.

13. original claim: p > 0.15

$\hat{p} = x/n = 149/880 = 0.1693$

H_o: p = 0.15
H_1: p > 0.15
$\alpha = 0.05$
C.V. $z = z_\alpha = 1.645$
calculations:

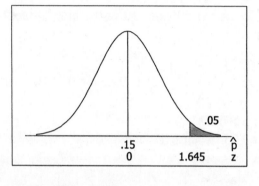

$z\hat{p} = (\hat{p} - \mu\hat{p})/\sigma\hat{p}$

$= (0.1693 − 0.15)/\sqrt{(0.15)(0.85)/880}$

$= 0.0193/0.0120 = 1.60$

P-value = P(z>1.60) = 1 − P(z<1.60) = 1 − 0.9452 = 0.0548
conclusion:

Do not reject H_o; there is not sufficient evidence to conclude that p > 0.15.

No. Technology is changing so rapidly that figures from 1997 are no longer valid today. By the time you read this, e-mail could be so common that essentially 100% of the households use it – or it could have been replaced by something newer so that no one still uses it!

15. original claim: p = 0.000340

$\hat{p} = x/n = 135/420,095 = 0.000321$

H_o: p = 0.000340
H_1: p ≠ 0.000340
$\alpha = 0.005$
C.V. $z = \pm z_{\alpha/2} = \pm 2.81$
calculations:

$z\hat{p} = (\hat{p} - \mu\hat{p})/\sigma\hat{p}$

$= (0.000321 − 0.000340)/\sqrt{\dfrac{(0.000340)(0.999660)}{420,095}}$

$= -0.000019/0.0000284 = -0.66$

P-value = 2·P(z<-0.66) = 2(0.2546) = 0.5092
conclusion:

Do not reject H_o; there is not sufficient evidence to reject the claim that p = 0.000340.

No. Based on these results, cell phone uses have no reason for such concern.

17. original claim: p = 0.01

   $\hat{p} = x/n = 20/1234 = 0.0162$

H_o: p = 0.01
H_1: p ≠ 0.01
α = 0.05
C.V. $z = \pm z_{\alpha/2} = \pm 1.96$
calculations:

$z\hat{p} = (\hat{p} - \mu\hat{p})/\sigma\hat{p}$

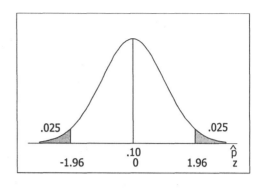

$= (0.0162 - 0.01)/\sqrt{(0.01)(0.99)/1234}$

$= 0.0062/0.00283 = 2.19$

P-value = 2·P(z>2.19) = 2[1 − P(z<2.19)] = 2[1 − 0.9857] = 2[0.0143] = 0.0286
conclusion:

Reject H_o; there is sufficient evidence to reject the claim that p = 0.01 and conclude that
p ≠ 0.01 (in fact, that p > 0.01).
No. Based on these results, consumers appear to be subjected to more overcharges than under
the old pre-scanner system.

19. original claim: p < 1/3

   $\hat{p} = x/n = 312/976 = 0.3197$

H_o: p = 1/3
H_1: p < 1/3
α = 0.05
C.V. $z = -z_\alpha = -1.645$
calculations:

$z\hat{p} = (\hat{p} - \mu\hat{p})/\sigma\hat{p}$

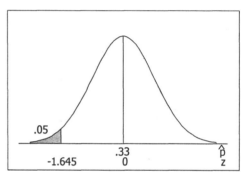

$= (0.3197 - 1/3)/\sqrt{(1/3)(2/3)/976}$

$= -0.0137/0.0151 = -0.91$

P-value = P(z<-0.91) = 0.1814
conclusion:

Do not reject H_o; there is not sufficient evidence to conclude that p < 1/3.
The question is worded appropriately, with nothing to encourage the respondents to answer
untruthfully. It is difficult, however, to trust people to give honest responses – even in matters
like height and weight, and especially regarding private behaviors. Here, a likely misstatement
would be for people who drink "every day" to respond with "a few times a week."

21. NOTE: The value for x is not given. In truth, any 524 ≤ x ≤ 534 rounds to the given
    $\hat{p} = x/1125 = 47\%$. All that can be said is that 0.4658 ≤ $\hat{p}$ ≤ 0.4747. For want of a more precise

    value, $\hat{p} = 0.47$ is used in the calculation of the test statistic $z\hat{p}$.

   original claim: p = 0.50
H_o: p = 0.50
H_1: p ≠ 0.50
α = 0.05
C.V. $z = \pm z_{\alpha/2} = \pm 1.96$
calculations:

$z\hat{p} = (\hat{p} - \mu\hat{p})/\sigma\hat{p}$

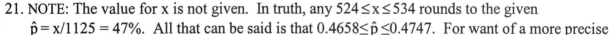

$= (0.47 - 0.50)/\sqrt{(0.50)(0.50)/1125}$

$= -0.03/0.0149 = -2.01$

P-value = $2 \cdot P(z < -2.01) = 2(0.0222) = 0.0444$
conclusion:
   Reject H_0; there is sufficient evidence to reject the claim that $p = 0.50$ and conclude that $p \neq 0.50$ (in fact, that $p < 0.50$).
Yes. It is difficult to trust people to give accurate responses – even in matters like height and weight, and especially regarding private behaviors. In this situation, what it means to fly "never or rarely" is not precisely defined and is open to private interpretation.

23. NOTE: The value for x is not given. In truth, either x=34 or x=35 rounds to the given $\hat{p} = x/1655 = 2.1\%$. All that can be said is that $\hat{p} = 0.02054$ or $\hat{p} = 0.02115$. For want of a more precise value, $\hat{p} = 0.021$ was used in the calculation of the test statistic $z\hat{p}$.
original claim: p > 0.012
H_0: p = 0.012
H_1: p > 0.012
$\alpha = 0.01$
C.V. $z = z_\alpha = 2.326$
calculations:

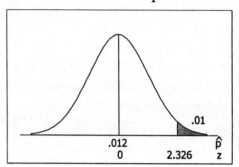

$z\hat{p} = (\hat{p} - \mu\hat{p})/\sigma\hat{p}$
   $= (0.021 - 0.012)/\sqrt{(0.012)(0.988)/1655}$
   $= 0.009/0.00268 = 3.36$
P-value = $P(z > 3.36) = 1 - P(z < 3.36) = 1 - 0.9994 = 0.0006$
conclusion:
   Reject H_0; there is sufficient evidence to conclude that p > 0.012.
Yes. It appears that fatigue is an adverse reaction of Clarinex.

25. There are 100 M&M's listed in Data Set 13, and 27 of them are blue.
original claim: p = 0.24
   $\hat{p} = x/n = 27/100 = 0.27$
H_0: p = 0.24
H_1: p ≠ 0.24
$\alpha = 0.05$ [assumed]
C.V. $z = \pm z_{\alpha/2} = \pm 1.96$
calculations:

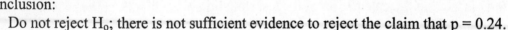

   $z\hat{p} = (\hat{p} - \mu\hat{p})/\sigma\hat{p}$
      $= (0.27 - 0.24)/\sqrt{(0.24)(0.76)/100}$
      $= 0.03/0.0427 = 0.70$
P-value = $2 \cdot P(z > 0.70) = 2[1 - P(z < 0.70)] = 2[1 - 0.7580] = 2[0.2420] = 0.4840$
conclusion:
   Do not reject H_0; there is not sufficient evidence to reject the claim that p = 0.24.

27. There are data for 35 days in Data Set 8. For 9 of those days the difference between the actual high and one day forecasted high is more than 2°, and so $\hat{p} = x/n = 9/35 = 0.2571$.

original claim: p < 0.50

H_0: p = 0.50

H_1: p < 0.50

$\alpha = 0.05$

C.V. $z = -z_\alpha = -1.645$

calculations:

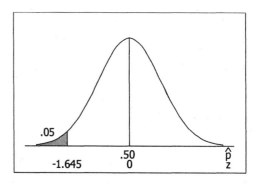

$$z\hat{p} = (\hat{p} - \mu\hat{p})/\sigma\hat{p}$$

$$= (0.2571 - 0.50)/\sqrt{(0.50)(0.50)/35}$$

$$= -0.2429/0.0845 = -2.87$$

P-value = P(z<-2.87) = 0.0021

conclusion:

Reject H_0; there is sufficient evidence to conclude that p < 0.50.

The result suggests that the one-day high forecasts are fairly accurate in that they are off by more than 2° less than 50% of the time.

29. original claim: p = 0.10

$\hat{p} = x/n = 119/1000 = 0.119$

H_0: p = 0.10

H_1: p ≠ 0.10

$\alpha = 0.05$

C.V. $z = \pm z_{\alpha/2} = \pm 1.96$

calculations:

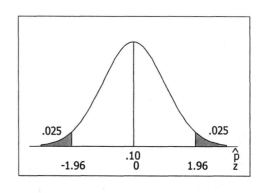

$$z\hat{p} = (\hat{p} - \mu\hat{p})/\sigma\hat{p}$$

$$= (0.119 - 0.10)/\sqrt{(0.10)(0.90)/1000}$$

$$= 0.019/0.00949 = 2.00$$

P-value = 2·P(z>2.00) = 2[1 − P(z<2.00)] = 2[1 − 0.9772] = 2[0.0228] = 0.0456

conclusion:

Reject H_0; there is sufficient evidence to reject the claim that p = 0.10 and conclude that p ≠ 0.10 (in fact, that p > 0.10).

a. As seen above, the traditional method leads to rejection of the claim that p=0.10 because the calculated z=2.00 is greater than the critical value of 1.96.

b. As seen above, the P-value method leads to rejection of the claim that p=0.10 because the calculated P-value=0.0456 is less than the level of significance of 0.05.

c. $\alpha = 0.05$ and $\hat{p} = x/n = 119/1000 = 0.1190$

$$\hat{p} \pm z_{\alpha/2}\sqrt{\hat{p}\hat{q}/n}$$

$$0.1190 \pm 1.96\sqrt{(0.1190)(0.8810)/1000}$$

0.1190 ± 0.0201

0.0989 < p < 0.139

Since 0.10 is inside the confidence interval, p=0.10 is a reasonable claim that should not be rejected.

d. The traditional method and the P-value method are mathematically equivalent and will always agree. As seen by this example, the confidence interval method does not always lead to the same conclusion as the other two methods.

31. a. The claim p>0.0025 results in the null hypothesis H_o: p = 0.0025.
   Since np = (80)(0.0025) = 0.2 < 5, the normal approximation to the binomial is not
   appropriate and the methods of this section cannot be used.

   b. binomial: n = 80 and p = 0.0025

   $P(x) = [n!/(n-x)!x!]p^x q^{n-x}$

   $P(x=0) = [80!/80!0!](0.0025)^0(0.9975)^{80} = 0.818525754280$
   $P(x=1) = [80!/79!1!](0.0025)^1(0.9975)^{79} = 0.164115439455$
   $P(x=2) = [80!/78!2!](0.0025)^2(0.9975)^{78} = 0.016247017189$
   $P(x=3) = [80!/77!3!](0.0025)^3(0.9975)^{77} = 0.001058702874$
   $P(x=4) = [80!/76!4!](0.0025)^4(0.9975)^{76} = 0.000051077770$
   $P(x=5) = [80!/75!5!](0.0025)^5(0.9975)^{75} = 0.000001945820$
   $P(x=6) = [80!/74!6!](0.0025)^6(0.9975)^{74} = \underline{0.000000060959}$

   0.999999998347

   $P(x \geq 7) = 1 - P(x \leq 6) = 1 - 0.999999998347 = 0.000000001653$

   c. Based on the result from part (b), getting 7 or more males with this color blindness if
   p = 0.0025 would be an extremely rare event. This leads one to reject the claim that
   p = 0.0025. This is essentially the P-value approach to hypothesis testing for

   H_o: p = 0.0025
   H_1: p > 0.0025
   $\alpha = 0.01$
   P-value = P(getting a result as extreme as or more extreme than x=7)
       $= P(x \geq 7)$
       = 0.000000001653

   NOTE: This problem may also be worked using the classical approach to hypothesis testing.
   Using the binomial probabilities in part (b),

   P(x > 0) = 1 - [P(x=0)]   = 1 - [0.818526]   = 0.181474 > 0.01
   P(x > 1) = 1 - [P(x=0)+P(x=1)]   = 1 - [0.818526+0.164115] = 0.017359 > 0.01
   P(x > 2) = 1 - [P(x=0)+P(x=1)+P(x=2)] = 1 - [0.81526+0.164115+0.016247] = 0.001112 ≤ 0.01

   To place 0.01 (or as close to it as possible, without going over) in the upper tail, the C.V.
   must be x > 2, and the classical test [accompanied by a non-appropriate normal-looking
   figure for illustrative purposes only] is as follows.

   H_o: p = 0.0025
   H_1: p > 0.0025
   $\alpha = 0.01$
   C.V. x = 2
   calculations:
       x = 7

   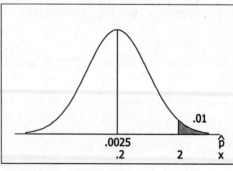

   conclusion:
       Reject H_o; there is sufficient evidence to conclude p > 0.0025.

33. The test of hypothesis is given below
and illustrated by the figure at the right.
H_o: $p = 0.40$
H_1: $p < 0.40$
$\alpha = 0.05$ [assumed]
C.V. $z = -z_\alpha = -1.645$
The c corresponding to $z = -1.645$ is
found by solving $z_c = (c-p)/\sqrt{pq/n}$
for c as follows:

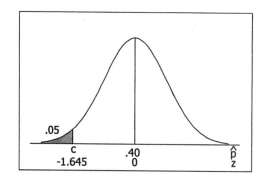

$$c = p + z_c \cdot \sqrt{pq/n}$$
$$= 0.40 + (-1.645)\sqrt{(.40)(.60)/50}$$
$$= 0.40 - (1.645)(0.0693)$$
$$= 0.2860$$

a. The power calculations are given below
and illustrated by the figure at the right.
power = P(rejecting H_o|H_o is false)
$$= P(\hat{p} < 0.2860 | p=0.25)$$
$$= P(z<0.59) = 0.7224$$
The z corresponding to c= 0.2860 is
found as follows:

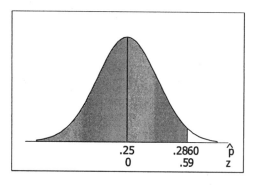

$$z_c = (c-p)/\sqrt{pq/n} = (0.2860-0.25)/\sqrt{(.25)(.75)/50} = 0.0360/0.0612 = 0.59$$

b. β = P(type II error) = P(not rejecting H_o|H_o is false)
$$= 1 - P(\text{rejecting } H_o | H_o \text{ is false})$$
$$= 1 - 0.7224 = 0.2776$$

c. No test of hypothesis can lead the researcher to the right conclusion 100% of the time. If H_o
is true, there is a 95% chance H_o will (correctly) not be rejected; if H_o is false and p=0.25 is
true, there is a 72% chance that H_o will (correctly) be rejected. These are reasonable
probabilities of success, and the test is reasonably effective.

8-4 Testing a Claim About a Mean: σ Known

1. No, a sample of size n>30 is not necessary to use the methods of this section. If the x's are
normally distributed, the $\bar{x}$'s are normally distributed with $\sigma_{\bar{x}} = \sigma/\sqrt{n}$ for any value of n. If
the x's are not normally distributed, then it must be true that n>30. In either case, the
underlying assumption is that the data represent a simple random sample.

3. A usual two-sided confidence interval at the 1-α level places α/2 in each tail of the distribution.
To correspond with a one-tailed test at the 0.01 level, a two-sided confidence interval should
be constructed using α = 0.02 – i.e., using the 98% level of confidence.

5. Yes. The methods of this section apply to this situation..

7. No. Since σ is unknown, the methods of this section do not apply.

9. a. $z_{\bar{x}} = (\bar{x} - \mu_{\bar{x}})/\sigma_{\bar{x}}$

   $= (\bar{x} - \mu)/(\sigma/\sqrt{n}) = (0.8560 - 0.8535)/(0.0565/\sqrt{27}) = 0.0025/0.0109 = 0.23$

   b. $z = \pm z_{\alpha/2} = \pm 1.96$

   c. P-value $= 2 \cdot P(z > .23) = 2[1 - P(z < .23)] = 2[1 - 0.5910] = 2[0.4090] = 0.8180$

   d. Fail to reject H_o.

   e. There is not sufficient evidence to reject the claim that the true mean weight of blue M&M's is 0.8535 grams.

11. original claim: $\mu > 30.0 \,°C$

    $H_o: \mu = 30.0 \,°C$

    $H_1: \mu > 30.0 \,°C$

    $\alpha = 0.05$

    C.V. $z > z_{.05} = 1.645$

    calculations:

    $z_{\bar{x}} = (\bar{x} - \mu_{\bar{x}})/\sigma_{\bar{x}} = 1.732$ [Excel]

    P-value $= 0.0416$ [Excel]

    conclusion:

    Reject H_o; there is sufficient evidence to conclude that $\mu > 30.0$.

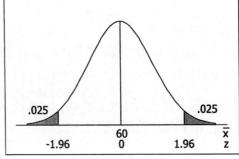

13. original claim: $\mu = 60 \, sec$

    $H_o: \mu = 60 \, sec$

    $H_1: \mu \neq 60 \, sec$

    $\alpha = 0.05$

    C.V. $z = \pm z_{\alpha/2} = \pm 1.96$

    calculations:

    $z_{\bar{x}} = (\bar{x} - \mu_{\bar{x}})/\sigma_{\bar{x}}$

    $= (58.3 - 60)/(9.5/\sqrt{40})$

    $= -1.7/1.502 = -1.13$

    P-value $= 2 \cdot P(z < -1.13) = 2(0.1292) = 0.2584$

    conclusion:

    Do not reject H_o; there is not sufficient evidence to reject the claim that $\mu = 60$.

    Yes. Based on this result, the overall perception of one minute appears to be reasonably accurate – although individual results might vary considerably.

15. original claim: $\mu < 0 \, lbs$

    $H_o: \mu = 0 \, lbs$

    $H_1: \mu < 0 \, lbs$

    $\alpha = 0.05$

    C.V. $z = -z_\alpha = -1.645$

    calculations:

    $z_{\bar{x}} = (\bar{x} - \mu_{\bar{x}})/\sigma_{\bar{x}}$

    $= (-2.1 - 0)/(4.8/\sqrt{40})$

    $= -2.1/0.7589 = -2.77$

    P-value $= P(z < -2.77) = 0.0028$

    conclusion:

    Reject H_o; there is sufficient evidence to conclude that $\mu < 0$.

    Yes. Based on this result, the diet is effective. If the goal of the diet is to lose weight, the mean weight loss of only 2.1 pounds after one full year is not worth the effort. If the goal of the diet is to not add weight and to be a generally healthier person, then it is worth the effort.

17. summary statistics: $n = 14$ $\Sigma x = 1875$ $\bar{x} = 133.93$
    original claim: $\mu < 140$ mmHg
    H_o: $\mu = 140$ mmHg
    H_1: $\mu < 140$ mmHg
    $\alpha = 0.05$
    C.V. $z = -z_\alpha = -1.645$
    calculations:

    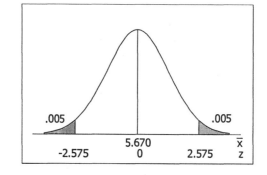

$$z_{\bar{x}} = (\bar{x} - \mu_{\bar{x}})/\sigma_{\bar{x}}$$
$$= (133.93 - 140)/(10/\sqrt{14})$$
$$= -6.07/2.673 = -2.27$$
$$\text{P-value} = P(z < -2.27) = 0.0116$$

conclusion:
    Reject H_o; there is sufficient evidence to conclude that $\mu < 140$.
    Yes. It can be concluded with 95% confidence that the person does not have hypertension.

19. summary statistics: $n = 40$ $\Sigma x = 225.5719$ $\bar{x} = 5.6393$
    original claim: $\mu = 5.670$
    H_o: $\mu = 5.670$
    H_1: $\mu \neq 5.670$
    $\alpha = 0.01$
    C.V. $z = \pm z_{\alpha/2} = \pm 2.575$
    calculations:

    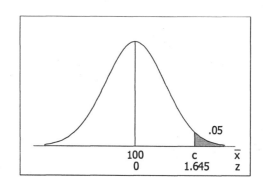

$$z_{\bar{x}} = (\bar{x} - \mu_{\bar{x}})/\sigma_{\bar{x}}$$
$$= (5.6393 - 5.670)/(0.068/\sqrt{40})$$
$$= -0.0307/0.01075 = -2.86$$
$$\text{P-value} = 2 \cdot P(z < -2.86) = 2(0.0021) = 0.0042$$

conclusion:
    Reject H_o; there is sufficient evidence to reject the claim that $\mu = 5.670$ and conclude that $\mu \neq 5.670$ (in fact, that $\mu < 5.670$). If the quarters were in uncirculated condition, they appear not to have been manufactured according to specifications – but if they were worn from use, all that can be said is that their mean original weight was at least 5.6393 grams.

21. The test of hypothesis is given below
    and illustrated by the figure at the right.
    H_o: $\mu = 100$
    H_1: $\mu > 100$
    $\alpha = 0.05$ [assumed]
    C.V. $z = z_\alpha = 1.645$
    The c corresponding to $z = 1.645$ is
    found by solving $z_c = (c - \mu)/(\sigma/\sqrt{n})$
    for c as follows.

$$c = \mu + z_c \cdot (\sigma/\sqrt{n})$$
$$= 100 + 1.645(15/\sqrt{40})$$
$$= 100 + 3.90$$
$$= 103.90$$

a. The power calculations are given below
   and illustrated by the figure at the right.

   power = P(rejecting $H_o | H_o$ is false)
      $= P(\overline{x} > 103.90 | \mu = 108)$
      $= P(z > -1.73)$
      $= 1 - P(z < -1.73)$
      $= 1 - 0.0418$
      $= 0.9582$

   The z corresponding to c= 103.90 is

   $z_c = (c - \mu)/(\sigma/\sqrt{n})$

   $= (103.90 - 108)/(15/\sqrt{40}) = -4.099/2.372 = -1.73$

The test is very effective in recognizing that the mean is greater than 100 when the mean is
actually equal to 108, for it will do so 95.8% of the time.

b. $\beta = P(\text{type II error}) = P(\text{not rejecting } H_o | H_o \text{ is false})$
      $= 1 - P(\text{rejecting } H_o | H_o \text{ is false})$
      $= 1 - 0.9582 = 0.0418$

8-5 Testing a Claim About a Mean: σ Not Known

1. When using $\overline{x}$ and s to make inferences about μ, the degrees of freedom is df = n-1. If a
   sample consists of n = 5 values, then df = n-1 = 4. Given $\overline{x} = (\Sigma x)/n$, one can make up n-1 of
   the x values – but the final x value has to be whatever must be added to those n-1 values to
   produce $\Sigma x = n \cdot \overline{x}$. And so in this instance, 4 of the 5 values can be made up before the last
   one is determined by the restriction that the mean is 20.0.

3. The claim that μ is less than 12 is supported by the sample data only when $\overline{x}$ is significantly
   less than 12. The sample data definitely do not support the claim that μ is less than 12 when $\overline{x}$
   is greater than or equal to 12.

5. Use t. When σ is unknown and the x's are normally distributed, use t.

7. Neither the z nor the t applies. When σ is unknown and the x's are not normally distributed,
   sample sizes n≤30 cannot be used with the techniques in this chapter.

NOTE: Exercises #9-#12 may be worked as follows.
   table: find the correct df row in Table A-3, and see what values surround the given t.
   TI-83+: use tcdf(lower bound, upper bound, df) to get the probability between 2 values
For the remainder of this manual, exact P-values for the t distribution USING THE UNROUNDED
VALUES of t will be given in TI-83+ calculator format as described above and used in exercises
#9-#12. While the input and/or output formats may be different for other technologies
[e.g., Minitab, Excel, Statdisk, etc], the final answers should be the same.

9. P-value = $P(t_6 > 3.500)$
      table for area in one tail: [3.707 > 3.500 > 3.143] 0.005 < P- value < 0.01
      TI-83+: tcdf(3.500,99,6) = 0.00641

11. P-value = $2 \cdot P(t_{20} > 9.883)$
      table for area in two tails: [9.833 > 2.845] P-value < 0.01
      TI-83+: $2 \cdot$tcdf(9.883,99,20) = $2 \cdot (1.93 \times 10^{-9}) = 3.85 \times 10^{-9} = 0.00000000385$

13. original claim: $\mu > 120$

    H_o: $\mu = 120$

    H_1: $\mu > 120$

    $\alpha = 0.05$ and df = 11

    C.V. $t = t_\alpha = 1.796$

    calculations:

        $t_{\bar{x}} = (\bar{x} - \mu)/s_{\bar{x}}$

          $= (132 - 120)/(12/\sqrt{12})$

          $= 12/3.464 = 3.464$

    P-value = $P(t_{11} > 3.464) = \text{tcdf}(3.464,99,11) = 0.0026$

    conclusion:

        Reject H_o; there is sufficient evidence to conclude that $\mu > 120$.

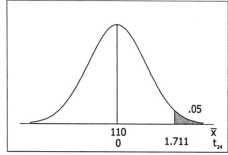

15. original claim: $\mu > 110$

    H_o: $\mu = 110$

    H_1: $\mu > 110$

    $\alpha = 0.05$ and df = 24

    C.V. $t = t_\alpha = 1.711$

    calculations:

        $t_{\bar{x}} = (\bar{x} - \mu)/s_{\bar{x}}$

          $= 3.74$ [Minitab]

        P-value = $P(t_{24} > 3.74) = 0.001$ [Minitab]

    conclusion:

        Reject H_o; there is sufficient evidence to conclude that $\mu > 110$.

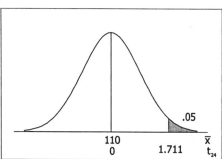

17. original claim: $\mu < 98.6$ °F

    H_o: $\mu = 98.6$ °F

    H_1: $\mu < 98.6$ °F

    $\alpha = 0.05$ and df = 105

    C.V. $t = -t_\alpha = -1.660$

    calculations:

        $t_{\bar{x}} = (\bar{x} - \mu)/s_{\bar{x}}$

          $= (98.20 - 98.6)/(0.62/\sqrt{106})$

          $= -0.4/0.0602 = -6.642$

    P-value = $P(t_{105} < -6.642) = \text{tcdf}(-99,-6.642,105) = 0.000000000702$

    conclusion:

        Reject H_o; there is sufficient evidence to conclude that $\mu < 98.6$.

    Yes. It appears that the commonly-used mean of 98.6°F is not correct.

19. original claim: $\mu < 3103$ g
    H_o: $\mu = 3103$ g
    H_1: $\mu < 3103$ g
    $\alpha = 0.01$ and df = 189
    C.V. $t = -t_\alpha = -2.345$
    calculations:

$$t_{\overline{x}} = (\overline{x} - \mu)/s_{\overline{x}}$$
$$= (2700 - 3103)/(645/\sqrt{190})$$
$$= -403/46.793 = -8.612$$

    P-value = $P(t_{189} < -8.612) = tcdf(-99,-8.612,189) = 0$
conclusion:
    Reject H_o; there is sufficient evidence to conclude that $\mu < 3103$.
Not necessarily. There is sufficient evidence to conclude that weights of babies born to
cocaine uses are less than weights of babies born to non-users, but the association is not
necessarily one of cause and effect. The birth weights might not be affected by the cocaine *per
se*, but by other factors associated with cocaine users – e.g., poor diet, poor pre-natal health
care, etc.

21. original claim: $\mu > 0$
    H_o: $\mu = 0$
    H_1: $\mu > 0$
    $\alpha = 0.01$ and df = 20
    C.V. $t = t_\alpha = 2.528$
    calculations:

$$t_{\overline{x}} = (\overline{x} - \mu)/s_{\overline{x}}$$
$$= (4.00 - 0)/(2.17/\sqrt{21})$$
$$= 4.00/0.4735 = 8.447$$

    P-value = $P(t_{20} > 8.447) = tcdf(8.447,99,20) = 0.0000000249$
conclusion:
    Reject H_o; there is sufficient evidence to conclude that $\mu > 0$.
Yes. The treatment appears to be effective.

23. original claim: $\mu > 63.6$ in
    H_o: $\mu = 63.6$ in
    H_1: $\mu > 63.6$ in
    $\alpha = 0.01$ and df = 8
    C.V. $t = t_\alpha = 2.896$
    calculations:

$$t_{\overline{x}} = (\overline{x} - \mu)/s_{\overline{x}}$$
$$= (70.2 - 63.6)/(1.5/\sqrt{9})$$
$$= 6.6/0.5 = 13.200$$

    P-value = $P(t_8 > 13.200) = tcdf(13.200,99,8) = 0.000000517$
conclusion:
    Reject H_o; there is sufficient evidence to conclude that $\mu > 63.6$.
Yes. Even though the sample was only n = 9, the conclusion is valid. The sample size is built
into the critical region – the smaller the sample size, the larger t must be in order to reject H_o.

25. summary statistics: n = 6, $\Sigma x = 9.23$, $\Sigma x^2 = 32.5197$, $\bar{x} = 1.538$, s = 1.914
    original claim: $\mu > 1.5$ µg/m^3
    H_0: $\mu = 1.5$ µg/m^3
    H_1: $\mu > 1.5$ µg/m^3
    $\alpha = 0.05$ and df = 5
    C.V. $t = t_\alpha = 2.015$
    calculations:

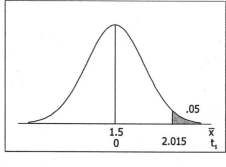

$$t_{\bar{x}} = (\bar{x} - \mu)/s_{\bar{x}}$$
$$= (1.538 - 1.5)/(1.914/\sqrt{6})$$
$$= 0.0383/0.78147 = 0.049$$

    P-value = $P(t_5 > 0.049)$ = tcdf(0.049,99,5) = 0.4814
    conclusion:
       Do not reject H_0; there is not sufficient evidence to conclude that $\mu > 1.5$.
    Yes. Since 5 of the 6 sample values are below the sample mean, there is reason to doubt that
    the population values are normally distributed.

27. summary statistics: n = 15, $\Sigma x = 25.7$, $\Sigma x^2 = 44.97$, $\bar{x} = 1.713$, s = 0.2588
    original claim: $\mu = 1.8$ g
    H_0: $\mu = 1.8$ g
    H_1: $\mu \neq 1.8$ g
    $\alpha = 0.05$ [assumed] and df = 14
    C.V. $t = \pm t_{\alpha/2} = \pm 2.145$
    calculations:

$$t_{\bar{x}} = (\bar{x} - \mu)/s_{\bar{x}}$$
$$= (1.713 - 1.8)/(0.2588/\sqrt{15})$$
$$= -0.087/0.0668 = -1.297$$

    P-value = $2 \cdot P(t_{14} < -1.297)$ = 2·tcdf(-99,-1.297,14) = 2(0.1078) = 0.2155
    conclusion:
       Do not reject H_0; there is not sufficient evidence to reject the claim that $\mu = 1.8$.

29. summary statistics from statistical software: n = 40, $\bar{x} = 5.63930$, s = 0.06194
    original claim: $\mu = 5.670$
    H_0: $\mu = 5.670$
    H_1: $\mu \neq 5.670$
    $\alpha = 0.05$ [assumed] and df = 39
    C.V. $t = \pm t_{\alpha/2} = \pm 2.024$
    calculations:

$$t_{\bar{x}} = (\bar{x} - \mu)/s_{\bar{x}}$$
$$= (5.63930 - 5.670)/(0.06194/\sqrt{40})$$
$$= -0.03070/0.00979 = -3.135$$

    P-value = $2 \cdot P(t_{39} < -3.135)$ = 2·tcdf(-99,-3.135,39) = 2(0.0016) = 0.0033
    conclusion:
       Reject H_0; there is sufficient evidence to reject the claim that $\mu = 5.670$ and conclude that
       $\mu \neq 5.670$ (in fact, that $\mu < 5.670$).
    NOTE: This does not mean that the quarters are not manufactured according to specifications,
    but only that they weigh less than the specified manufacturing weight when they are sampled.

31. summary statistics from statistical software: $n = 40$, $\bar{x} = 69.4000$, $s = 11.2974$
    original claim: $\mu > 60$ beats/min
    H_o: $\mu = 60$ beats/min
    H_1: $\mu > 60$ beats/min
    $\alpha = 0.05$ [assumed] and df = 39
    C.V. $t = t_\alpha = 1.686$
    calculations:

    $$t_{\bar{x}} = (\bar{x} - \mu)/s_{\bar{x}}$$

    $$= (69.40 - 60)/(11.2974/\sqrt{40})$$

    $$= 9.40/1.7863 = 5.262$$

    P-value $= P(t_{39} > 5.262) = tcdf(5.262, 99, 39) = 0.00000272$
    conclusion:

    Reject H_o; there is sufficient evidence to conclude that $\mu > 60$.

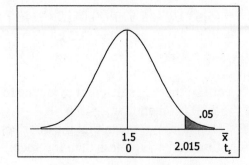

33. The two methods may be compared as follows.

method of this section	alternative method
H_o: $\mu = 100$	H_o: $\mu = 100$
H_1: $\mu \neq 100$	H_1: $\mu \neq 100$
$\alpha = 0.05$ and df = 31	$\alpha = 0.05$
C.V. $t = \pm t_{\alpha/2} = \pm 2.040$	C.V. $z = \pm z_{\alpha/2} = \pm 1.960$
calculations:	calculations:

    method of this section:
    $$t_{\bar{x}} = (\bar{x} - \mu)/s_{\bar{x}}$$
    $$= (105.3 - 100)/(15.0/\sqrt{32})$$
    $$= 5.3/2.6517 = 1.999$$

    alternative method:
    $$z_{\bar{x}} = (\bar{x} - \mu)/\sigma_{\bar{x}}$$
    $$= (105.3 - 100)/(15.0/\sqrt{32})$$
    $$= 5.3/2.6517 = 1.999$$

    conclusion:
    Do not reject H_o; there is not enough evidence to reject the claim that $\mu = 100$.

    conclusion:
    Reject H_o; there is enough evidence to reject the claim that $\mu = 100$ and conclude that $\mu \neq 100$ (in fact, that $\mu > 100$).

    The two methods lead to different conclusions – because of the unwarranted artificial precision created in the alternative method by assuming that the s value could be used for σ.

35. The new test of hypothesis is given below. In this instance the presence of a drastic outlier drastically changed the summary statistics but did not have much of an impact on the overall test of hypothesis. In general, however, the effects of an outlier can be substantial. One could argue that the effect is not so great in this exercise because according to the $\bar{x} \pm 2s$ standard the original test already involved an outlier – so much larger than the others that it was the only value above $\bar{x}$.
    summary statistics: $n = 6$, $\Sigma x = 543.83$, $\Sigma x^2 = 291{,}603.3597$, $\bar{x} = 90.638$, $s = 220.142$
    original claim: $\mu > 1.5$
    H_o: $\mu = 1.5$
    H_1: $\mu > 1.5$
    $\alpha = 0.05$ and df = 5
    C.V. $t = t_\alpha = 2.015$
    calculations:

    $$t_{\bar{x}} = (\bar{x} - \mu)/s_{\bar{x}}$$

    $$= (90.638 - 1.5)/(220.142/\sqrt{6})$$

    $$= 89.138/89.872 = 0.992$$

    P-value $= P(t_5 > 0.992) = tcdf(0.992, 99, 5) = 0.1834$

conclusion:
   Do not reject H_o; there is not sufficient evidence to conclude that $\mu > 1.5$.

37. The set up for the basic test of hypothesis is as follows.
   summary statistics: $n = 15$, $\Sigma x = 25.7$, $\Sigma x^2 = 44.97$, $\bar{x} = 1.713$, $s = 0.2588$
   original claim: $\mu < 1.8$
   H_o: $\mu = 1.8$
   H_1: $\mu < 1.8$
   $\alpha = 0.05$ [assumed] and df $= 14$
   C.V. $t = -t_\alpha = -1.761$

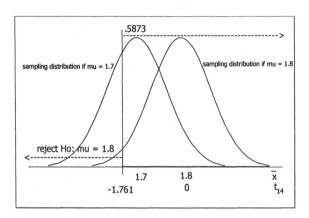

   The set up for the test indicates that
   $\alpha = P(\text{type I error})$
      $= P(\text{rejecting } H_o | H_o \text{ is true}) = 0.05$.
   In general,
   $\beta = P(\text{type II error})$
      $= P(\text{not rejecting } H_o | H_o \text{ is false})$
      $= \text{depends on how false } H_o \text{ really is.}$
   In particular, Minitab has determined that
   $\beta = P(\text{type II error})$
      $= P(\text{not rejecting } H_o | \mu = 1.7) = 0.5873$.

   a. power of the test $= 1 - \beta = 1 - 0.5873 = 0.4127$

   b. P(type II error) $= \beta = 0.5873$

   c. No. The power of the test when $\mu = 1.7$ is not very high. The probability of properly
      concluding that μ is less than 1.8 when it is really 1.7 is only 0.4127. This is less than 50%,
      and a test should have a power of at least 80% to be considered effective.

8-6 Testing a Claim About a Standard Deviation or Variance

   1. If a test is *not robust* against departures from normality, then the original population must be
      very close to a normal distribution in order for the statistical theory and methods to apply.
      Since the chi-square test of this section is not robust, a stricter set of guidelines should be
      applied than in previous sections when checking whether the original population is normal.

   3. No. The chi-square test of this section is not robust against departures from normality, and the
      original population of x values from a fair die follows a uniform distribution.

NOTE: Following the pattern used with the z and t distributions, this manual uses the closest entry
from Table A-4 for the χ^2 as if it were the precise value necessary and does not use interpolation.
This sacrifices little accuracy – and even interpolation does not yield precise values. When more
accuracy is needed, refer either to more complete tables or to computer-produced values.
The P-value portion of exercises #5-#8 may be worked as follows.
   table: find the correct df row in Table A-4, and see what values surround the calculated χ^2
   TI-83+: use χ^2cdf(lower bound, upper bound, df) to get the probability between 2 values
For the remainder of this manual, exact P-values for the χ^2 distribution USING THE UNROUNDED
VALUES of χ^2 will be given in TI-83+ calculator format as described above and used in exercises
#5-#8. While the input and/or output formats may be different for other technologies [e.g., Excel,
Minitab, Statdisk, etc], the final answers should be the same.

5. a. test statistic: $\chi^2 = (n-1)s^2/\sigma^2 = (9)(3.00)^2/(2.00)^2 = 20.250$

   b. critical values for $\alpha = 0.05$ and df = 9: $\chi^2 = \chi^2_{1-\alpha/2} = 2.700$

$$\chi^2 = \chi^2_{\alpha/2} = 19.023$$

   c. P-value limits: $19.023 < 20.250 < 21.666$

$$0.02 < \text{P-value} < 0.05$$

       P-value exact: $2 \cdot \chi^2 \text{cdf}(20.250, 999, 9) = 2(0.0164) = 0.0329$

   d. conclusion: Reject H_o; there is sufficient to conclude that $\sigma \neq 2.00$ (in fact, that $\sigma > 2.00$).

NOTE: As with the t distribution a subscript (df) may be used to identify which χ^2 distribution to use in the tables. As before, it is also acceptable to use the actual numerical values in place of the generic α, $\alpha/2$ or $1-\alpha/2$. In exercise #5, for example, one could indicate

     $\chi^2_{9,0.975} = 2.700$ [the χ^2 value with df = 9 and $1-\alpha/2 = 0.975$ in the upper tail]

     $\chi^2_{9,0.025} = 19.023$ [the χ^2 value with df − 9 and $\alpha/2 = 0.025$ in the upper tail]

7. a. test statistic: $\chi^2 = (n-1)s^2/\sigma^2 = (20)(10)^2/(15)^2 = 8.889$

   b. critical value for $\alpha = 0.01$ and df = 20: $\chi^2 = \chi^2_{1-\alpha} = 8.260$

   c. P-value limits: $8.260 < 8.889 < 9.591$

$$0.01 < \text{P-value} < 0.025$$

       P-value exact: $\chi^2 \text{cdf}(0, 8.889, 20) = 0.0158$

   d. conclusion: Do not reject H_o; there is not sufficient evidence to conclude that $\sigma < 15$.

NOTE: The formula $\chi^2 = (n-1)s^2/\sigma^2$ used in the calculations contains df = n-1. When the exact df needed in the problem does not appear in the table and the closest χ^2 value is used to determine the critical region, some instructors recommend using the same df in the calculations that were used to determine the C.V. This manual typically uses the closest entry to determine the C.V. and the n from the problem in the calculations – even though this introduces a slight discrepancy.

9. original claim: $\sigma \neq 696$ g

   H_o: $\sigma = 696$ g

   H_1: $\sigma \neq 696$ g

   $\alpha = 0.05$ and df = 189

   C.V. $\chi^2 = \chi^2_{1-\alpha/2} = 152.8222$

       $\chi^2 = \chi^2_{\alpha/2} = 228.9638$

   calculations:

$$\chi^2 = (n-1)s^2/\sigma^2$$
$$= (189)(645)^2/(696)^2$$
$$= 162.317$$

   P-value $= 2 \cdot \chi^2 \text{cdf}(0, 162.317, 189) = 2(0.0795) = 0.1591$

conclusion:

   Do not reject H_o; there is not sufficient evidence to conclude that $\sigma \neq 696$.

Based on this result, the variation in the weights of babies born to mothers who use cocaine is not significantly different from the variation in the weights for mothers who don't use cocaine.

GRAPHICS NOTE: While the χ^2 distribution is neither symmetric nor centered at zero, there are some important guidelines that help to keep χ^2 values in perspective. Loosely speaking, it "bunches up" around its df (which is actually its expected value) – and so the lower critical value will be less than df and the upper critical will be greater than df. Similarly, a calculated χ^2 value less than df falls in the lower tail and a calculated χ^2 value greater than df falls in the upper tail.

To illustrate χ^2 tests of hypotheses, this manual uses a "generic" figure resembling a χ^2 distribution with df=4. Actually χ^2 distributions with df=1 and df=2 have no upper limit and approach the y axis asymptotically, while χ^2 distributions with df>30 are essentially symmetric and normal-looking. Because the χ^2 distribution is positively skewed, the expected value df is noted slightly to the right of the figure's peak.

11. original claim: $\sigma > 0.056$ g
    H_0: $\sigma = 0.056$ g
    H_1: $\sigma > 0.056$ g
    $\alpha = 0.01$ and df = 40
    C.V. $\chi^2 = \chi^2_\alpha = 63.691$

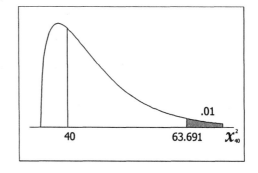

    calculations:
      $\chi^2 = (n-1)s^2/\sigma^2$
        $= (40)(0.31)^2/(0.056)^2$
        $= 1225.765$
      P-value $= \chi^2$cdf(1225.765,9999, 40) = 0
    conclusion:
      Reject H_0; there is sufficient evidence to conclude that $\sigma > 0.056$.
The peanut candy weights should have a larger standard deviation because adding a peanut introduces more variability – peanuts do not all have the same weight.

13. preliminary values: n = 16, $\Sigma x = 10565$, $\Sigma x^2 = 7114147$, $\bar{x} = 660.3$, $s^2 = 9196.3625$, s = 95.898
    original claim: $\sigma \neq 83$
    H_0: $\sigma = 83$
    H_1: $\sigma \neq 83$
    $\alpha = 0.05$ and df = 15
    C.V. $\chi^2 = \chi^2_{1-\alpha/2} = 6.262$
        $\chi^2 = \chi^2_{\alpha/2} = 27.488$

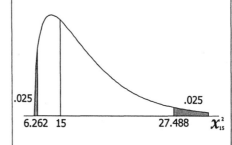

    calculations:
      $\chi^2 = (n-1)s^2/\sigma^2$
        $= (15)(9196.3625)/(83)^2$
        $= 20.024$
      P-value $= 2 \cdot \chi^2$cdf(20.024,999,15) = 2(0.1710) = 0.3420
    conclusion:
      Do not reject H_0; there is not sufficient evidence to conclude that $\sigma \neq 83$.
No. Based on these results, applicants from the new branch do not appear to have credit ratings that vary more than those at the main bank.

15. preliminary values: n = 9, Σx = 1089, Σx^2 = 132223, $\bar{x}$ = 121.0, s^2 = 56.75, s = 7.533

original claim: σ < 29 lbs

H_o: σ = 29 lbs

H_1: σ < 29 lbs

α = 0.01 and df = 8

C.V. $\chi^2 = \chi^2_{1-\alpha}$ = 1.646

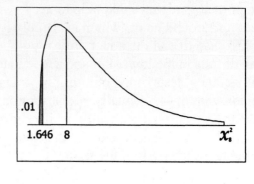

calculations:

$\chi^2 = (n-1)s^2/\sigma^2$

$= (8)(56.75)/(29)^2$

$= 0.540$

P-value = χ^2cdf(0,0.540,8) = 0.0002

conclusion:

Reject H_o; there is sufficient evidence to conclude that σ < 29.

17. preliminary values: n = 6, Σx = 9.23, Σx^2 = 32.5197, $\bar{x}$ = 1.538, s^2 = 3.664, s = 1.914

original claim: σ > 0.4 $\mu g/m^3$

H_o: σ = 0.4 $\mu g/m^3$

H_1: σ > 0.4 $\mu g/m^3$

α = 0.05 and df = 5

C.V. $\chi^2 = \chi^2_\alpha$ = 11.071

calculations:

$\chi^2 = (n-1)s^2/\sigma^2$

$= (5)(3.664)/(0.4)^2$

$= 114.506$

P-value = χ^2cdf(114.506,999,5) = 4.57x10^{-23} $\approx$ 0

conclusion:

Reject H_o; there is sufficient evidence to conclude that σ > 0.4.

Yes. Since 5 of the n=6 sample values are below the sample mean, the data do not appear to come from a population with a normal distribution. Considering that this test is non-robust, the methods of this section should not be used on these data.

19. summary statistics from statistical software: n = 40, $\bar{x}$ = 5.63930, s = 0.06194

original claim: σ = 0.068 g

H_o: σ = 0.068 g

H_1: σ $\neq$ 0.068 g

α = 0.05 and df = 39

C.V. $\chi^2 = \chi^2_{1-\alpha/2}$ = 24.433

$\chi^2 = \chi^2_{\alpha/2}$ = 59.342

calculations:

$\chi^2 = (n-1)s^2/\sigma^2$

$= (39)(0.06194)^2/(0.068)^2$

$= 32.359$

P-value = $2\cdot\chi^2$cdf(0,32.359,39) = 2(0.2349) = 0.4698

conclusion:

Do not reject H_o; there is not sufficient evidence to reject the claim that σ = 0.068.

Yes. The variation of the weights appears to be as desired.

21. lower $\chi^2 = \frac{1}{2}(-z_{\alpha/2} + \sqrt{2 \cdot df - 1})^2$

$\qquad = \frac{1}{2}(-1.96 + \sqrt{2 \cdot (189) - 1})^2$

$\qquad = \frac{1}{2}(-1.96 + \sqrt{377})^2$

$\qquad = \frac{1}{2}(304.7290)$

$\qquad = 152.3645$, which is close to the STATDISK value of 152.8222

upper $\chi^2 = \frac{1}{2}(z_{\alpha/2} + \sqrt{2 \cdot df - 1})^2$

$\qquad = \frac{1}{2}(1.96 + \sqrt{2 \cdot (189) - 1})^2$

$\qquad = \frac{1}{2}(1.96 + \sqrt{377})^2$

$\qquad = \frac{1}{2}(456.9542)$

$\qquad = 228.4771$, which is close to the STATDISK value of 228.9638

23. Yes, although the effect may not be "dramatic" if the sample is extremely large. Because this section tests the "spread" of the scores, and by definition an outlier is one that is "spread" away from the others, an outlier directly affects the property being measured and tested.

Statistical Literacy and Critical Thinking

1. Yes; the mean weight loss is statistically significant. No; the mean weight loss does not suggest that this diet is practical. No; the diet should not be recommended for weight loss. NOTE: This illustrates the difference between statistical significance and practical significance. In one sense, every null hypothesis is false. Consider, for example, the claim that $\mu = 50$. No matter what the story problem, it is unlikely that the true population mean is <u>exactly</u> 50 (i.e., to 100 decimal places) – and the true mean in the population might be, for example, 50.000013. With a large enough sample that difference can be discovered and declared statistically significant – even though such a difference is likely of no practical consequence.

Mathematically in a test about a population mean, $t = (\bar{x} - \mu)/(s/\sqrt{n}) = \sqrt{n} \cdot [(\bar{x} - \mu)/s]$ and increasing n by a factor of 100, for example, multiplies the t by 10. The same $\bar{x}$ and s that give an insignificant $t = 1.00$ for a particular value of n give a significant $t = 10.00$ for a sample 100 times as large. In general, taking a large enough sample can make any result statistically significant. This is the phenomenon behind the given weight loss scenario.

2. The data gathered applies only to those who had access to the survey and cared enough to respond. The population is precisely those who responded. The "sample" is not representative of any larger group. Since the data comes from a voluntary-response sample it cannot be used to make inferences using the techniques presented in the text.

3. The smaller the P-value, the less likely it is that the observed results occurred by chance alone. Prefer the smallest given P-value of 0.001.

4. The null and alternative hypotheses are as follows.

$\qquad H_o: \mu = 12$

$\qquad H_1: \mu > 12$

$\qquad \alpha = 0.05$ [assumed]

Since P-value = 0.250 > 0.05, the correct conclusion is as follows.

$\qquad$ Do not reject H_o; there is not sufficient evident to conclude that $\mu > 12$.

The P-value indicates that if the null hypothesis is true and the population mean is really 12, one expects 25% of the time to get a sample mean as large as or larger than the one obtained.

Review Exercises

1. a. H_1: $\mu < 10{,}000$
     sampling distribution: t, with df = 749
   b. H_1: $p > 0.50$
     sampling distribution: z
   c. H_1: $\mu \neq 100$
     sampling distribution: z
   d. H_1: $\sigma > 1.8$
     sampling distribution: none of the methods in this chapter apply
   e. H_1: $\sigma < 15$
     sampling distribution: χ^2, with df = 23

2. original claim: $p < 0.50$
   H_o: $p = 0.50$
   H_1: $p < 0.50$
   $\alpha = 0.05$
   C.V. $z = -z_\alpha = -1.645$
   calculations:

   $z\hat{p} = (\hat{p} - \mu\hat{p})/\sigma\hat{p}$

   $= (0.44 - 0.50)/\sqrt{(0.50)(0.50)/150}$

   $= -0.06/0.0408 = -1.47$

   P-value = P(z<-1.47) = 0.0708

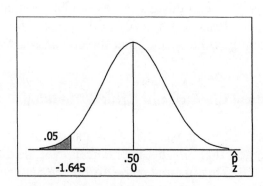

conclusion:
   Do not reject H_o; there is not sufficient evidence to conclude that $p < 0.50$.

3. NOTE: The value for x is not given. In truth, any $1488 \leq x \leq 1512$ rounds to the given
   $\hat{p} = x/2500 = 60\%$. All that can be said is that $0.5952 \leq \hat{p} \leq 0.6048$. For want of a more
   precise value, $\hat{p} = 0.60$ was used in the calculation of the test statistic $z\hat{p}$.
   original claim: $p < 0.62$
   H_o: $p = 0.62$
   H_1: $p < 0.62$
   $\alpha = 0.01$
   C.V. $z = -z_\alpha = -2.326$
   calculations:

   $z\hat{p} = (\hat{p} - \mu\hat{p})/\sigma\hat{p}$

   $= (0.60 - 0.62)/\sqrt{(0.62)(0.38)/2500}$

   $= -0.02/0.009708 = -2.06$

   P-value = P(z<-2.06) = 0.0197

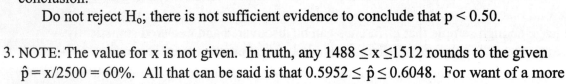

conclusion:
   Do not reject H_o; there is not sufficient evidence to conclude that $p < 0.62$.
If the results were obtained from a voluntary-response survey, then they cannot be used with
the methods of this chapter to make inferences.

4. original claim: $\mu = 3.5$ g

H_o: $\mu = 3.5$ g

H_1: $\mu \neq 3.5$ g

$\alpha = 0.05$ [assumed] and df = 69

C.V. $t = \pm t_{\alpha/2} = \pm 1.994$

calculations:

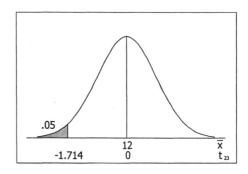

$\quad t_{\bar{x}} = (\bar{x} - \mu)/s_{\bar{x}}$

$\quad\quad = (3.586 - 3.5)/(0.074/\sqrt{70}\,)$

$\quad\quad = 0.086/0.00884 = 9.723$

P-value = $2 \cdot P(t_{69} > 9.723) = 2 \cdot tcdf(9.723,999,69) = 2(7.32 \times 10^{-15}) = 1.46 \times 10^{-14}$

conclusion:

Reject H_o; there is sufficient evidence to reject the claim that $\mu = 3.5$ and conclude that $\mu \neq 3.5$ (in fact, that $\mu > 3.5$).

No. The extra 0.086 grams amounts to $0.086/3.5 = 2.5\%$ more than the stated amount – which is not large enough to create a health hazard for unsuspecting consumers.

5. original claim: $\mu < 12$ oz

H_o: $\mu = 12$ oz

H_1: $\mu < 12$ oz

$\alpha = 0.05$ [assumed] and df = 23

C.V. $t = -t_{\alpha} = -1.714$

calculations:

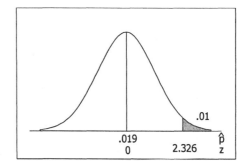

$\quad t_{\bar{x}} = (\bar{x} - \mu)/s_{\bar{x}}$

$\quad\quad = (11.4 - 12)/(0.62/\sqrt{24}\,)$

$\quad\quad = -0.6/0.1266 = -4.741$

P-value = $P(t_{23} < -4.741) = tcdf(-99,-4.741,23) = 0.0000444$

conclusion:

Reject H_o; there is sufficient evidence to conclude that $\mu < 12$.

No. The claim that the sample is too small has no validity. The test incorporates the sample size into the decision-making by requiring a larger test statistic t for the smaller degrees of freedom associated with the smaller values of n.

6. original claim: $p > 0.019$

$\quad \hat{p} = x/n = 19/863 = 0.022$

H_o: $p = 0.019$

H_1: $p > 0.019$

$\alpha = 0.01$

C.V. $z = z_{\alpha} = 2.326$

calculations:

$\quad z\hat{p} = (\hat{p} - \mu\hat{p})/\sigma\hat{p}$

$\quad\quad = (0.022 - 0.019)/\sqrt{(0.019)(0.981)/863}$

$\quad\quad = 0.003/0.004647 = 0.65$

P-value = $P(z>0.65) = 1 - P(z<0.65) = 1 - 0.7422 = 0.2578$

conclusion:

Do not reject H_o; there is not sufficient evidence to conclude that $p > 0.019$.

No. There is not sufficient evidence to say flu symptoms are an adverse treatment reaction.

7. original claim: $\sigma = 15$
    H_o: $\sigma = 15$
    H_1: $\sigma \neq 15$
    $\alpha = 0.05$ and df = 12
    C.V. $\chi^2 = \chi^2_{1-\alpha/2} = 4.404$
         $\chi^2 = \chi^2_{\alpha/2} = 23.337$
    calculations:

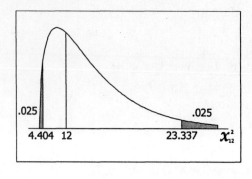

$\chi^2 = (n\text{-}1)s^2/\sigma^2$
    $= (12)(7.2)^2/(15)^2$
    $= 2.7648$
    P-value $= 2 \cdot \chi^2 cdf(0,2.7648,12) = 2(0.0030) = 0.0060$
conclusion:
    Reject H_o; there is sufficient evidence to reject the claim that $\sigma = 15$ and conclude that $\sigma \neq 15$ (in fact, that $\sigma > 15$).
Based on this result, we conclude that the standard deviation of IQ scores for statistics professors is less than that of the general population.

8. original claim: $\mu = 100$
    H_o: $\mu = 100$
    H_1: $\mu \neq 100$
    $\alpha = 0.10$
    C.V. $z = \pm z_{\alpha/2} = \pm 1.645$
    calculations:

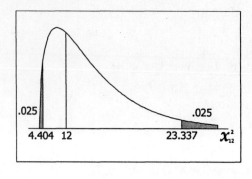

$z_{\bar{x}} = (\bar{x} - \mu_{\bar{x}})/\sigma_{\bar{x}}$
        $= (98.4 - 100)/(15/\sqrt{50})$
        $= -1.6/2.1213 = -0.75$
    P-value $= 2 \cdot P(z < -0.75) = 2(0.2266) = 0.4532$
conclusion:
    Do not reject H_o; there is not sufficient evidence to reject the claim that $\mu = 100$.
Based on these results, it appears that the random number generator is working correctly.

9. preliminary values: n = 12, $\Sigma x = 1180.7$, $\Sigma x^2 = 116174.19$, $\bar{x} = 98.3917$, s = 0.5351
    original claim: $\mu < 98.6$ °F
    H_o: $\mu = 98.6$ °F
    H_1: $\mu < 98.6$ °F
    $\alpha = 0.05$ and df = 11
    C.V. $t = -t_{\alpha} = -1.796$
    calculations:

$t_{\bar{x}} = (\bar{x} - \mu)/s_{\bar{x}}$
        $= (98.2917 - 98.6)/(0.5351/\sqrt{12})$
        $= -0.2083/0.1544 = -1.349$

    P-value $= P(t_{11} < -1.349) = tcdf(-99,-1.349,11) = 0.1023$
conclusion:
    Do not reject H_o; there is not sufficient evidence to conclude that $\mu < 98.6$.

10. original claim: $\sigma = 2.52$ in

   H_0: $\sigma = 2.52$ in
   H_1: $\sigma \neq 2.52$ in
   $\alpha = 0.05$ and df $= 39$
   C.V. $\chi^2 = \chi^2_{1-\alpha/2} = 24.433$
   $\chi^2 = \chi^2_{\alpha/2} = 59.342$

   calculations:
   $\chi^2 = (n-1)s^2/\sigma^2$
   $= (39)(2.74)^2/(2.52)^2$
   $= 46.107$
   P-value $= 2 \cdot \chi^2$cdf$(46.107,999,39) = 2(0.2019) = 0.4038$

   conclusion:

   Do not reject H_0; there is not sufficient evidence to reject the claim that $\sigma = 2.52$. Equipment is generally designed to accommodate all but a certain extreme percentage (which depends on the particular equipment) of the population. If the population has more variability than is planned for, then the portion of the population not accommodated by the equipment will be larger than proposed.

11. original claim: $p > 0.610$

   $\hat{p} = x/n = 2231/3581 = 0.623$

   H_0: $p = 0.610$
   H_1: $p > 0.610$
   $\alpha = 0.05$
   C.V. $z = z_\alpha = 1.645$

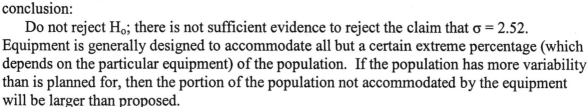

   calculations:
   $z\hat{p} = (\hat{p} - \mu\hat{p})/\sigma\hat{p}$
   $= (0.623 - 0.610)/\sqrt{(0.610)(0.390)/3581}$
   $= 0.013/0.008151 = 1.60$
   P-value $= P(z>1.60) = 1 - P(z<1.60) = 1 - 0.9452 = 0.0548$

   conclusion:

   Do not reject H_0; there is not sufficient evidence to conclude that $p > 0.610$.

12. preliminary values: $n = 10$, $\Sigma x = 433.0$, $\Sigma x^2 = 18878.90$, $\overline{x} = 43.30$, $s = 3.8006$

   original claim: $\mu > 40$ mg
   H_0: $\mu = 40$ mg
   H_1: $\mu > 40$ mg
   $\alpha = 0.01$ and df $= 9$
   C.V. $t = t_\alpha = 2.821$

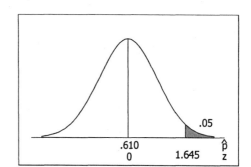

   calculations:
   $t_{\overline{x}} = (\overline{x} - \mu)/s_{\overline{x}}$
   $= (43.30 - 40)/(3.8006/\sqrt{10})$
   $= 3.30/1.2019 = 2.746$
   P-value $= P(t_9 > 2.746) = $ tcdf$(2.746,99,9) = 0.0113$

   conclusion:

   Do not reject H_0; there is not sufficient evidence to conclude that $\mu > 40$.

Cumulative Review Exercises

1. scores in order: 67 69 69.5 70 70 70.5 71 71 71.5 72 72 72 73 74 74 74.5 75
   summary statistics: n = 17, Σx = 1216.0, Σx^2 = 87053.00
   a. $\bar{x} = (\Sigma x)/n = (1216.0)/17 = 71.53$ in
   b. $\tilde{x} = 71.5$ in
   c. $s^2 = [n(\Sigma x^2) - (\Sigma x)^2]/[n(n-1)] = [17(87053.00) - (1216.0)^2]/[17(16)] = 4.577$
      s = 2.139, rounded to 2.1 in
   d. $s^2 = 4.577$, rounded to 4.6 in^2
   e. R = 75 – 67 = 8 in
   f. σ unknown and n ≤ 30: assuming the data come from a normal population, use t
      α = 0.05 and df = 16

   $$\bar{x} \pm t_{\alpha/2} \cdot s/\sqrt{n}$$

   $71.53 \pm 2.120 \cdot (2.139)/\sqrt{17}$
   71.53 ± 1.10
   $70.43 < \mu < 72.63$ (inches)

   g. original claim: μ > 69.0 in

   H_o: μ = 69 in
   H_1: μ > 69 in
   α = 0.05 and df = 16
   C.V. $t = t_\alpha = 1.746$
   calculations:

   $$t_{\bar{x}} = (\bar{x} - \mu)/s_{\bar{x}}$$
   $$= (71.53 - 69)/(2.139/\sqrt{17})$$
   $$= 2.53/0.5189 = 4.875$$

   P-value = $P(t_{16} > 4.875)$ = tcdf(4.875,99,16) = 0.0001
   conclusion:

   Reject H_o; there is sufficient evidence to conclude that μ > 69.
   Yes. Presidents appear to be taller than the typical man.

2. The normal quantile plot is given below, to the right of the columns used in the calculations.
   As in section 6-7, x = the original scores arranged in order
   cp = the cumulative probability values 1/21, 3/2n, 5/2n,...,(2n-1)/2n
   z = the normal z scores that have cp to their left

x	cp	z
67	0.029	-1.89
69	0.088	-1.35
69.5	0.147	-1.05
70	0.206	-0.82
70	0.265	-0.63
70.5	0.324	-0.46
71	0.382	-0.30
71	0.441	-0.15
71.5	0.500	0.00
72	0.559	0.15
72	0.618	0.30
72	0.676	0.46
73	0.735	0.63
74	0.794	0.82
74	0.853	1.05
74.5	0.912	1.35
75	0.971	1.89

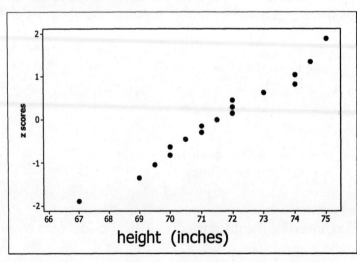

Yes. Based on the normal quantile plot, which approximates a straight line, the heights appear to come from a normal distribution.

3. a. normal distribution

$\mu = 496$

$\sigma = 108$

$P(x>500)$

$\quad = 1 - P(x<500)$

$\quad = 1 - P(z<0.04)$

$\quad = 1 - 0.5160$

$\quad = 0.4840$

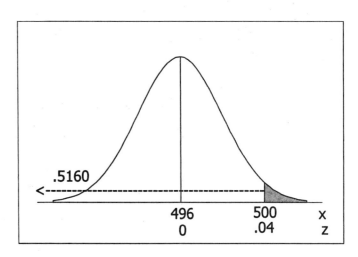

b. let A = a selected score is above 500

$P(A) = 0.4840$, for each selection

$P(A_1 \text{ and } A_2 \text{ and } A_3 \text{ and } A_4 \text{ and } A_5)$

$\quad = P(A_1){\cdot}P(A_2){\cdot}P(A_3){\cdot}(A_4){\cdot}P(A_5)$

$\quad = (0.4840)^5$

$\quad = 0.0266$

c. normal distribution, since the original distribution is so

$\mu_{\overline{x}} = \mu = 496$

$\sigma_{\overline{x}} = \sigma/\sqrt{n} = 108/\sqrt{5} = 48.39$

$P(\overline{x}>500)$

$\quad = 1 - P(\overline{x}<500)$

$\quad = 1 - P(z<0.08)$

$\quad = 1 - 0.5319$

$\quad = 0.4681$

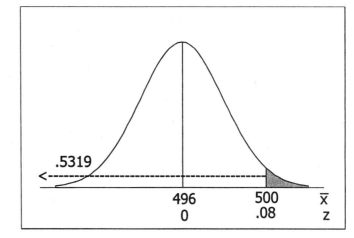

d. For P_{90}, A = 0.9000 [0.8997] and z = 1.28. [or 1.282 from the "large" df row of the t table]

$x = \mu + z{\cdot}\sigma$

$\quad = 496 + (1.282)(108)$

$\quad = 496 + 138$

$\quad = 634$

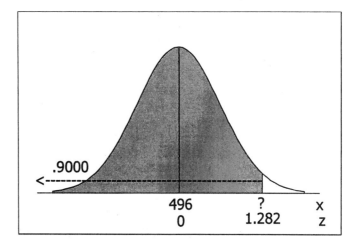

Chapter 9

Inferences from Two Samples

9-2 Inferences About Two Proportions

1. The pooled proportion for two samples is given by the symbol $\bar{p}$. It is the total number of successes from both samples divided by the total number of trials from both samples. It is literally the result of conceptually pooling both samples into one large sample.

3. We are 95% confident that the interval from 0.200 to 0.300 contains the true difference $p_1 - p_2$, where p_1 and p_2 are the true proportions in two populations.

5. $x = (0.23)(3250) = 747.50 \approx 748$ [NOTE: Any $712 \leq x \leq 763$ rounds to $x/3250 = 23\%$.]

7. $x = (0.07)(976) = 68.32 \approx 68$ [NOTE: Any $64 \leq x \leq 73$ rounds to $x/976 = 7\%$.]

9. $\hat{p}_1 = x_1/n_1 = 100/500 = 0.200$

   $\hat{p}_2 = x_1/n_1 = 50/400 = 0.125$

   $\hat{p}_1 - \hat{p}_2 = 0.200 - 0.125 = 0.075$

  a. $\bar{p} = (x_1+x_2)/(n_1+n_2)$

     $= (100+50)/(500+400) = 150/900 = 0.167$

  b. $z_{\hat{p}_1-\hat{p}_2} = (\hat{p}_1-\hat{p}_2 - \mu_{\hat{p}_1-\hat{p}_2})/\sigma_{\hat{p}_1-\hat{p}_2}$

     $= (0.075 - 0)/\sqrt{(0.167)(0.833)/500 + (0.167)(0.833)/400} = 0.075/0.025 = 3.00$

  c. for $\alpha = 0.05$, the critical values are $z = \pm z_{\alpha/2} = \pm 1.96$

  d. P-value $= 2 \cdot P(z>3.00) = 2(0.0013) = 0.0026$

NOTE: While the intermediate step for calculating $\sigma_{\hat{p}_1-\hat{p}_2}$ shows decimals with a fixed number of decimal places, this manual always uses the exact values – i.e., the original fractions recalled from the calculator memory.

11. Let the basketball games be group 1.
   original claim: $p_1-p_2 = 0$
   H_o: $p_1-p_2 = 0$
   H_1: $p_1-p_2 \neq 0$
   $\alpha = 0.05$ [assumed]
   C.V. $z = \pm z_{\alpha/2} = \pm 1.96$
   calculations:

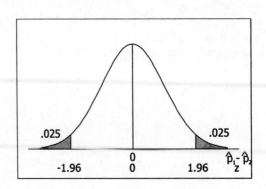

     $z_{\hat{p}_1-\hat{p}_2} = (\hat{p}_1-\hat{p}_2 - \mu_{\hat{p}_1-\hat{p}_2})/\sigma_{\hat{p}_1-\hat{p}_2}$
       $= 1.099$ [TI-83/84+]
     P-value $= 0.2719$ [TI-83/84+]
   conclusion:
     Do not reject H_o; there is not sufficient evidence to reject the claim that $p_1-p_2 = 0$.
   No. There does not appear to be a significant difference between the proportions of home wins for basketball and football. Since $\bar{p} = 0.620$ [TI-83/84+], the home field advantage seems to be real.

13. Let the XSORT babies be group 1.

$\hat{p}_1 = x_1/n_1 = 295/325 = 0.9077$

$\hat{p}_2 = x_2/n_2 = 39/51 = 0.7647$

$\hat{p}_1 - \hat{p}_2 = 0.9077 - 0.7647 = 0.1430$

a. $\alpha = 0.05$

$(\hat{p}_1 - \hat{p}_2) \pm z_{\alpha/2}\sigma_{\hat{p}_1 - \hat{p}_2}$

$0.1430 \pm 1.96\sqrt{(0.9077)(0.0923)/325 + (0.7647)(0.2353)/51}$

0.1430 ± 0.1206

$0.0224 < p_1-p_2 < 0.264$

b. Yes. Since the confidence interval does not include 0, there does appear to be a difference. While both the XSORT and YSORT methods appear to be effective, the XSORT method appears to be more effective than the YSORT method.

15. Let those contacted for the standard survey be group 1.

$\hat{p}_1 = x_1/n_1 = 720/1720 = 0.419$

$\hat{p}_2 = x_2/n_2 = 429/1640 = 0.262$

$\hat{p}_1 - \hat{p}_2 = 0.419 - 0.262 = 0.157$

$\bar{p} = (720+429)/(1720+1640) = 1149/3360 = 0.342$

original claim: $p_1-p_2 > 0$

H_0: $p_1-p_2 = 0$

H_1: $p_1-p_2 > 0$

$\alpha = 0.01$

C.V. $z = z_\alpha = 2.326$

calculations:

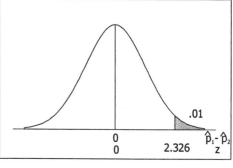

$z_{\hat{p}_1 - \hat{p}_2} = (\hat{p}_1 - \hat{p}_2 - \mu_{\hat{p}_1 - \hat{p}_2})/\sigma_{\hat{p}_1 - \hat{p}_2}$

$= (0.157 - 0)/\sqrt{(0.342)(0.658)/1720 + (0.342)(0.658)/1640} = 0.157/0.0164 = 9.59$

P-value = $P(z>9.59) = 1 - P(z<9.59) = 1 - 0.9999 = 0.0001$

conclusion:

Reject H_0; there is sufficient evidence to conclude that $p_1-p_2 > 0$.

Yes. Since the rigorous survey has a lower refusal weight, it is more likely to produce the better results.

17. Let the workers be group 1.

$\hat{p}_1 = x_1/n_1 = 192/436 = 0.4404$

$\hat{p}_2 = x_2/n_2 = 40/121 = 0.3306$

$\hat{p}_1 - \hat{p}_2 = 0.440 - 0.331 = 0.1098$

$\bar{p} = (192+40)/(436+121) = 232/557 = 0.4165$

original claim: $p_1-p_2 > 0$

H_0: $p_1-p_2 = 0$

H_1: $p_1-p_2 > 0$

$\alpha = 0.05$

C.V. $z = z_\alpha = 1.645$

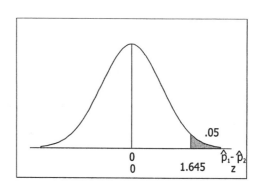

calculations:

$$z_{\hat{p}_1-\hat{p}_2} = (\hat{p}_1-\hat{p}_2 - \mu_{\hat{p}_1-\hat{p}_2})/\sigma_{\hat{p}_1-\hat{p}_2}$$

$$= (0.1098 - 0)/\sqrt{(0.4165)(0.5835)/436 + (0.4165)(0.5835)/121} = 0.1098/0.0507 = 2.17$$

P-value = $P(z>2.17) = 1 - P(z<2.17) = 1 - 0.9850 = 0.0150$

conclusion:

Reject H_o; there is sufficient evidence to conclude that $p_1-p_2 > 0$.

19. Let the employees of hospitals with the smoking ban be group 1.

$\hat{p}_1 = x_1/n_1 = 56/843 = 0.0664$

$\hat{p}_2 = x_2/n_2 = 27/703 = 0.0384$

$\hat{p}_1 - \hat{p}_2 = 0.0664 - 0.0384 = 0.0280$

$\bar{p} = (56+27)/(843+703) = 83/1546 = 0.0537$

original claim: $p_1-p_2 \neq 0$

H_o: $p_1-p_2 = 0$

H_1: $p_1-p_2 \neq 0$

$\alpha = 0.05$

C.V. $z = \pm z_{\alpha/2} = \pm1.96$

calculations:

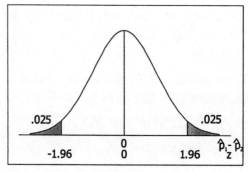

$$z_{\hat{p}_1-\hat{p}_2} = (\hat{p}_1-\hat{p}_2 - \mu_{\hat{p}_1-\hat{p}_2})/\sigma_{\hat{p}_1-\hat{p}_2}$$

$$= (0.0280 - 0)/\sqrt{(0.0537)(0.9463)/843 + (0.0537)(0.9463)/703} = 0.0280/0.0115 = 2.43$$

P-value = $2 \cdot P(z>2.43) = 2[1 - P(z<2.43)] = 2[1 - 0.9925] = 2[0.0075] = 0.0150$

conclusion:

Reject H_o; there is sufficient evidence to conclude that $p_1-p_2 \neq 0$ (in fact, that $p_1-p_2 > 0$). Since P-value = 0.0150 > 0.01, the difference is not significant at the 0.01 level. Even though one cannot be 99% certain, it appears that the hospital employees with the smoking ban had a higher quit smoking rate – BUT, the difference in quit smoking rates could be a result of the working place (hospital vs. non-hospital) or the policy (ban vs. no-ban) or a combination.

NOTE: The information given for exercises #21, #22, #25, #26, #27, #28 and #35 is not sufficient to determine the exact values of x_1 and x_2. As a consequence, fractions cannot be used and the usual desired level of accuracy cannot be achieved.

21. Let the workers be group 1.

$\hat{p}_1 = x_1/n_1 = x_1/1655 = 0.021$ [as any $34 \leq x_1 \leq 35$ gives $\hat{p}_1 = 0.021$, use 0.021 in all calculations]

$\hat{p}_2 = x_2/n_2 = x_2/1652 = 0.012$ [as any $19 \leq x_2 \leq 20$ gives $\hat{p}_2 = 0.012$, use 0.012 in all calculations]

$\hat{p}_1 - \hat{p}_2 = 0.021 - 0.012 = 0.009$

$\bar{p} = [(0.021)(1655)+(0.012)(1652)]/[1655+1652]$

$= 54.579/3307 = 0.0165$

original claim: $p_1-p_2 > 0$

H_o: $p_1-p_2 = 0$

H_1: $p_1-p_2 > 0$

$\alpha = 0.05$

C.V. $z = z_\alpha = 1.645$

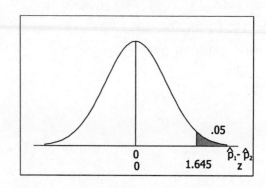

calculations:

$$z_{\hat{p}_1-\hat{p}_2} = (\hat{p}_1-\hat{p}_2 - \mu_{\hat{p}_1-\hat{p}_2})/\sigma_{\hat{p}_1-\hat{p}_2}$$

$$= (0.009 - 0)/\sqrt{(0.0165)(0.9835)/1655 + (0.0165)(0.9835)/1652} = 0.009/0.00443 = 2.03$$

P-value = $P(z>2.03) = 1 - P(z<2.03) = 1 - 0.9788 = 0.0212$

conclusion:

Reject H_o; there is sufficient evidence to conclude that $p_1-p_2 > 0$.

While those who use Clarinex are more likely to experience fatigue, a difference of 0.009 = 0.9% does not pose a "major concern."

23. Let the homeowners be group 1.

$\hat{p}_1 = x_1/n_1 = 880/1068 = 0.824$

$\hat{p}_2 = x_2/n_2 = 725/1064 = 0.681$

$\hat{p}_1 - \hat{p}_2 = 0.824 - 0.681 = 0.143$

$\alpha = 0.05$

$(\hat{p}_1 - \hat{p}_2) \pm z_{\alpha/2}\,\sigma_{\hat{p}_1-\hat{p}_2}$

$\quad 0.143 \pm 1.96\sqrt{(0.824)(0.176)/1068 + (0.681)(0.319)/1064}$

$\quad 0.143 \pm 0.036$

$\quad 0.106 < p_1-p_2 < 0.179$

Yes. Since the confidence interval does not include 0, there seems to be a significant difference between the proportions. One major factor might be income levels: homeowners are more likely to have the higher income necessary to maintain a vehicle for commuting – which often necessitates at least two vehicles in the household. Another major factor might be the availability of public transportation: renters are more likely to live in the more densely populated areas typically served by public transportation – homeowners in suburban/rural areas might not have convenient access to public transportation even if they wanted to use it.

25. Let the males be group 1.

$\hat{p}_1 = x_1/n_1 = x_1/2200 = 0.72$ [as any $1574 \le x_1 \le 1594$ gives $\hat{p}_1 = 0.72$, use 0.72 in all calculations]

$\hat{p}_2 = x_2/n_2 = x_2/2380 = 0.84$ [as any $1988 \le x_2 \le 2011$ gives $\hat{p}_2 = 0.84$, use 0.84 in all calculations]

$\hat{p}_1 - \hat{p}_2 = 0.72 - 0.84 = -0.12$

$\bar{p} = [(0.84)(2200) + (0.72)(2380)]/[2200 + 2380]$

$\quad = 3583.2/4580 = 0.782$

original claim: $p_1-p_2 = 0$

H_o: $p_1-p_2 = 0$

H_1: $p_1-p_2 \ne 0$

$\alpha = 0.05$

C.V. $z = \pm z_{\alpha/2} = \pm 1.96$

calculations:

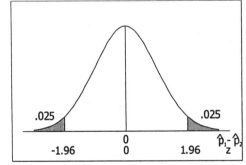

$$z_{\hat{p}_1-\hat{p}_2} = (\hat{p}_1-\hat{p}_2 - \mu_{\hat{p}_1-\hat{p}_2})/\sigma_{\hat{p}_1-\hat{p}_2}$$

$$= (-0.12 - 0)/\sqrt{(0.782)(0.218)/2200 + (0.782)(0.218)/2380} = -0.12/0.0122 = -9.83$$

P-value = $2 \cdot P(z<-9.83) = 2(0.0001) = 0.0002$

conclusion:

Reject H_o; there is sufficient evidence to reject the claim that $p_1-p_2 = 0$ and conclude that $p_1-p_2 \ne 0$ (in fact, that $p_1-p_2 < 0$).

Yes. There does appear to be a gender gap in that males have a lower seat belt usage rate.

27. Let the single women be group 1.

$\hat{p}_1 = x_1/n_1 = x_1/205 = 0.24$ [as any $49 \le x_1 \le 50$ gives $\hat{p}_1 = 0.24$, use 0.24 in all calculations]

$\hat{p}_2 = x_2/n_2 = x_2/260 = 0.27$ [as any $70 \le x_2 \le 71$ gives $\hat{p}_2 = 0.27$, use 0.27 in all calculations]

$\hat{p}_1 - \hat{p}_2 = 0.24 - 0.27 = -0.03$

$\alpha = 0.01$

$(\hat{p}_1 - \hat{p}_2) \pm z_{\alpha/2}\, \sigma_{\hat{p}_1 - \hat{p}_2}$

$-0.0300 \pm 2.575 \sqrt{(0.24)(0.76)/205 + (0.27)(0.73)/260}$

-0.0300 ± 0.1045

$-0.135 < p_1 - p_2 < 0.0745$ [See the accuracy comments at the beginning of the exercise.]

No. Since the confidence interval contains 0, there does not appear to be a gender gap on this issue.

29. Cut and paste to place the 261 Mon-Fri values into one column and the 104 Sat-Sun values into another column. Then sort the two resulting columns in descending order to quickly count the numbers of days with precipitation. In Minitab, for example, this can be done as follows.

    STACK C1-C5 C8
    STACK C6-67 C9
    SORT C8 C8
    SORT C9 C9

Let the weekdays be group 1.

$\hat{p}_1 = x_1/n_1 = 92/261 = 0.352$

$\hat{p}_2 = x_2/n_2 = 35/104 = 0.337$

$\hat{p}_1 - \hat{p}_2 = 0.352 - 0.337 = 0.016$

$\bar{p} = (92+35)/(261+104) = 127/365 = 0.348$

original claim: $p_1 - p_2 < 0$

H_o: $p_1 - p_2 = 0$

H_1: $p_1 - p_2 < 0$

$\alpha = 0.05$ [assumed]

C.V. $z = -z_\alpha = -1.645$

calculations:

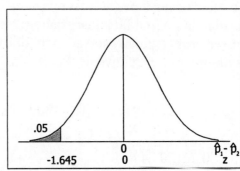

$z_{\hat{p}_1 - \hat{p}_2} = (\hat{p}_1 - \hat{p}_2 - \mu_{\hat{p}_1 - \hat{p}_2})/\sigma_{\hat{p}_1 - \hat{p}_2}$

$= (0.016 - 0)/\sqrt{(0.348)(0.652)/261 + (0.348)(0.652)/104} = 0.016/0.0552 = 0.29$

P-value $= P(z < 0.29) = 0.6141$

conclusion:

   Do not reject H_o; there is not sufficient evidence to conclude that $p_1 - p_2 < 0$.

No. The Boston data do not support the claim that is rains more often on weekends.

31. Let the movies tested for alcohol be group 1.

$\hat{p}_1 = x_1/n_1 = 25/50 = 0.50$

$\hat{p}_2 = x_2/n_2 = 28/50 = 0.56$

$\hat{p}_1 - \hat{p}_2 = 0.50 - 0.56 = -0.06$

$\overline{p} = (25+28)/(50+50) = 53/100 = 0.53$

original claim: $p_1-p_2 < 0$

H_o: $p_1-p_2 = 0$

H_1: $p_1-p_2 < 0$

$\alpha = 0.05$ [assumed]

C.V. $z = -z_\alpha = -1.645$

calculations:

$z_{\hat{p}_1-\hat{p}_2} = (\hat{p}_1-\hat{p}_2 - \mu_{\hat{p}_1-\hat{p}_2})/\sigma_{\hat{p}_1-\hat{p}_2}$

$= (-0.06 - 0)/\sqrt{(0.53)(0.47)/50 + (0.53)(0.47)/50} = -0.06/0.0998 = -0.60$

P-value = $P(z<-0.60) = 0.2743$

conclusion:

Do not reject H_o; there is not sufficient evidence to conclude that $p_1-p_2 < 0$.

No. One requirement for this section is that the two samples be independent. The data in Data Set 5 are not from independent samples – but from the same sample. This is analogous to testing whether the proportion of people who declare bankruptcy is different from the proportion of people who have more than \$10,000 credit card debt – and using the same 50 people.

33. For all parts of this exercise

$\hat{p}_1 = x_1/n_1 = 112/200 = 0.560$

$\hat{p}_2 = x_2/n_2 = 88/200 = 0.440$

$\hat{p}_1 - \hat{p}_2 = 0.560 - 0.440 = 0.120$

$\overline{p} = (112+88)/(200+200) = 200/400 = 0.500$

Use with $\alpha = 0.05$ for all parts.

a. $(\hat{p}_1 - \hat{p}_2) \pm z_{\alpha/2}\, \sigma_{\hat{p}_1-\hat{p}_2}$

$0.1200 \pm 1.96 \sqrt{(0.560)(0.440)/200 + (0.440)(0.560)/200}$

0.1200 ± 0.0973

$0.0227 < p_1-p_2 < 0.217$

Since the interval does not include 0, conclude that p_1 and p_2 are different.

Since the interval lies entirely above 0, conclude that $p_1 - p_2 > 0$.

b. for group 1   for group 2

$\hat{p} \pm z_{\alpha/2} \sqrt{\hat{p}\hat{q}/n}$ $\hat{p} \pm z_{\alpha/2} \sqrt{\hat{p}\hat{q}/n}$

$0.560 \pm 1.96 \sqrt{(0.56)(0.44)/200}$ $0.440 \pm 1.96 \sqrt{(0.44)(0.56)/200}$

0.560 ± 0.069 0.440 ± 0.069

$0.491 < p < 0.629$ $0.371 < p < 0.509$

Since the intervals overlap, the implication is that p_1 and p_2 could have the same value.

c. original claim: $p_1-p_2 = 0$
   H_o: $p_1-p_2 = 0$
   H_1: $p_1-p_2 \neq 0$
   $\alpha = 0.05$
   C.V. $z = \pm z_\alpha = \pm 1.96$

calculations:

$$z_{\hat{p}_1-\hat{p}_2} = (\hat{p}_1-\hat{p}_2 - \mu_{\hat{p}_1-\hat{p}_2})/\sigma_{\hat{p}_1-\hat{p}_2}$$

$$= (0.12 - 0)/\sqrt{(0.5625)(0.4375)/40 + (0.5625)(0.4375)/40} = 0.12/0.05 = 2.40$$

P-value $= 2 \cdot P(z>2.40) = 2[1 - P(z<2.40)] = 2[1 - 0.9918] = 2[0.0082] = 0.0164$
conclusion:

Reject H_o; there is sufficient evidence to reject the claim that $p_1-p_2 = 0$ and conclude that $p_1-p_2 \neq 0$ (in fact, that $p_1-p_2 > 0$)

d. Based on the preceding results, conclude that p_1 and p_2 are unequal and that $p_1 > p_2$. The overlapping interval method of part (b) appears to be the least effective method for comparing two populations.

35. Let the vinyl gloves be group 1.
   $\hat{p}_1 = x_1/n_1 = x_1/240 = 0.63$ [as any $151 \leq x_1 \leq 152$ gives $\hat{p}_1 = 0.63$, use 0.63 in all calculations]
   $\hat{p}_2 = x_2/n_2 = x_2/240 = 0.07$ [as any $16 \leq x_2 \leq 17$ gives $\hat{p}_2 = 0.07$, use 0.07 in all calculations]
   $\hat{p}_1 - \hat{p}_2 = 0.63 - 0.07 = 0.56$
   $\bar{p} = [(0.63)(240) + (0.07)(240)]/[240 + 240]$
   $\quad = 168/480 = 0.35$
   original claim: $p_1-p_2 = 0.50$
   H_o: $p_1-p_2 = 0.50$
   H_1: $p_1-p_2 \neq 0.50$
   $\alpha = 0.05$
   C.V. $z = \pm z_{\alpha/2} = \pm 1.96$
   calculations:

$$z_{\hat{p}_1-\hat{p}_2} = (\hat{p}_1-\hat{p}_2 - \mu_{\hat{p}_1-\hat{p}_2})/\sigma_{\hat{p}_1-\hat{p}_2}$$

$$= (0.56 - 0.50)/\sqrt{(0.63)(0.37)/240 + (0.07)(0.93)/240} = 0.06/0.0352 = 1.70$$

   [See the accuracy comments at the beginning of the exercise.]
   P-value $= 2 \cdot P(z>1.70) = 2[1 - P(z<1.70)] = 2[1 - 0.9554] = 2[0.0446] = 0.0892$
conclusion:

Do not reject H_o; there is not sufficient evidence to reject the claim that $p_1-p_2 = 0.50$.

37. a.

#	sample	$\hat{p}$	$\hat{p}-\mu_{\hat{p}}$	$(\hat{p}-\mu_{\hat{p}})^2$
1	HH	1.0	0.5	0.25
2	HT	0.5	0.0	0.00
3	TH	0.5	0.0	0.00
4	TT	0.0	-0.5	0.25
		2.0	0.0	0.50

$\mu_{\hat{p}} = (\Sigma \hat{p})/N = 2.0/5 = 0.5$

$\sigma_{\hat{p}} = \Sigma(\hat{p}-\mu_{\hat{p}})^2/N = 0.50/4 = 0.125$

b. The 16 possible $\hat{p}_D - \hat{p}_Q$ values are given at the right, identified by the sample #'s used to generate them.

$\mu_{D-Q} = \Sigma(\hat{p}_D - \hat{p}_Q)/N$

$\quad = 0/16 = 0$

$\sigma^2_{D-Q} = \Sigma[(\hat{p}_D - \hat{p}_Q) - \mu]^2/N$

$\quad = \Sigma[(\hat{p}_D - \hat{p}_Q) - 0]^2/N$

$\quad = \Sigma(\hat{p}_D - \hat{p}_Q)^2/N$

$\quad = 4.00/16 = 0.25$

c. $\sigma^2_{D-Q} = \sigma^2_D + \sigma^2_Q$

$\quad = 0.125 + 0.125$

$\quad = 0.25$

sample	$\hat{p}_D - \hat{p}_Q$	$(\hat{p}_D - \hat{p}_Q)^2$
1-1	0	0
1-2	0.5	0.25
1-3	0.5	0.25
1-4	1.0	1.00
2-1	-0.5	0.25
2-2	0	0
2-3	0	0
2-4	0.5	0.25
3-1	-0.5	0.25
3-2	0	0
3-3	0	0
3-4	0.5	0.25
4-1	-1.0	1.00
4-2	-0.5	0.25
4-3	-0.5	0.25
4-4	0	0
	0	4.00

9-3 Inferences About Two Means: Independent Samples

1. The requirements of this section are not satisfied. The age of 80 would be an outlier for the sample data and suggest that the original population is not normal, but positively skewed.

3. The critical value t = 1.717 is "more conservative" than t = 1.682 because it requires a larger difference between the two means before rejecting the hypothesis that they are equal.
NOTE: Formula 9-1 yields a df value between $\min(df_1, df_2)$ and $(df_1 + df_2)$. The smaller the df, the more variability in the t distribution, and the larger (i.e., the farther from 0) the t score for any level of significance. A larger critical t score requires a greater deviation from the null hypothesis before rejecting the null hypothesis – making the test more conservative. Choosing $df = \min(df_1, df_2)$ selects the smallest reasonable df and uses the most conservative standard.

5. Independent samples, since there is no connection between the two groups – i.e., the two groups are selected and analyzed separately.

7. Matched pairs, since each pair of before and after weights is gathered from the same subject.

NOTE: To be consistent with the previous notation, reinforcing patterns and concepts presented in those sections, the manual uses the usual t formula written to apply to $\bar{x}_1 - \bar{x}_2$'s

$$t_{\bar{x}_1 - \bar{x}_2} = (\bar{x}_1 - \bar{x}_2 - \mu_{\bar{x}_1 - \bar{x}_2})/s_{\bar{x}_1 - \bar{x}_2}, \text{ with } \mu_{\bar{x}_1 - \bar{x}_2} = \mu_1 - \mu_2 \text{ and } s_{\bar{x}_1 - \bar{x}_2} = \sqrt{s_1^2/n_1 + s_2^2/n_2}.$$

9. Let the children given the echinacea treatment be group 1.

$\bar{x}_1 - \bar{x}_2 = 0.81 - 0.64 = 0.17$

original claim: $\mu_1 - \mu_2 \neq 0$ days

H_0: $\mu_1 - \mu_2 = 0$ days

H_1: $\mu_1 - \mu_2 \neq 0$ days

$\alpha = 0.05$ and df = 336

C.V. $t = \pm t_{\alpha/2} = \pm 1.968$

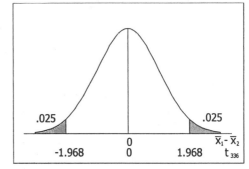

calculations:

$$t_{\overline{x}_1-\overline{x}_2} = (\overline{x}_1-\overline{x}_2 - \mu_{\overline{x}_1-\overline{x}_2})/s_{\overline{x}_1-\overline{x}_2}$$

$$= (0.17 - 0)/\sqrt{(1.50)^2/337 + (1.16)^2/370}$$

$$= 0.17/0.1016 = 1.674$$

P-value = 2·tcdf(1.674,99,336) = 2(0.0475) = 0.0951

conclusion:

Do not reject H_o; there is not sufficient evidence to conclude that $\mu_1 - \mu_2 \neq 0$.

No. Based on these results, echinacea does not appear to have an effect.

11. Let the children with low birth weights be group 1.

$$\overline{x}_1-\overline{x}_2 = 95.5 - 104.9 = -9.4$$

$\alpha = 0.05$ and df = 219

$$(\overline{x}_1-\overline{x}_2) \pm t_{\alpha/2} s_{\overline{x}_1-\overline{x}_2}$$

$$-9.4 \pm 1.972 \sqrt{(16.0)^2/258 + (14.1)^2/220}$$

$$-9.4 \pm 2.7$$

$$-12.1 < \mu_1 - \mu_2 < -6.7$$

Yes. Since the confidence interval is entirely below 0, it appears that children with extremely low birth weights do have lower IQ's at age 8.

13. Let the light users be group 1.

$$\overline{x}_1-\overline{x}_2 = 53.3 - 51.3 = 2.0$$

original claim: $\mu_1 - \mu_2 > 0$ items

H_o: $\mu_1 - \mu_2 = 0$ items

H_1: $\mu_1 - \mu_2 > 0$ items

$\alpha = 0.01$ and df = 63

C.V. $t = t_\alpha = 2.385$

calculations:

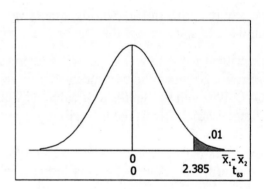

$$t_{\overline{x}_1-\overline{x}_2} = (\overline{x}_1-\overline{x}_2 - \mu_{\overline{x}_1-\overline{x}_2})/s_{\overline{x}_1-\overline{x}_2}$$

$$= (2.0 - 0)/\sqrt{(3.6)^2/64 + (4.5)^2/65}$$

$$= 2.0/0.7170 = 2.790$$

P-value = tcdf(2.790,99,63) = 0.0034

conclusion:

Reject H_o; there is sufficient evidence to conclude that $\mu_1 - \mu_2 > 0$.

Yes. Based on these results, heavy marijuana use is associated with poor performance.

15. Let the placebo users be group 1.

$$\overline{x}_1-\overline{x}_2 = 21.57 - 20.38 = 1.19$$

$\alpha = 0.05$ and df = 32

$$(\overline{x}_1-\overline{x}_2) \pm t_{\alpha/2} s_{\overline{x}_1-\overline{x}_2}$$

$$1.19 \pm 2.037 \sqrt{(3.87)^2/43 + (3.91)^2/33}$$

$$1.19 \pm 1.84$$

$$-0.65 < \mu_1 - \mu_2 < 3.03$$

No. Since the confidence interval includes 0, we cannot be 95% certain that the two populations have different means. There is not enough evidence to make this the generally recommended treatment for bipolar depression.

17. Let those receiving the magnet treatment be group 1.

$\overline{x}_1 - \overline{x}_2 = 0.49 - 0.44 = 0.05$

original claim: $\mu_1 - \mu_2 > 0$
H_o: $\mu_1 - \mu_2 = 0$
H_1: $\mu_1 - \mu_2 > 0$
$\alpha = 0.05$ and df = 19
C.V. $t = t_\alpha = 1.729$
calculations:

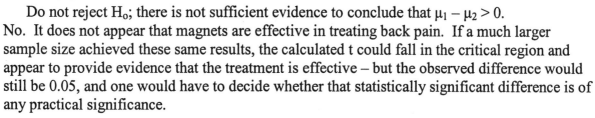

$$t_{\overline{x}_1 - \overline{x}_2} = (\overline{x}_1 - \overline{x}_2 - \mu_{\overline{x}_1 - \overline{x}_2}) / s_{\overline{x}_1 - \overline{x}_2}$$

$$= (0.05 - 0)/ \sqrt{(0.96)^2/20 + (1.4)^2/20}$$

$$= 0.05/0.3796 = 0.132$$

P-value = tcdf(0.132,99,19) = 0.4483

conclusion:

Do not reject H_o; there is not sufficient evidence to conclude that $\mu_1 - \mu_2 > 0$.
No. It does not appear that magnets are effective in treating back pain. If a much larger sample size achieved these same results, the calculated t could fall in the critical region and appear to provide evidence that the treatment is effective – but the observed difference would still be 0.05, and one would have to decide whether that statistically significant difference is of any practical significance.

19. Let the control subjects be group 1.

$\overline{x}_1 - \overline{x}_2 = 0.45 - 0.34 = 0.11$

$\alpha = 0.01$ and df = 9

$(\overline{x}_1 - \overline{x}_2) \pm t_{\alpha/2} s_{\overline{x}_1 - \overline{x}_2}$

$$0.11 \pm 3.250 \sqrt{(.08)^2/10 + (.08)^2/10}$$

$$0.11 \pm 0.12$$

$$-0.01 < \mu_1 - \mu_2 < 0.23 \text{ (mL)}$$

The confidence interval suggests that the two samples could come from populations with the same mean. Based on this result, one cannot be 99% certain that there is such a biological basis for the disorders.

21. Let the treatment subjects be group 1.

$\overline{x}_1 - \overline{x}_2 = 4.20 - 1.71 = 2.49$

$\alpha = 0.05$ and df = 21

$(\overline{x}_1 - \overline{x}_2) \pm t_{\alpha/2} s_{\overline{x}_1 - \overline{x}_2}$

$$2.49 \pm 2.080 \sqrt{(2.20)^2/22 + (.72)^2/22}$$

$$2.49 \pm 1.03$$

$$1.46 < \mu_1 - \mu_2 < 3.52 \text{ (errors)}$$

Yes. Since high scores indicate the presence of more errors and the confidence interval falls entirely above 0, the results support the common belief that drinking alcohol is hazardous for the operators of passenger vehicles – and for their passengers.

23. Let the oceanfront homes be group 1.
    <u>group 1: front</u> (n=5)   <u>group 2: side</u> (n=5)
    $\Sigma x = 12871$ $\Sigma x = 5184$
    $\Sigma x^2 = 35,458,531$ $\Sigma x^2 = 6,014,756$
    $\overline{x} = 2574.2$ $\overline{x} = 1036.8$
    $s^2 = 581500.7$ (s=762.56) $s^2 = 159996.2$ (s=400.00)
$\overline{x}_1 - \overline{x}_2 = 2574.2 - 1036.8 = 1537.4$

original claim: $\mu_1 - \mu_2 > 0$ ($1000's)
H_o: $\mu_1 - \mu_2 = 0$ ($1000's)
H_1: $\mu_1 - \mu_2 > 0$ ($1000's)
$\alpha = 0.05$ and df = 4
C.V. $t = t_\alpha = 2.132$
calculations:

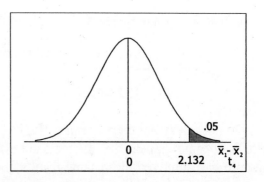

$$t_{\overline{x}_1 - \overline{x}_2} = (\overline{x}_1 - \overline{x}_2 - \mu_{\overline{x}_1 - \overline{x}_2}) / s_{\overline{x}_1 - \overline{x}_2}$$
$$= (1537.4 - 0) / \sqrt{581500.7/5 + 159996.2/5}$$
$$= 1537.4/385.097 = 3.992$$
P-value = tcdf(3.992,99,4) = 0.0081
conclusion:
    Reject H_o; there is sufficient evidence to conclude that $\mu_1 - \mu_2 > 0$.
Yes. The fact that there are only five values in each sample is built into the critical region of the test. When the test indicates there is sufficient evidence to reject the null hypothesis, that conclusion is valid regardless of the sample size.

25. Let the filtered cigarettes be group 1.
    <u>group 1: filtered</u> (n=21)   <u>group 2: unfiltered</u> (n=8)
    $\Sigma x = 279$ $\Sigma x = 192$
    $\Sigma x^2 = 3987$ $\Sigma x^2 = 4628$
    $\overline{x} = 13.286$ $\overline{x} = 24.000$
    $s^2 = 14.0143$ (s=3.744) $s^2 = 2.8571$ (s=1.690)
$\overline{x}_1 - \overline{x}_2 = 13.286 - 24.000 = -10.714$

original claim: $\mu_1 - \mu_2 < 0$ mg
H_o: $\mu_1 - \mu_2 = 0$ mg
H_1: $\mu_1 - \mu_2 < 0$ mg
$\alpha = 0.05$ and df = 7
C.V. $t = -t_\alpha = -1.895$
calculations:

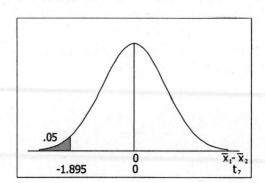

$$t_{\overline{x}_1 - \overline{x}_2} = (\overline{x}_1 - \overline{x}_2 - \mu_{\overline{x}_1 - \overline{x}_2}) / s_{\overline{x}_1 - \overline{x}_2}$$
$$= (-10.714 - 0) / \sqrt{14.0143/21 + 2.8571/8}$$
$$= -10.714/1.0122 = -10.585$$
P-value = tcdf(-99,-10.585,7) = 0.00000734
conclusion:
    Reject H_o; there is sufficient evidence to conclude that $\mu_1 - \mu_2 < 0$.

27. Let the pre-1964 quarters be group 1. Statistical software yields the following summary.
    group 1 (pre-1964):   n = 40 $\overline{x} = 6.192672$ s = 0.086995
    group 2 (post-1964):   n = 40 $\overline{x} = 5.639298$ s = 0.061937
    $\overline{x}_1 - \overline{x}_2 = 6.192672 - 5.639298 = 0.553374$

original claim: $\mu_1 - \mu_2 = 0$ g
H_0: $\mu_1 - \mu_2 = 0$ g
H_1: $\mu_1 - \mu_2 \neq 0$ g
$\alpha = 0.05$ [assumed] and df = 39
C.V. $t = \pm t_{\alpha/2} = \pm 2.024$

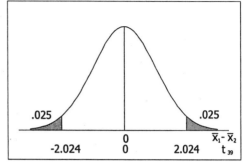

calculations:
$$t_{\overline{x}_1 - \overline{x}_2} = (\overline{x}_1 - \overline{x}_2 - \mu_{\overline{x}_1 - \overline{x}_2})/s_{\overline{x}_1 - \overline{x}_2}$$
$$= (0.553374 - 0)/\sqrt{(0.086995)^2/40 + (0.061937)^2/40}$$
$$= 0.553374/0.016885 = 32.773$$
P-value = $2 \cdot \text{tcdf}(32.773, 99, 39) = 2(2.68 \times 10^{-30}) = 5.37 \times 10^{-30} \approx 0$

conclusion:

    Reject H_0; there is sufficient evidence to reject the claim that $\mu_1 - \mu_2 = 0$ and conclude that $\mu_1 - \mu_2 \neq 0$ (in fact, that $\mu_1 - \mu_2 > 0$).

Yes. The conclusion is valid even though the population sizes are extremely large. If the requirements for using the test are satisfied, the size of the original population is irrelevant – except in cases where n>.05N and the finite population factor should be employed.

29. Let the people using the Weight Watchers diet be group 1.

    $\overline{x}_1 - \overline{x}_2 = 3.0 - 2.1 = 0.9$ $s_p^2 = [(df_1)s_1^2 + (df_2)s_2^2]/[df_1 + df_2]$

    df = $df_1 + df_2 = 39 + 39 = 78$ $= [(39)(4.9)^2 + (39)(4.8)^2]/[39 + 39] = 1834.95/78 = 23.525$

    $\alpha = 0.05$ and df = 78

    $(\overline{x}_1 - \overline{x}_2) \pm t_{\alpha/2} s_{\overline{x}_1 - \overline{x}_2}$

        $0.9 \pm 1.990 \sqrt{23.525/40 + 23.525/40}$

        0.9 ± 2.2

        $-1.3 < \mu_1 - \mu_2 < 3.1$ (lbs)

To one decimal accuracy, the results are the same as in exercise #12. In this case the results are not affected by the assumption of equal standard deviations.

31. Let those receiving the magnet treatment be group 1.

    $\overline{x}_1 - \overline{x}_2 = 0.49 - 0.44 = 0.05$ $s_p^2 = [(df_1)s_1^2 + (df_2)s_2^2]/[df_1 + df_2]$

    df = $df_1 + df_2 = 19 + 19 = 38$ $= [(19)(0.96)^2 + (19)(1.4)^2]/[19+19] = 54.7504/38 = 1.4408$

    original claim: $\mu_1 - \mu_2 > 0$
    H_0: $\mu_1 - \mu_2 = 0$
    H_1: $\mu_1 - \mu_2 > 0$
    $\alpha = 0.05$ and df = 38
    C.V. $t = t_{\alpha} = 1.686$
    calculations:

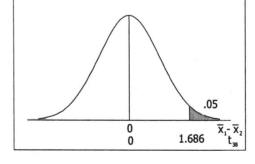

        $t_{\overline{x}_1 - \overline{x}_2} = (\overline{x}_1 - \overline{x}_2 - \mu_{\overline{x}_1 - \overline{x}_2})/s_{\overline{x}_1 - \overline{x}_2}$

          $= (0.05 - 0)/\sqrt{1.4408/20 + 1.4408/20}$

          $= 0.05/0.3796 = 0.132$

    P-value = tcdf(0.132, 99, 38) = 0.4479

conclusion:

    Do not reject H_0; there is not sufficient evidence to conclude that $\mu_1 - \mu_2 > 0$.

The conclusion is the same as in exercise #17, and the values in the problem changed very little. The calculated t statistic is exactly the same, as will be the case whenever $n_1 = n_2$.

33. Let the cars be group 1.

group 1: cars (n=28)

$\Sigma x = 200$

$\Sigma x^2 = 3724$

$\bar{x} = 7.143$

$s^2 = 85.0159$ (s=9.220)

group 2: taxis (n=20)

$\Sigma x = 117$

$\Sigma x^2 = 837$

$\bar{x} = 5.850$

$s^2 = 8.0289$ (s=2.834)

$\bar{x}_1 - \bar{x}_2 = 7.143 - 5.850 = 1.293$

original claim: $\mu_1 - \mu_2 \neq 0$ yrs

H_o: $\mu_1 - \mu_2 = 0$ yrs

H_1: $\mu_1 - \mu_2 \neq 0$ yrs

$\alpha = 0.05$ and df = 19

C.V. $t = \pm t_{\alpha/2} = \pm 2.093$

calculations:

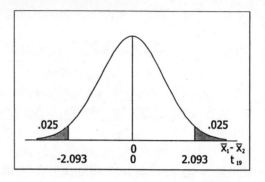

$$t_{\bar{x}_1 - \bar{x}_2} = (\bar{x}_1 - \bar{x}_2 - \mu_{\bar{x}_1 - \bar{x}_2})/s_{\bar{x}_1 - \bar{x}_2}$$

$$= (1.293 - 0)/\sqrt{85.0159/28 + 8.0289/20}$$

$$= 1.293/1.854 = 0.697$$

P-value = $2 \cdot \text{tcdf}(0.697, 99, 19) = 2(0.2470) = 0.4941$

conclusion:

Do not reject H_o; there is not sufficient evidence to conclude that $\mu_1 - \mu_2 \neq 0$.

The conclusion is the same. The large change in $\bar{x}_1$ does not affect the problem because of the corresponding increase in the variability in the problem.

35. a. x = 5,10,15

$\mu = (\Sigma x)/N = 30/3 = 10$

$\sigma^2 = \Sigma(x-\mu)^2/N$

$\quad = [(-5)^2 + (0)^2 + (5)^2]/3 = 50/3$

b. y = 1,2,3

$\mu = (\Sigma y)/N = 6/3 = 2$

$\sigma^2 = \Sigma(y-\mu)^2/N$

$\quad = [(-1)^2 + (0)^2 + (1)^2]/3 = 2/3$

c. z = x-y = 4,3,2,9,8,7,14,13,12

$\mu = (\Sigma z)/N = 72/9 = 8$

$\sigma^2 = \Sigma(z-\mu)^2/N$

$\quad = [(-4)^2 + (-5)^2 + (-6)^2 + (1)^2 + (0)^2 + (-1)^2 + (6)^2 + (5)^2 + (4)^2]/9 = 156/9 = 52/3$

d. $\sigma^2_{x-y} = \sigma^2_x + \sigma^2_y$

$52/3 = 50/3 + 2/3$

$52/3 = 52/3$

The variance of the differences is equal to the sum of the variances.

e. Let R stand for range.

$R_{x-y} = \text{highest}_{x-y} - \text{lowest}_{x-y}$

$\quad = (\text{highest x} - \text{lowest y}) - (\text{lowest x} - \text{highest y})$

$\quad = (\text{highest x} - \text{lowest x}) + (\text{highest y} - \text{lowest y})$

$\quad = R_x + R_y$

Just as the variance of the differences is equal to the sum of the variances, so is the range of the differences equal to the sum of the ranges.

NOTE: This entire problem refers to all possible x-y difference (where n_x and n_y might even be unequal), and not to x-y differences for paired data.

37. $A = s_1^2/n_1 = 0.0064/10 = 0.00064$

$B = s_2^2/n_2 = 0.0064/10 = 0.00064$

$df = (A+B)^2/(A^2/df_1 + B^2/df_2)$
$= (0.00064 + 0.00064)^2/(0.00064^2/9 + 0.00064^2/9)$
$= 0.000001638/0.000000091$
$= 18$

When $s_1^2 = s_2^2$ and $n_1 = n_2$, Formula 9-1 yields $df = df_1 + df_2$.

In exercises #19 and #20, for which $\alpha = 0.01$, the tabled t value changes from $t_{9,\alpha/2} = 3.250$ to $t_{18,\alpha/2} = 2.878$. In general, the larger df signifies a "tighter" t distribution that is closer to the z distribution. In the test of hypothesis, the sampling distribution will be "tighter" and the critical t will be smaller – and since the calculated t is not affected, the P-value will be smaller. In the confidence interval, the tabled t will be smaller – and since the other values are not affected, the interval will be narrower. Using $df = \min(df_1, df_2)$ is more conservative in that such action will not reject H_o as often and will lead to wider confidence intervals – i.e., it allows for a wider range of possible values for the parameter.

9-4 Inferences from Matched Pairs

1. No. The methods of this section apply to the differences between values that are matched in some way – but the difference between values with different units is not a meaningful concept.

3. When n d values represent the differences (in some specified order) between the two values in n matched pair of values, then $\overline{d}$ denotes the mean of those differences and s_d denotes the standard deviation of those differences.

NOTE: To be consistent with the notation of the previous sections, thus reinforcing patterns and concepts presented in those sections, the manual uses the usual t formula written to apply to $\overline{d}$'s
$$t_{\overline{d}} = (\overline{d} - \mu_{\overline{d}})/s_{\overline{d}}, \text{ with } \mu_{\overline{d}} = \mu_d \text{ and } s_{\overline{d}} = s_d/\sqrt{n}$$

5. $d = x - y$: 0 2 -2 -3 1
summary: $n = 5$ $\Sigma d = -2$ $\Sigma d^2 = 18$
   a. $\overline{d} = (\Sigma d)/n = -2/5 = -0.4$
   b. $s_d^2 = [n(\Sigma d^2) - (\Sigma d)^2]/[n(n-1)]$
      $= [5(18) - (-2)^2]/[5(4)] = 86/20 = 4.3$
   $s_d = 2.07$
   c. $t_{\overline{d}} = (\overline{d} - \mu_{\overline{d}})/s_{\overline{d}}$
      $= (-0.4 - 0)/(2.07/\sqrt{5}) = -0.4/0.927 = -0.431$
   d. with df=4 and α=0.05, the critical values are $t = \pm t_{\alpha/2} = \pm 2.776$

7. $\alpha = 0.05$ and $df = 4$
   $\overline{d} \pm t_{\alpha/2} \cdot s_{\overline{d}}$
   $-0.4 \pm 2.776 \cdot (2.07/\sqrt{5})$
   -0.4 ± 2.6
   $-3.0 < \mu_d < 2.2$

9. a. original claim: $\mu_d \neq 0$
   H_o: $\mu_d = 0$
   H_1: $\mu_d \neq 0$
   $\alpha = 0.05$ and df $=9$
   C.V. $t = \pm t_{\alpha/2} = \pm 2.262$
   calculations:

   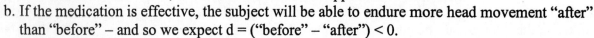

$$t_{\bar{d}} = (\bar{d} - \mu_{\bar{d}})/s_{\bar{d}}$$
$$= -0.41 \text{ [Minitab]}$$
      P-value = 0.6910 [Minitab]
   conclusion:
      Do not reject H_o; there is not sufficient evidence to conclude that $\mu_d \neq 0$.
   No. Based on this result, the medication does not appear to be effective.

   b. If the medication is effective, the subject will be able to endure more head movement "after"
      than "before" – and so we expect d = ("before" – "after") < 0.
      original claim: $\mu_d < 0$
      H_o: $\mu_d = 0$
      H_1: $\mu_d < 0$
      $\alpha = 0.05$ and df $= 9$
      C.V. $t = -t_{\alpha} = -1.833$
      calculations:

   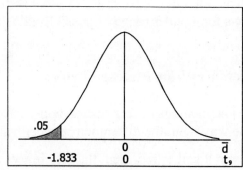

$$t_{\bar{d}} = (\bar{d} - \mu_{\bar{d}})/s_{\bar{d}}$$
$$= -0.41 \text{ [Minitab]}$$
         P-value = ½(0.6910) = 0.3455 [Minitab]
      conclusion:
         Do not reject H_o; there is not sufficient evidence to conclude that $\mu_d < 0$.

11. d = actual – forecast: 2 2 0 0 -3
    $n = 5$ $\Sigma d = 1$ $\Sigma d^2 = 17$ $\bar{d} = 0.2$ $s_d = 2.049$
    a. original claim: $\mu_d = 0$ °F
       H_o: $\mu_d = 0$ °F
       H_1: $\mu_d \neq 0$ °F
       $\alpha = 0.05$ and df $= 4$
       C.V. $t = \pm t_{\alpha/2} = \pm 2.776$
       calculations

$$t_{\bar{d}} = (\bar{d} - \mu_{\bar{d}})/s_{\bar{d}}$$
$$= (0.2 - 0)/(2.049/\sqrt{5})$$
$$= 0.2/0.917 = 0.218$$
          P-value = 2·tcdf(0.218,99,4) = 2(0.4190) = 0.8379
       conclusion:
          Do not reject H_o; there is not sufficient evidence to reject the claim that $\mu_d = 0$.
       The results suggest that, on the average, the one-day forecasts are on target.
    b. $\alpha = 0.05$ and df $= 4$
       $$\bar{d} \pm t_{\alpha/2} \cdot s_{\bar{d}}$$

       $0.2 \pm 2.776 \cdot (2.049/\sqrt{5})$
       0.2 ± 2.5
       $-2.3 < \mu_d < 2.7$ (°F)
       Since the confidence interval contains 0, there is no reason to reject the notion that the one-
    day predictions are correct on the average.

13. $d = day1 - day2$:   22   10   10   30   0   -10   -4   12   10   14   3   0   0   12

   $n = 14$ $\Sigma d = 109$ $\Sigma d^2 = 2293$ $\overline{d} = 7.786$ $s_d = 10.541$

   $\alpha = 0.05$ and df = 13

   $\overline{d} \pm t_{\alpha/2} \cdot s_{\overline{d}}$

   $7.8 \pm 2.160 \cdot (10.541/\sqrt{14})$

   7.8 ± 6.1

   $1.7 < \mu_d < 13.9$ (mmHg)

   The results suggest either that blood pressure is not a stable and/or consistent quantity, or that it cannot (with present technology) be measured with accuracy and/or consistency.

15. $d = second - first$:   -20   -10   0   -20   30   20   -40   20   30

   $n = 9$ $\Sigma d = -10$ $\Sigma d^2 = 5100$ $\overline{d} = -1.111$ $s_d = 25.221$

   original claim: $\mu_d = 0$

   H_o: $\mu_d = 0$

   H_1: $\mu_d \neq 0$

   $\alpha = 0.05$ [assumed] and df = 8

   C.V. $t = \pm t_{\alpha/2} = \pm 2.306$

   calculations

   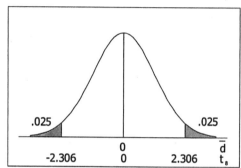

   $t_{\overline{d}} = (\overline{d} - \mu_{\overline{d}})/s_{\overline{d}}$

   $= (-1.111 - 0)/(25.221/\sqrt{9})$

   $= -1.111/8.404 = -0.132$

   P-value = $2 \cdot tcdf(-99, -0.132, 8) = 2(0.4491) = 0.8981$

   conclusion:

   Do not reject H_o; there is not sufficient evidence to reject the claim that $\mu_d = 0$.

   The results suggest that there is not much change between the first and second scores of those repeating the SAT with no intermediate preparatory course.

17. $d = before - after$:   -0.2   4.1   1.6   1.8   3.2   2.0   2.9   9.6

   $n = 8$ $\Sigma d = 25.0$ $\Sigma d^2 = 137.46$ $\overline{d} = 3.125$ $s_d = 2.9114$

   a. $\alpha = 0.05$ and df = 7

   $\overline{d} \pm t_{\alpha/2} \cdot s_{\overline{d}}$

   $3.125 \pm 2.365 \cdot (2.9114/\sqrt{8})$

   3.125 ± 2.434

   $0.69 < \mu_d < 5.56$ (cm)

   b. original claim: $\mu_d > 0$ cm

   H_o: $\mu_d = 0$ cm

   H_1: $\mu_d > 0$ cm

   $\alpha = 0.05$ and df = 7

   C.V. $t = t_{\alpha} = 1.895$

   calculations

   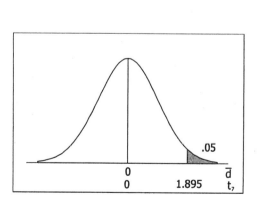

   $t_{\overline{d}} = (\overline{d} - \mu_{\overline{d}})/s_{\overline{d}}$

   $= (3.125 - 0)/(2.9114/\sqrt{8})$

   $= 3.125/1.029 = 3.036$

   P-value = $tcdf(3.036, 99, 7) = 0.0095$

   conclusion:

   Reject H_o; there is sufficient evidence to conclude that $\mu_d > 0$.

c. Yes, hypnotism appears to be effective in reducing pain.

19. d = regular – kiln: -106 20 -101 33 -72 36 -62 -38 70 -127 -24

   $n = 11$ $\Sigma d = -371$ $\Sigma d^2 = 56299$ $\bar{d} = -33.73$ $s_d = 66.171$

   a. original claim: $\mu_d = 0$ lbs/acre

   H_o: $\mu_d = 0$ lbs/acre

   H_1: $\mu_d \neq 0$ lbs/acre

   $\alpha = 0.05$ and df = 10

   C.V. $t = \pm t_{\alpha/2} = \pm 2.228$

   calculations

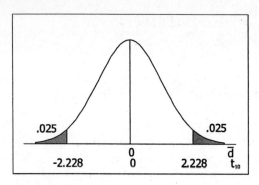

   $$t_{\bar{d}} = (\bar{d} - \mu_{\bar{d}})/s_{\bar{d}}$$

   $$= (-33.73 - 0)/(66.171/\sqrt{11})$$

   $$= -33.73/19.951 = -1.690$$

   P-value = 2·tcdf(-99,-1.690,10) = 2(0.0609) = 0.1218

   conclusion:

   Do not reject H_o; there is not sufficient evidence to reject the claim that $\mu_d = 0$.

   b. $\alpha = 0.05$ and df = 10

   $$\bar{d} \pm t_{\alpha/2} \cdot s_{\bar{d}}$$

   $-33.73 \pm 2.228 \cdot (66.171/\sqrt{11})$

   -33.73 ± 44.45

   $-78.2 < \mu_d < 10.7$ (lbs/acre)

c. We cannot be 95% certain that either type of seed is better. If there are no differences in cost, or any other considerations, choose the kiln dried – even though we can't be sure that it's generally better, it did have the higher yield in this particular trial.

21. The Minitab commands...let c11=c2-c8

                       name c11 'LA-L5'

                       ttest c11

produce the output below.

   Test of mu = 0 vs not = 0

   Variable N   Mean   StDev SE Mean   95% CI   T   p

   LA-L5   35 0.142857 5.440279 0.919575 (-1.725944, 2.011658) 0.16 0.877

   a. original claim: $\mu_d = 0$ °F

   H_o: $\mu_d = 0$ °F

   H_1: $\mu_d \neq 0$ °F

   $\alpha = 0.05$ and df = 34

   C.V. $t = \pm t_{\alpha/2} = \pm 2.032$

   calculations

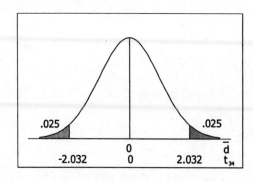

   $$t_{\bar{d}} = (\bar{d} - \mu_{\bar{d}})/s_{\bar{d}}$$

   $$= (0.142857 - 0)/(5.440279/\sqrt{35})$$

   $$= 0.142857/0.919575$$

   $$= 0.16 \text{ [Minitab]}$$

   P-value = 2·P(t_{34}>0.16) = 0.877 [Minitab]

   conclusion:

   Do not reject H_o; there is not sufficient evidence to reject the claim that $\mu_d = 0$.

b. $\alpha = 0.05$ and df $= 34$

$\overline{d} \pm t_{\alpha/2} \cdot s_{\overline{d}}$

$0.14 \pm 2.032 \cdot (5.440279/\sqrt{35})$

0.14 ± 1.87

$-1.7 < \mu_d < 2.0$ (°F)

c. The results are the same as in the example. It appears that, on the average, the forecast low temperatures are accurate.

23. The Minitab commands…let c10=c1-c2

    name c10 'S-L'

    ttest c10

produce the output below.

Test of mu = 0 vs not = 0

Variable	N	Mean	StDev	SE Mean	95% CI	T	p
S-L	40	-7265.00	8582.32	1356.98	(-10009.76, -4520.24)	-5.35	0.000

a. original claim: $\mu_d = \$0$

H$_0$: $\mu_d = \$0$

H$_1$: $\mu_d \neq \$0$

$\alpha = 0.05$ and df $= 39$

C.V. $t = \pm t_{\alpha/2} = \pm 2.024$

calculations

$t_{\overline{d}} = (\overline{d} - \mu_{\overline{d}})/s_{\overline{d}}$

$= (-7265 - 0)/(8582.32/\sqrt{40})$

$= -7265/1356.98$

$= -5.35$ [Minitab]

P-value $= 2 \cdot P(t_{39} < -5.35) = 0.000$ [Minitab]

conclusion:

Reject H$_0$; there is sufficient evidence to reject the claim that $\mu_d = 0$ and conclude that $\mu_d \neq 0$ (in fact, that $\mu_d < 0$ – i.e., that the selling price is less than the list price).

b. $\alpha = 0.05$ and df $= 39$

$\overline{d} \pm t_{\alpha/2} \cdot s_{\overline{d}}$

$-7265.0 \pm 2.024 \cdot (8582.32/\sqrt{40})$

-7265.0 ± 2746.5

$-10,011.5 < \mu_d < -4518.5$ ($) [or $-10,009.76 < \mu_d < -4520.24$ ($), from Minitab]

25. The points on the graph are as given below.

    before: 460 470 490 490 510 510 600 610 620

    after: 480 510 500 610 590 630 630 690 660

d = after – before: 20   40   10 120   80 120   30   80   40

$n = 9$ $\Sigma d = 540$ $\Sigma d^2 = 46200$ $\overline{d} = 60.0$ $s_d = 41.533$

original claim: $\mu_d > 0$
H_o: $\mu_d = 0$
H_1: $\mu_d > 0$
$\alpha = 0.05$ [assumed] and df = 8
C.V. $t = t_\alpha = 1.860$
calculations

$t_{\bar{d}} = (\bar{d} - \mu_{\bar{d}})/s_{\bar{d}}$

$= (60.0 - 0)/(41.533/\sqrt{9})$

$= 60.0/13.844 = 4.334$

P-value $= tcdf(4.334,99,8) = 0.0012$

conclusion:
    Reject H_o; there is sufficient evidence to conclude that $\mu_d > 0$.
NOTE: If one defines the difference in terms of "before – after", then the original claim is
"$\mu_d < 0$" and the C.V. and calculated t are the negative of the above values. The choice is
arbitrary.

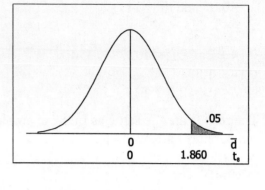

27. The following table of summary statistics applies to all parts of this exercise.

	values	n	Σv	Σv^2	$\bar{v}$	s_v
x:	1 3 2 2 1 2 3 3 2 1	10	20	46	2.0	0.816
y:	1 2 1 2 1 2 1 2 1 2	10	15	25	1.5	0.527
d = x-y:	0 1 1 0 0 0 2 1 1 -1	10	5	9	0.5	0.850

a. original claim: $\mu_d > 0$
   H_o: $\mu_d = 0$
   H_1: $\mu_d > 0$
   $\alpha = 0.05$ and df = 9
   C.V. $t = t_\alpha = 1.833$
   calculations

   $t_{\bar{d}} = (\bar{d} - \mu_{\bar{d}})/s_{\bar{d}}$

   $= (0.5 - 0)/(0.850/\sqrt{10})$

   $= 0.5/0.2687 = 1.861$

   P-value $= tcdf(1.861,99,9) = 0.0479$

   conclusion:
       Reject H_o; there is sufficient evidence to conclude that $\mu_d > 0$.

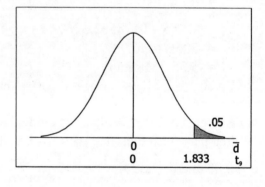

b. $\bar{x}_1 - \bar{x}_2 = 2.0 - 1.5 = 0.5$
   original claim: $\mu_1 - \mu_2 > 0$
   H_o: $\mu_1 - \mu_2 = 0$
   H_1: $\mu_1 - \mu_2 > 0$
   $\alpha = 0.05$ and df = 9
   C.V. $t = t_\alpha = 1.833$
   calculations:

   $t_{\bar{x}_1 - \bar{x}_2} = (\bar{x}_1 - \bar{x}_2 - \mu_{\bar{x}_1 - \bar{x}_2})/s_{\bar{x}_1 - \bar{x}_2}$

   $= (0.5 - 0)/\sqrt{(0.816)^2/10 + (0.527)^2/10}$

   $= 0.5/0.3073 = 1.627$

   P-value $= tcdf(1.627,99,9) = 0.0691$

conclusion:

Do not reject H_o; there is not sufficient evidence to conclude that $\mu_1 - \mu_2 > 0$.

c. Parts (a) and (b) do not have the same conclusion. Since different methods can give different results, it is important that the correct method be used.

9-5 Comparing Variation in Two Samples

1. The F test described in this section is not robust – i.e., it is very sensitive to departures from normality. Two alternatives that perform well even when the original populations are not normal are (1) the count five test and (2) the Levene-Brown-Forsythe test.

3. The F distribution is the sampling distribution formed by the ratio of the sample variances from two independent normal distributions with equal variances.

NOTE: The following conventions are used in this manual regarding the F test.
• The set of scores with the larger sample variance is designated group 1.
• Even though designating the scores with the larger sample variance as group 1 makes lower tail critical values unnecessary in two-tailed tests, the lower critical vales are calculated [see exercise #23] and included in the accompanying graphs for completeness and consistency with the other tests. The F distribution always "bunches up" around 1.0 regardless of the df values.
• Since the F value depends on two degrees of freedom, the df for group 1 (numerator) and group 2 (denominator) may be used with the F as a superscript and a subscript respectively to clarify which F distribution is being used.
• If the desired df does not appear in Table A-5, the closest entry is used. For a desired entry exactly halfway between two tabled values, the conservative approach of using the smaller df is used – and 120 is used for all df larger than 120.
• F is statistically defined to be $F = (s_1^2/\sigma_1^2)/(s_2^2/\sigma_2^2)$. Since all hypotheses in the text question the equality of σ_1^2 and σ_2^2, the calculation of F may be shortened to $F = s_1^2/s_2^2$.
• Some problems are stated in terms of variance, and some are stated in terms of standard deviation – and some are stated simply in terms of "variation." Since $\sigma_1 = \sigma_2$ is equivalent to $\sigma_1^2 = \sigma_2^2$, this manual simply states all claims and hypotheses and conclusions using the variance.

5. Let the treatment group be group 1.

original claim: $\sigma_1^2 \neq \sigma_2^2$

$H_0: \sigma_1^2 = \sigma_2^2$

$H_1: \sigma_1^2 \neq \sigma_2^2$

$\alpha = 0.05$ and $df_1 = 15$, $df_2 = 40$

C.V. $F = F_{\alpha/2} = 2.1819$

calculations:

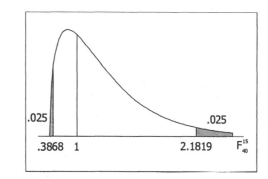

$$F = s_1^2/s_2^2$$

$$= (0.80)^2/(0.40)^2 = 4.000$$

P-value $= 2 \cdot Fcdf(4.000,99,15,40) = 2(0.0002) = 0.0005$

conclusion:

Reject H_o; there is sufficient evidence to conclude that $\sigma_1^2 \neq \sigma_2^2$ (in fact, that $\sigma_1^2 > \sigma_2^2$).

7. Let the regular Coke be group 1.

original claim: $\sigma_1^2 \neq \sigma_2^2$

$H_0 : \sigma_1^2 = \sigma_2^2$

$H_1 : \sigma_1^2 \neq \sigma_2^2$

$\alpha = 0.05$ and $df_1 = 35$, $df_2 = 35$

C.V. $F = F_{\alpha/2} = 2.0739$

calculations:

    $F = s_1^2 / s_2^2$

       $= (0.0075)^2 / (0.0044)^2 = 2.9233$ [TI-83/84+]

    P-value $= 0.0021$ [TI-83/84+]

conclusion:

Reject H_0; there is sufficient evidence to conclude that $\sigma_1^2 \neq \sigma_2^2$ (in fact, that $\sigma_1^2 > \sigma_2^2$). The addition of sugar to the regular Coke may account for the larger variation. In general, each additional ingredient contributes a certain amount of its own variability – in particular, it would be difficult to ensure there is precisely the same amount of sugar in each can.

9. Let the diet Coke be group 1.

original claim: $\sigma_1^2 \neq \sigma_2^2$

$H_0 : \sigma_1^2 = \sigma_2^2$

$H_1 : \sigma_1^2 \neq \sigma_2^2$

$\alpha = 0.05$ and $df_1 = 35$, $df_2 = 35$

C.V. $F = F_{\alpha/2} = 2.0739$

calculations:

    $F = s_1^2 / s_2^2$

       $= (0.004391)^2 / (0.004362)^2 = 1.0133$

    P-value $= 2 \cdot \text{Fcdf}(1.0133, 99, 35, 35) = 2(0.4845) = 0.9690$

conclusion:

Do not reject H_0; there is not sufficient evidence to conclude that $\sigma_1^2 \neq \sigma_2^2$.

11. Let the sham group be group 1.

original claim: $\sigma_1^2 > \sigma_2^2$

$H_0 : \sigma_1^2 = \sigma_2^2$

$H_1 : \sigma_1^2 > \sigma_2^2$

$\alpha = 0.05$ and $df_1 = 19$, $df_2 = 19$

C.V. $F = F_{\alpha} = 2.1555$

calculations:

    $F = s_1^2 / s_2^2$

       $= (1.4)^2 / (0.96)^2 = 2.1267$

    P-value $= \text{Fcdf}(2.1267, 99, 19, 19) = 0.0543$

conclusion:

Do not reject H_0; there is not sufficient evidence to conclude that $\sigma_1^2 > \sigma_2^2$.

13. Let the treatment group be group 1.

original claim: $\sigma_1^2 > \sigma_2^2$

$H_0: \sigma_1^2 = \sigma_2^2$

$H_1: \sigma_1^2 > \sigma_2^2$

$\alpha = 0.05$ and $df_1 = 21$, $df_2 = 21$

C.V. $F = F_\alpha = 2.0960$

calculations:

$\quad F = s_1^2/s_2^2$

$\qquad = (2.20)^2/(0.72)^2 = 9.3364$

$\quad$ P-value = Fcdf(9.3364,99,21,21) = 0.00000168

conclusion:

$\quad$ Reject H_0; there is sufficient evidence to conclude that $\sigma_1^2 > \sigma_2^2$.

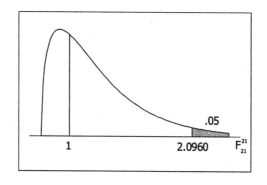

15. Let the pre-1964 quarters be group 1.

original claim: $\sigma_1^2 = \sigma_2^2$

$H_0: \sigma_1^2 = \sigma_2^2$

$H_1: \sigma_1^2 \neq \sigma_2^2$

$\alpha = 0.05$ and $df_1 = 39$, $df_2 = 39$

C.V. $F = F_{\alpha/2} = 1.8752$

calculations:

$\quad F = s_1^2/s_2^2$

$\qquad = (0.08700)^2/(0.06194)^2 = 1.9729$

$\quad$ P-value = 2·Fcdf(1.9729,99,39,39) = 2(0.0184) = 0.0368

conclusion:

$\quad$ Reject H_0; there is sufficient evidence to conclude that $\sigma_1^2 \neq \sigma_2^2$ (in fact, that $\sigma_1^2 > \sigma_2^2$).

No. It appears that the new quarters are <u>less</u> variable than the old ones, and so machines that accept quarters will not need adjustments to handle extra variability.

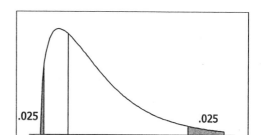

17. Let the easy to difficult group be group 1.

original claim: $\sigma_1^2 = \sigma_2^2$

$H_0: \sigma_1^2 = \sigma_2^2$

$H_1: \sigma_1^2 \neq \sigma_2^2$

$\alpha = 0.05$ and $df_1 = 24$, $df_2 = 15$

C.V. $F = F_{\alpha/2} = 2.7006$

calculations:

$\quad F = s_1^2/s_2^2$

$\qquad = 47.020/18.150 = 2.5906$

$\quad$ P-value = 2·Fcdf(2.5906,99,24,15) = 2(0.0299) = 0.0599

conclusion:

$\quad$ Do not reject H_0; there is not sufficient evidence to reject the claim that $\sigma_1^2 = \sigma_2^2$.

19. data summary from statistical software

variable	n	$\overline{X}$	s
WED	53	0.0508	0.1346
SUN	52	0.0667	0.2000

a. Let the Sunday amounts be group 1.

original claim: $\sigma_1^2 = \sigma_2^2$

$H_0: \sigma_1^2 = \sigma_2^2$

$H_1: \sigma_1^2 \neq \sigma_2^2$

$\alpha = 0.05$ and $df_1 = 51$, $df_2 = 52$

C.V. $F = F_{\alpha/2} = 1.6668$

calculations:

$F = s_1^2/s_2^2$

$= (0.2000)^2/(0.1346)^2 = 2.2079$

P-value $= 2 \cdot Fcdf(2.2079,99,51,52) = 2(0.0026) = 0.0052$

conclusion:

Reject H_0; there is sufficient evidence to reject the claim that $\sigma_1^2 = \sigma_2^2$ and conclude

that $\sigma_1^2 \neq \sigma_2^2$ (in fact, that $\sigma_1^2 > \sigma_2^2$).

b. For Sundays, 37 of the 52 observations are 0; for Wednesdays, 37 of the 53 observations are 0. The data for both days are very positively skewed. The rainfall amounts do not come from populations with normal distributions.

c. Because the original populations are not normally distributed, the test in part (a) is not valid – which is good news, since there is no reason why rainfall amounts should be more variable on Sundays than on Wednesdays.

21. The Minitab commands…let c9 = abs(c1 - 0.81682)

let c10 = abs(c5 - 0.82410)

sort c9 c11

sort c10 c12

indicate that the largest Coke MAD 0.02672 is the only value larger than the largest Pepsi MAD 0.01600 – and so $c_1 = 1$ and $c_2 = 0$.

original claim: Coke and Pepsi have equal variation

H_0: Coke and Pepsi have equal variation

H_1: Coke and Pepsi do not have equal variation

$\alpha = 0.05$

C.V. c = 5.

calculations:

$c_1 = 1$

$c_2 = 0$

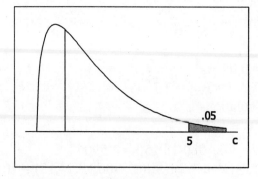

conclusion:

Do not reject H_0; there is not sufficient evidence to reject the claim that Coke and Pepsi have equal variation.

23. a. $\alpha = 0.05$ $F_L = F^9_{9,1-\alpha/2} = 1/F^9_{9,\alpha/2} = 1/4.0260 = 0.2484$ $F_R = F^9_{9,\alpha/2} = 4.0260$

b. $\alpha = 0.05$ $F_L = F^9_{6,1-\alpha/2} = 1/F^6_{9,\alpha/2} = 1/4.3197 = 0.2315$ $F_R = F^9_{6,\alpha/2} = 5.5234$

c. $\alpha = 0.05$ $F_L = F^6_{9,1-\alpha/2} = 1/F^9_{6,\alpha/2} = 1/5.5234 = 0.1810$ $F_R = F^6_{9,\alpha/2} = 4.3197$

Statistical Literacy and Critical Thinking

1. The statement "he is preferred by more male voters than female voters" is not clear.
   • If the statement intends to say "most of the voters who prefer him are male", then the claim is $p > 0.5$ and the techniques of section 8.3 are appropriate.
   • If the statement intends to say "the proportion of voters who prefer him is larger among males than among females", then the claim is $p_1 - p_2 > 0$ and the techniques of section 9.2 are appropriate.

2. Yes. Since every group of 200 creditors has the same chance of being selected, this satisfies the requirement for being a simple random sample of creditors.
   NOTE: There are two ambiguities in this problem that need to be resolved before proceeding.

   • Is the intent to compare the debt of men who live in the state to the debt of women who live in the state – no matter in what state the money is owed? Or is the intent to compare the debt owed within the state by men to the debt owed within the state by women – no matter in what state the debtors reside?
   • The interest appears to be in the debtors, but the simple random sample was taken from among the creditors. Since a single debtor can owe money to more than one creditor, debtors with multiple creditors are more likely to appear in the survey – and a simple random sample of creditors will not produce a simple random sample of debtors.

3. No, this procedure is not acceptable. The procedure described counts the mean income from each state equally, but the mean income from a large population state has a greater influence on the national mean than does the mean income from a small population state. The researcher should use weighted means.

4. If two samples are independent, then the inclusion of one observation in one of the samples in no way affects or determines the inclusion of any observation in the other sample.

Review Exercises

1. Let the black drivers be group 1.

$\hat{p}_1 = x_1/n_1 = 24/200 = 0.120$

$\hat{p}_2 = x_2/n_2 = 147/1400 = 0.105$

$\hat{p}_1 - \hat{p}_2 = 0.120 - 0.105 = 0.015$

$\bar{p} = (24+147)/(200+1400) = 171/1600 = 0.107$

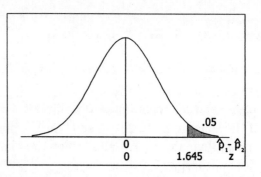

a. original claim: $p_1 - p_2 > 0$

   $H_o: p_1 - p_2 = 0$

   $H_1: p_1 - p_2 > 0$

   $\alpha = 0.05$

   C.V. $z = z_\alpha = 1.645$

   calculations:

   $z_{\hat{p}_1 - \hat{p}_2} = (\hat{p}_1 - \hat{p}_2 - \mu_{\hat{p}_1 - \hat{p}_2})/\sigma_{\hat{p}_1 - \hat{p}_2}$

   $= (0.015 - 0)/\sqrt{(0.107)(0.893)/200 + (0.107)(0.893)/1400}$

   $= 0.015/0.02335 = 0.64$

   P-value $= P(z>0.64) = 1 - P(z<0.64) = 1 - 0.7389 = 0.2611$

   conclusion:

   Do not reject H_o; there is not sufficient evidence to conclude that $p_1 - p_2 > 0$.

b. $\alpha = 0.10$

   $(\hat{p}_1 - \hat{p}_2) \pm z_{\alpha/2}\, \sigma_{\hat{p}_1 - \hat{p}_2}$

   $0.015 \pm 1.645 \sqrt{(0.120)(0.880)/200 + (0.105)(0.895)/1400}$

   0.015 ± 0.040

   $-0.025 < p_1 - p_2 < 0.055$

   Since the interval includes 0, there is not sufficient evidence to conclude that $p_1 - p_2 > 0$.

2. d = reported − measured: 1.2   0.1   0.4   1.8   0.1   1.6   0.1   1.4   -0.9   0.4   1.2

   $n = 11$ $\Sigma d = 7.4$ $\Sigma d^2 = 11.80$ $\bar{d} = 0.673$ $s_d = 0.8259$

   original claim: $\mu_d > 0$ in

   $H_o: \mu_d = 0$ in

   $H_1: \mu_d > 0$ in

   $\alpha = 0.05$ [assumed] and df = 10

   C.V. $t = t_\alpha = 1.812$

   calculations

   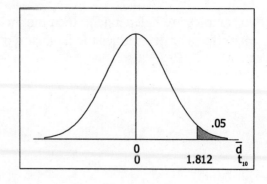

   $t_{\bar{d}} = (\bar{d} - \mu_{\bar{d}})/s_{\bar{d}}$

   $= (0.673 - 0)/(0.8259/\sqrt{11})$

   $= 0.673/0.2490 = 2.701$

   P-value $= \text{tcdf}(2.701,99,10) = 0.0111$

   conclusion:

   Reject H_o; there is sufficient evidence to conclude that $\mu_d > 0$.

3. Let the Rowling scores be group 1.

group 1: Rowling (n=12)
$\Sigma x = 969.0$
$\Sigma x^2 = 78487.82$
$\bar{x} = 80.75$
$s^2 = 21.915$ (s=4.681)

group 2: Tolstoy (n=12)
$\Sigma x = 793.8$
$\Sigma x^2 = 53189.10$
$\bar{x} = 66.15$
$s^2 = 61.748$ (s=7.858)

$\bar{x}_1 - \bar{x}_2 = 80.75 - 66.15 = 14.60$

original claim: $\mu_1 - \mu_2 > 0$
H_o: $\mu_1 - \mu_2 = 0$
H_1: $\mu_1 - \mu_2 > 0$
$\alpha = 0.05$ and $df = 11$
C.V. $t = t_\alpha = 1.796$
calculations:

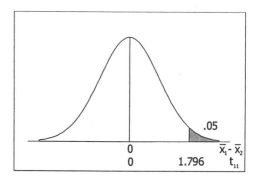

$$t_{\bar{x}_1-\bar{x}_2} = (\bar{x}_1-\bar{x}_2 - \mu_{\bar{x}_1-\bar{x}_2})/s_{\bar{x}_1-\bar{x}_2}$$
$$= (14.60 - 0)/\sqrt{21.915/12 + 61.748/12}$$
$$= 14.60/2.6404 = 5.529$$

P-value = tcdf(5.529,99,11) = 0.0001
conclusion:

Reject H_o; there is sufficient evidence to conclude that $\mu_1 - \mu_2 > 0$.
Yes. It was expected that the Harry Potter book would be easier to read than *War and Peace*.

4. Refer to the summary calculations in exercise #3.

Let the Tolstoy scores be group 1.

original claim: $\sigma_1^2 = \sigma_2^2$

H_o: $\sigma_1^2 = \sigma_2^2$

H_1: $\sigma_1^2 \neq \sigma_2^2$

$\alpha = 0.05$ and $df_1 = 11$, $df_2 = 11$
C.V. $F = F_{\alpha/2} = 3.5257$
calculations:

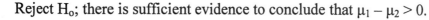

$F = s_1^2/s_2^2$
$$= 61.748/21.915 = 2.8176$$

P-value = 2·Fcdf(2.8176,99,11,11) = 2(0.0500) = 0.1000
conclusion:

Do not reject H_o; there is not sufficient evidence to reject the claim that $\sigma_1^2 = \sigma_2^2$.

5. Let those that were warmed be group 1.

$\hat{p}_1 = x_1/n_1 = 6/104 = 0.0577$

$\hat{p}_2 = x_2/n_2 = 18/96 = 0.1875$

$\hat{p}_1 - \hat{p}_2 = 0.0577 - 0.1875 = -0.1298$

$\bar{p} = (6+18)/(104+96) = 24/200 = 0.12$

a. original claim: $p_1-p_2 < 0$

H_o: $p_1-p_2 = 0$
H_1: $p_1-p_2 < 0$
$\alpha = 0.05$
C.V. $z = -z_\alpha = -1.645$

calculations:

$z_{\hat{p}_1-\hat{p}_2} = (\hat{p}_1-\hat{p}_2 - \mu_{\hat{p}_1-\hat{p}_2})/\sigma_{\hat{p}_1-\hat{p}_2}$

$= (-0.1298 - 0)/ \sqrt{(0.12)(0.88)/104 + (0.12)(0.88)/96}$

$= -0.1298/0.0460 = -2.82$

P-value $= P(z<-2.82) = 0.0024$

conclusion:

Reject H_o; there is sufficient evidence to conclude that $p_1-p_2 < 0$.

Yes. If these results are verified, surgical patients should be routinely warmed.

b. The test in part (a) places 0.05 in the lower tail. The appropriate two-sided interval for testing the claim in part (a) places 0.05 in each tail – i.e., is a 90% confidence interval.

c. $\alpha = 0.10$

$(\hat{p}_1 - \hat{p}_2) \pm z_{\alpha/2}\, \sigma_{\hat{p}_1-\hat{p}_2}$

$-0.1298 \pm 1.645 \sqrt{(0.0577)(0.9423)/104 + (0.1875)(0.8125)/96}$

-0.1298 ± 0.0756

$-0.205 < p_1-p_2 < -0.054$

d. Since the test of hypothesis and the confidence interval use different estimates for $\sigma_{\hat{p}_1-\hat{p}_2}$, they are not mathematically equivalent – and so it is possible that they may lead to different conclusions.

6. Let the obsessive-compulsive patients be group 1.

$\bar{x}_1-\bar{x}_2 = 1390.03 - 1268.41 = 121.62$

a. $\alpha = 0.05$ and df = 9

$(\bar{x}_1-\bar{x}_2) \pm t_{\alpha/2}\, s_{\bar{x}_1-\bar{x}_2}$

$121.62 \pm 2.262 \sqrt{(156.84)^2/10 + (137.97)^2/10}$

121.62 ± 149.42

$-27.80 < \mu_1 - \mu_2 < 271.04$

b. original claim: $\mu_1 - \mu_2 = 0$

H_o: $\mu_1 - \mu_2 = 0$
H_1: $\mu_1 - \mu_2 \neq 0$
$\alpha = 0.05$ and df = 9
C.V. $t = \pm t_{\alpha/2} = \pm 2.262$

calculations:

$t_{\bar{x}_1-\bar{x}_2} = (\bar{x}_1-\bar{x}_2 - \mu_{\bar{x}_1-\bar{x}_2})/s_{\bar{x}_1-\bar{x}_2}$

$= (121.62 - 0)/ \sqrt{(156.84)^2/10 + (137.97)^2/10}$

$= 121.62/66.056 = 1.841$

P-value $= 2 \cdot \text{tcdf}(1.841,99,9) = 2(0.0494) = 0.0987$

conclusion:

     Do not reject H_o; there is not sufficient evidence to reject the claim that $\mu_1 - \mu_2 = 0$.

c. No. It does not appear that the total brain volume can be used as a reliable indicator.

7. Let the obsessive-compulsive patients be group 1.

   original claim: $\sigma_1^2 \neq \sigma_2^2$

   $H_o: \sigma_1^2 = \sigma_2^2$

   $H_1: \sigma_1^2 \neq \sigma_2^2$

   $\alpha = 0.05$ and $df_1 = 9$, $df_2 = 9$

   C.V. $F = F_{\alpha/2} = 4.0260$

   calculations:

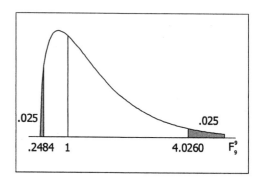

      $F = s_1^2/s_2^2$

        $= (156.84)^2/(137.97)^2 = 1.2922$

      P-value $= 2 \cdot \text{Fcdf}(1.2922,99,9,9) = 2(0.3544) = 0.7087$

   conclusion:

     Do not reject H_o; there is not sufficient evidence to conclude that $\sigma_1^2 \neq \sigma_2^2$.

8. d = regular – kiln: -5.75 -1.25 -1.00 -5.00 0.00 0.25 2.25 -0.50 0.75 -1.50 -0.25

   $n = 11$ $\Sigma d = -12.00$ $\Sigma d^2 = 68.8750$ $\bar{d} = -1.091$ $s_d = 2.3619$

  a. original claim: $\mu_d = 0$ cwt/acre

     $H_o: \mu_d = 0$ cwt/acre

     $H_1: \mu_d \neq 0$ cwt/acre

     $\alpha = 0.05$ and $df = 10$

     C.V. $t = \pm t_{\alpha/2} = \pm 2.228$

     calculations

        $t_{\bar{d}} = (\bar{d} - \mu_{\bar{d}})/s_{\bar{d}}$

           $= (-1.091 - 0)/(2.3619/\sqrt{11})$

           $= -1.091/0.7121 = -1.532$

     P-value $= 2 \cdot \text{tcdf}(-99,-1.532,10) = 2(0.0783) = 0.1565$

     conclusion:

       Do not reject H_o; there is not sufficient evidence to reject the claim that $\mu_d = 0$.

  b. $\alpha = 0.05$ and $df = 10$

     $\bar{d} \pm t_{\alpha/2} \cdot s_{\bar{d}}$

    $-1.09 \pm 2.228 \cdot (2.3619/\sqrt{11})$

    -1.09 ± 1.587

    $-2.67 < \mu_d < 0.50$

  c. We cannot be 95% certain that either type of seed is better. If there are no differences in cost, or any other considerations, choose the kiln dried – even though we can't be sure that it's generally better, it did have the higher yield in this particular trial.

Cumulative Review Exercises

1. Let the north speeds be group 1.

group 1: north (n=20)	group 2: south (n=20)	combined (n=40)
$\Sigma x = 1390$	$\Sigma x = 1345$	$\Sigma x = 2735$
$\Sigma x^2 = 96826$	$\Sigma x^2 = 91433$	$\Sigma x^2 = 188259$
$\overline{x} = 69.50$	$\overline{x} = 67.25$	$\overline{x} = 68.375$
$s^2 = 11.6316$	$s^2 = 51.6711$	$s^2 = 32.1378$
$s = 3.411$	$s = 7.188$	$s = 5.669$

$\overline{x}_1 - \overline{x}_2 = 69.50 - 27.25 = 2.25$

a. The ordered northbound speeds are:
   65 65 65 65 66 66 68 68 68 68 71 71 72 72 73 73 73 73 74 74
   $\overline{x} = 69.5$ mph
   $\tilde{x} = (68+71)/2 = 69.5$ mph
   $s = 3.4$ mph
   $s^2 = 11.63$ mi^2/hr^2
   $R = 74 - 65 = 9$ mph

b. original claim: $\mu > 65$ mph
   H_o: $\mu = 65$ mph
   H_1: $\mu > 65$ mph
   $\alpha = 0.05$ [assumed] and df = 39
   C.V. $t = t_\alpha = 1.686$
   calculations:

   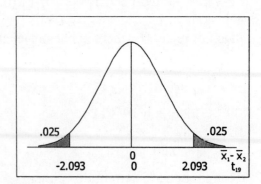

   $$t_{\overline{x}} = (\overline{x} - \mu)/s_{\overline{x}}$$
   $$= (68.375 - 65)/(5.669/\sqrt{40})$$
   $$= 3.375/0.8964 = 3.765$$
   P-value = tcdf(3.765,99,39) = 0.0003
   conclusion:
   Reject H_o; there is sufficient evidence to conclude that $\mu > 65$.

c. No. The northbound speeds in part (a) appear to be uniformly distributed from 65 to 74.

d. original claim: $\mu_1 - \mu_2 = 0$
   H_o: $\mu_1 - \mu_2 = 0$
   H_1: $\mu_1 - \mu_2 \neq 0$
   $\alpha = 0.05$ [assumed] and df = 19
   C.V. $t = \pm t_{\alpha/2} = \pm 2.093$
   calculations:

   $$t_{\overline{x}_1 - \overline{x}_2} = (\overline{x}_1 - \overline{x}_2 - \mu_{\overline{x}_1 - \overline{x}_2})/s_{\overline{x}_1 - \overline{x}_2}$$

   $$= (2.25 - 0)/\sqrt{11.6316/20 + 51.6711/20}$$
   $$= 2.25/1.7791$$
   $$= 1.265$$
   P-value = 2·tcdf(1.265,99,19) = 2(0.1106) = 0.2213
   conclusion:
   Do not reject H_o; there is not sufficient evidence to reject the claim that $\mu_1 - \mu_2 = 0$.
   Yes, the hypothesis test is likely to be valid despite the departure from normality noted in part (c) – since the uniform distribution is symmetric with no outliers, and the sample size is reasonable.

2. a. P(all 9 heads) = P(H_1 and H_2 and...H_9)

      = P(H_1)· P(H_2)·...· P(H_9)

      = (½)(½)...(½) = (½)^9

      = 1/512 = 0.00195

The probability of obtaining all 9 heads by chance alone is so small that it is reasonable to conclude that factors other than chance are involved – whether psychic abilities or deceitful procedures.

   b. Independent. The outcomes are not matched or paired in any way, nor does the outcome for one coin affect the outcome for the other.

   c. original claim: p > 0.50

      $\hat{p}$ = x/n = 14/18 = 0.778

    H_o: p = 0.50

    H_1: p > 0.50

    α = 0.01

    C.V. z = z_α = 2.326

    calculations:

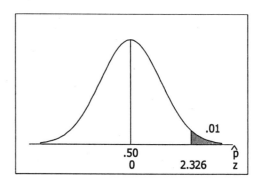

      $z\hat{p}$ = ($\hat{p} - \mu\hat{p}$)/σ$\hat{p}$

        = (0.778 – 0.50)/ $\sqrt{(0.50)(0.50)/18}$

        = 0.278/0.1179 = 2.36

    P-value = P(z>2.36) = 1 – P(z<2.36) = 1 – 0.9909 = 0.0091

    conclusion:

      Reject H_o; there is sufficient evidence to conclude that p > 0.50.

    NOTE: The conclusion is that the probability the given coins come up heads is greater than ½ – not that the person has such psychic powers. Being the eternal skeptic, the writer wonders whether the quarter might be an altered coin provided by the illusionist (either to come up heads almost all the time, or possibly even a two-headed quarter) and the penny might be an ordinary coin supplied from the audience – and as such the results can reasonably be expected to support the claims of the illusionist.

3. There is a problem with the reported results, and no statistical analysis would be appropriate. Since there were 100 drivers in each group, and the number in each group owning a cell phone must be a whole number between 0 and 100 inclusive, the sample proportion for each group must be a whole percent. The reported values of 13.7% and 10.6% are not mathematical possibilities for the sample success rates of groups of 100.

Chapter 10

Correlation and Regression

10-2 Correlation

1. The first requirement listed in this section is that the (x,y) pairs be a random sample of independent quantitative data. Since social security numbers are not quantitative data, it would not be appropriate to use the methods of this section.

3. Correlation is the existence of a relationship between two variables – so that knowing the value of one of the variables allows a researcher to make a reasonable inference about the value of the other. A lurking variable is a third variable that has a direct influence on the two variables being studied – and that may even be responsible for the correlation between them. There may be a correlation between the value of a house and the number of telephones in a house, for example, but those variables really don't affect each other directly – instead they are both a function of a lurking variable, the income of the household.

5. a. From Table A-6 for n = 8, C.V. = ±0.707. Therefore r = 0.993 indicates a significant (positive) linear correlation.

   b. The proportion of the variation in weights that can be explained by the variation in chest sizes is $r^2 = (0.993)^2 = 0.986$, or 98.6%.

7. a. From Table A-6 for n = 21, C.V. = ±0.444. Therefore r = -0.133 does not indicate a significant linear correlation.

   b. The proportion of the variation in Super Bowl points that can be explained by the variation in DJIA values is $r^2 = (-0.133)^2 = 0.017$, or 1.7%.

NOTE: In addition to the value of n, calculation of r requires five sums: Σx, Σy, Σx^2, Σy^2 and Σxy. The next problem gives the chart to find these sums. As the sums can usually be found conveniently using a calculator and without constructing the chart, subsequent problems typically give only the values of the sums and do not show a chart.
In addition, calculation of r involves three subcalculations.
   (1) $n(\Sigma xy) - (\Sigma x)(\Sigma y)$ determines the sign of r. If large values of x are associated with large values of y, it will be positive. If large values of x are associated with small values of y, it will be negative. If not, a mistake has been made.
   (2) $n(\Sigma x^2) - (\Sigma x)^2$ cannot be negative. If it is, a mistake had been made.
   (3) $n(\Sigma y^2) - (\Sigma y)^2$ cannot be negative. If it is, a mistake had been made.
Finally, r must be between -1 and 1 inclusive. If not, a mistake has been made. If this or any of the previous mistakes occurs, stop immediately and find the error – continuing is a waste of effort.

9.

x	y	xy	x^2	y^2
1	3	3	1	9
0	1	0	0	1
5	15	75	25	225
2	6	12	4	36
3	8	24	9	64
11	33	114	39	335

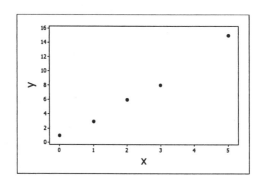

$n(\Sigma xy) - (\Sigma x)(\Sigma y) = 5(114) - (11)(33) = 207$
$n(\Sigma x^2) - (\Sigma x)^2 = 5(39) - (11)^2 = 74$
$n(\Sigma y^2) - (\Sigma y)^2 = 5(335) - (33)^2 = 586$

a. From the scatterplot, it appears that the points are very close to being on a straight line with positive slope. We expect a correlation very close to $r = +1.00$.

b. $r = [n(\Sigma xy) - (\Sigma x)(\Sigma y)]/[\sqrt{n(\Sigma x^2) - (\Sigma x)^2} \sqrt{n(\Sigma y^2) - (\Sigma y)^2}]$

$= 207/[\sqrt{74}\sqrt{586}] = 0.994$

From Table A-6 for n = 5, assuming $\alpha = 0.05$, C.V. $= \pm 0.878$. Therefore r = 0.994 indicates a significant (positive) linear correlation. This agrees with the interpretation of the scatterplot.

11. The following table and summary statistics apply to all parts of this exercise.

    x: 1 1 1 2 2 2 3 3 3 10
    y: 1 2 3 1 2 3 1 2 3 10
    using all the points: n =10 $\Sigma x = 28$ $\Sigma y = 28$ $\Sigma xy = 136$ $\Sigma x^2 = 142$ $\Sigma y^2 = 142$
    without the outlier: n = 9 $\Sigma x = 18$ $\Sigma y = 18$ $\Sigma xy = 36$ $\Sigma x^2 = 42$ $\Sigma y^2 = 42$

a. There appears to be a strong positive linear correlation, with r close to 1.

b. $n(\Sigma xy) - (\Sigma x)(\Sigma y) = 10(136) - (28)(28) = 576$
   $n(\Sigma x^2) - (\Sigma x)^2 = 10(142) - (28)^2 = 636$
   $n(\Sigma y^2) - (\Sigma y)^2 = 10(142) - (28)^2 = 636$

   $r = [n(\Sigma xy) - (\Sigma x)(\Sigma y)]/[\sqrt{n(\Sigma x^2) - (\Sigma x)^2} \sqrt{n(\Sigma y^2) - (\Sigma y)^2}]$

   $= 576/[\sqrt{636}\sqrt{636}] = 0.906$

From Table A-6 for n = 10, assuming $\alpha = 0.05$, C.V. $= \pm 0.632$. Therefore r = 0.906 indicates a significant (positive) linear correlation. This agrees with the interpretation of the scatterplot.

c. There appears to be no linear correlation, with r close to 0.
   $n(\Sigma xy) - (\Sigma x)(\Sigma y) = 9(36) - (18)(18) = 0$
   $n(\Sigma x^2) - (\Sigma x)^2 = 9(42) - (18)^2 = 54$
   $n(\Sigma y^2) - (\Sigma y)^2 = 9(42) - (18)^2 = 54$

   $r = [n(\Sigma xy) - (\Sigma x)(\Sigma y)]/[\sqrt{n(\Sigma x^2) - (\Sigma x)^2} \sqrt{n(\Sigma y^2) - (\Sigma y)^2}]$

   $= 0/[\sqrt{54}\sqrt{54}] = 0$

From Table A-6 for n = 9 assuming $\alpha = 0.05$, C.V. $= \pm 0.666$. Therefore r = 0 does not indicate a significant linear correlation. This agrees with the interpretation of the scatterplot.

d. The effect of a single pair of values can be dramatic, changing the conclusion entirely.

NOTE: In each of exercises #13-#28 the first variable listed is designated x, and the second variable listed is designated y. In correlation problems the designation of x and y is arbitrary – so long as a person remains consistent after making the designation. In each test of hypothesis, the C.V. and test statistic are given in terms of t using Method 1. The usual t formula written for r is

$$t_r = (r - \mu_r)/s_r, \text{ where } \mu_r = \rho = 0 \text{ and } s_r = \sqrt{(1-r^2)/(n-2)} \text{ and } df = n-2.$$

One advantage of Method 1 is that performing the test using the t statistic allows the calculation of exact P-values. For those using Method 2, the C.V. in terms of r is given in brackets and indicated on the accompanying graph – and the test statistic is simply r. The two methods are mathematically equivalent and always agree.

13. a. n = 8
    $\Sigma x = 1030$
    $\Sigma y = 688$
    $\Sigma xy = 88885$
    $\Sigma x^2 = 133850$
    $\Sigma y^2 = 60204$

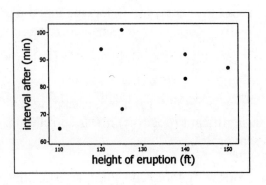

  b. $n(\Sigma xy) - (\Sigma x)(\Sigma y) = 8(88885) - (1030)(688) = 2440$
    $n(\Sigma x^2) - (\Sigma x)^2 = 8(133850) - (1030)^2 = 9900$
    $n(\Sigma y^2) - (\Sigma y)^2 = 8(60204) - (688)^2 = 8288$

    $r = [n(\Sigma xy) - (\Sigma x)(\Sigma y)]/[\sqrt{n(\Sigma x^2) - (\Sigma x)^2} \sqrt{n(\Sigma y^2) - (\Sigma y)^2}]$

      $= 2440/[\sqrt{9900}\sqrt{8288}] = 0.269$

  c. $H_o: \rho = 0$
    $H_1: \rho \neq 0$
    $\alpha = 0.05$ and df = 6
    C.V. $t = \pm t_{\alpha/2} = \pm 2.447$ [or $r = \pm 0.707$]
    calculations:
      $t_r = (r - \mu_r)/s_r$

       $= (0.269 - 0)/\sqrt{[1 - (0.269)^2]/6}$

       $= 0.269/0.393 = 0.685$

    P-value $= 2 \cdot tcdf(0.685,99,6) = 2(0.2594) = 0.5188$
  conclusion:

    Do not reject H_o; there is not sufficient evidence to conclude that $\rho \neq 0$.
No. In general, higher eruptions signify a greater release of pent up energy and are followed by longer intervals of "recovery" required to build up enough energy for another eruption – but the relationship is not strong enough to be statistically significant in a sample of size n = 8.

15. a. n = 7
    Σx = 627
    Σy = 1028
    Σxy = 153215
    Σx² = 73769
    Σy² = 398600

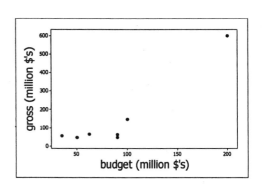

b. n(Σxy) – (Σx)(Σy) = 7(153215) – (627)(1028) = 427949
   n(Σx²) – (Σx)² = 7(73769) – (627)² = 123254
   n(Σy²) – (Σy)² = 7(398600) – (1028)² = 1733416

   $r = [n(\Sigma xy) - (\Sigma x)(\Sigma y)]/[\sqrt{n(\Sigma x^2) - (\Sigma x)^2}\sqrt{n(\Sigma y^2) - (\Sigma y)^2}]$

   $= 427949/[\sqrt{123254}\sqrt{1733416}] = 0.926$

c. $H_0: \rho = 0$
   $H_1: \rho \neq 0$
   α = 0.05 and df = 5
   C.V. t = ±$t_{\alpha/2}$ = ±2.571 [or r = ±0.754]
   calculations:

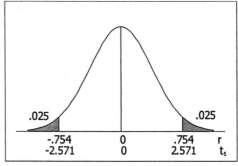

   $t_r = (r - \mu_r)/s_r$

   $= (0.926 - 0)/\sqrt{[1 - (0.926)^2]/5}$

   $= 0.926/0.169 = 5.478$

   P-value = 2·tcdf(5.478,99,5) = 2(0.0014) = 0.0028
conclusion:

   Reject H_0; there is sufficient evidence to conclude that $\rho \neq 0$ (in fact, that $\rho > 0$).
Yes; there does appear to be a significant positive correlation between budget and gross – but
the scatterplot indicates that this could be due to the effect of an outlier. Other important
factors that affect the amount a movie earns are its quality, its publicity and its stars.

17. a. n = 10
    Σx = 409
    Σy = 2544
    Σxy = 112504
    Σx² = 17411
    Σy² = 755448

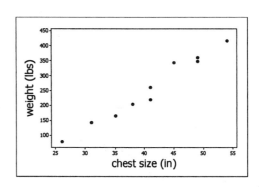

b. n(Σxy) – (Σx)(Σy) = 10(112504) – (409)(2544) = 84544
   n(Σx²) – (Σx)² = 10(17411) – (409)² = 6829
   n(Σy²) – (Σy)² = 10(755448) – (2544)² = 1082544

   $r = [n(\Sigma xy) - (\Sigma x)(\Sigma y)]/[\sqrt{n(\Sigma x^2) - (\Sigma x)^2}\sqrt{n(\Sigma y^2) - (\Sigma y)^2}]$

   $= 84544/[\sqrt{6829}\sqrt{1082544}] = 0.983$

c. H_o: $\rho = 0$
  H_1: $\rho \neq 0$
  $\alpha = 0.05$ and df = 8
  C.V. $t = \pm t_{\alpha/2} = \pm 2.306$ [or r = ± 0.632]
  calculations:
    $t_r = (r - \mu_r)/s_r$
      $= (0.983 - 0)/\sqrt{[1 - (0.983)^2]/8}$
      $= 0.983/0.0644$
      $= 15.277$
    P-value = $2 \cdot tcdf(15.277,99,8) = 2(1.67 \cdot 10^{-7}) = 3.34 \cdot 10^{-7} \approx 0$
  conclusion:
    Reject H_o; there is sufficient evidence to conclude that $\rho \neq 0$ (in fact, that $\rho > 0$).
  Yes. Bear chest sizes may be used to estimate bear weights – without having to maneuver
  anesthetized bears or scales in the field.

19. a. n = 14
    $\Sigma x = 1875$
    $\Sigma y = 1241$
    $\Sigma xy = 167023$
    $\Sigma x^2 = 252179$
    $\Sigma y^2 = 111459$

  b. $n(\Sigma xy) - (\Sigma x)(\Sigma y) = 14(167023) - (1875)(1241) = 11447$
     $n(\Sigma x^2) - (\Sigma x)^2 = 14(252179) - (1875)^2 = 14881$
     $n(\Sigma y^2) - (\Sigma y)^2 = 14(111459) - (1241)^2 = 20345$

     $r = [n(\Sigma xy) - (\Sigma x)(\Sigma y)]/[\sqrt{n(\Sigma x^2) - (\Sigma x)^2} \sqrt{n(\Sigma y^2) - (\Sigma y)^2}]$

       $= 11447/[\sqrt{14881}\sqrt{20345}] = 0.658$

  c. H_o: $\rho = 0$
     H_1: $\rho \neq 0$ and df = 12
     $\alpha = 0.05$
     C.V. $t = \pm t_{\alpha/2} = \pm 2.179$ [or r = ± 0.532]
     calculations:
       $t_r = (r - \mu_r)/s_r$
         $= (0.658 - 0)/\sqrt{[1 - (0.658)^2]/12}$
         $= 0.658/0.217$
         $= 3.026$
       P-value = $2 \cdot tcdf(3.026,99,12) = 2(0.0053) = 0.0105$
     conclusion:

       Reject H_o; there is sufficient evidence to conclude that $\rho \neq 0$ (in fact, that $\rho > 0$).
     Yes; high systolic readings are associated with high diastolic readings. One important issue
     suggested by the data is the variability of the scores – is it due to differences among the
     students, differences among the equipment, or natural fluctuations within the patient?

21. a. n = 8
$\Sigma x = 188.2$
$\Sigma y = 52.0$
$\Sigma xy = 1141.22$
$\Sigma x^2 = 11850.04$
$\Sigma y^2 = 402.90$

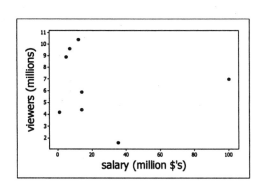

b. $n(\Sigma xy) - (\Sigma x)(\Sigma y) = 8(1141.22) - (188.2)(52.0) = -656.64$
$n(\Sigma x^2) - (\Sigma x)^2 = 8(11850.04) - (188.2)^2 = 59381.08$
$n(\Sigma y^2) - (\Sigma y)^2 = 8(402.90) - (52.0)^2 = 519.20$

$r = [n(\Sigma xy) - (\Sigma x)(\Sigma y)]/[\sqrt{n(\Sigma x^2) - (\Sigma x)^2} \sqrt{n(\Sigma y^2) - (\Sigma y)^2}]$

$= -656.64/[\sqrt{59381.08}\sqrt{519.20}] = -0.118$

c. $H_0: \rho = 0$
$H_1: \rho \neq 0$ and df = 6
$\alpha = 0.05$
C.V. $t = \pm t_{\alpha/2} = \pm 2.447$ [or $r = \pm 0.707$]
calculations:

$t_r = (r - \mu_r)/s_r$

$= (-0.118 - 0)/\sqrt{[1 - (-0.118)^2]/6}$

$= -0.118/0.405 = -0.292$

P-value = $2 \cdot \text{tcdf}(-99, -0.292, 6) = 2(0.3902) = 0.7803$

conclusion:

Do not reject H_0; there is not sufficient evidence to conclude that $\rho \neq 0$.
No; there is not a significant correlation between the salaries of stars and the number of viewers they attract. Susan Lucci has the lowest cost per viewer (23.81 cents), and Kelsey Grammer has the highest cost per viewer ($22.00).

23. a. n = 8
$\Sigma x = 478$
$\Sigma y = 1176.617$
$\Sigma xy = 70318.990$
$\Sigma x^2 = 29070$
$\Sigma y^2 = 173068.661845$

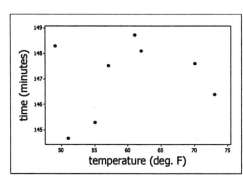

b. $n(\Sigma xy) - (\Sigma x)(\Sigma y) = 8(70318.990) - (478)(1176.617) = 128.994$
$n(\Sigma x^2) - (\Sigma x)^2 = 8(29070) - (478)^2 = 4076$
$n(\Sigma y^2) - (\Sigma y)^2 = 8(173068.661845) - (70318.990)^2 = 121.730071$

$r = [n(\Sigma xy) - (\Sigma x)(\Sigma y)]/[\sqrt{n(\Sigma x^2) - (\Sigma x)^2} \sqrt{n(\Sigma y^2) - (\Sigma y)^2}]$

$= 128.994/[\sqrt{4076}\sqrt{121.730071}] = 0.183$

c. H_o: $\rho = 0$
   H_1: $\rho \neq 0$
   $\alpha = 0.05$ and df $= 6$
   C.V. $t = \pm t_{\alpha/2} = \pm 2.447$ [or $r = \pm 0.707$]
   calculations:

$$t_r = (r - \mu_r)/s_r$$

$$= (0.183 - 0)/\sqrt{[1 - (0.183)^2]/6}$$

$$= 0.183/0.401$$

$$= 0.456$$

P-value $= 2 \cdot \text{tcdf}(0.456, 99, 6) = 2(0.3321) = 0.6642$

conclusion:

Do not reject H_o; there is not sufficient evidence to conclude that $\rho \neq 0$.

No; it does not appear that the winning time is affected by temperature. One factor complicating such a study is that winning times in most events tend to decrease over the years.

25. a. $n = 8$
    $\Sigma x = 8130$
    $\Sigma y = 646.0$
    $\Sigma xy = 663245.4$
    $\Sigma x^2 = 8391204$
    $\Sigma y^2 = 52626.60$

b. $n(\Sigma xy) - (\Sigma x)(\Sigma y) = 8(663245.4) - (8130)(646.0) = 53983.2$
   $n(\Sigma x^2) - (\Sigma x)^2 = 8(8391204) - (8130)^2 = 1032732$
   $n(\Sigma y^2) - (\Sigma y)^2 = 8(52626.60) - (646.0)^2 = 3696.8$

$$r = [n(\Sigma xy) - (\Sigma x)(\Sigma y)]/[\sqrt{n(\Sigma x^2) - (\Sigma x)^2}\sqrt{n(\Sigma y^2) - (\Sigma y)^2}]$$

$$= 53983.2/[\sqrt{1032732}\sqrt{3696.8}] = 0.874$$

c. H_o: $\rho = 0$
   H_1: $\rho \neq 0$
   $\alpha = 0.05$ and df $= 6$
   C.V. $t = \pm t_{\alpha/2} = \pm 2.447$ [or $r = \pm 0.707$]
   calculations:

$$t_r = (r - \mu_r)/s_r$$

$$= (0.874 - 0)/\sqrt{[1 - (0.874)^2]/6}$$

$$= 0.874/0.199$$

$$= 4.399$$

P-value $= 2 \cdot \text{tcdf}(4.399, 99, 6) = 2(0.0023) = 0.0046$

conclusion:

Reject H_o; there is sufficient evidence to conclude that $\rho \neq 0$ (in fact, that $\rho > 0$).

Yes; the higher the temperature, the more cricket chirps per minute.

27. a. n = 12
$\Sigma x = 762.4$
$\Sigma y = 872$
$\Sigma xy = 55392.4$
$\Sigma x^2 = 48514.38$
$\Sigma y^2 = 64032$

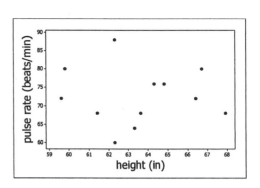

b. $n(\Sigma xy) - (\Sigma x)(\Sigma y) = 12(55392.4) - (762.4)(872) = -104.0$
$n(\Sigma x^2) - (\Sigma x)^2 = 12(48514.38) - (762.4)^2 = 918.8$
$n(\Sigma y^2) - (\Sigma y)^2 = 12(64032) - (872)^2 = 8000$

$r = [n(\Sigma xy) - (\Sigma x)(\Sigma y)]/[\sqrt{n(\Sigma x^2) - (\Sigma x)^2}\sqrt{n(\Sigma y^2) - (\Sigma y)^2}]$

$= -104.0/[\sqrt{918.8}\sqrt{8000}] = -0.038$

c. $H_o: \rho = 0$
$H_1: \rho \neq 0$
$\alpha = 0.05$ and df = 10
C.V. $t = \pm t_{\alpha/2} = \pm 2.228$ [or $r = \pm 0.576$]
calculations:

$t_r = (r - \mu_r)/s_r$

$= (-0.038 - 0)/\sqrt{[1 - (-0.038)^2]/10}$

$= -0.038/0.316 = -0.121$

P-value = $2 \cdot tcdf(-99, -0.121, 10) = 2(0.4529) = 0.9058$
conclusion:

Do not reject H_o; there is not sufficient evidence to conclude that $\rho \neq 0$.
No; the medical student's hypothesis does not appear to be correct.

29. For the n = 40 sample observations, the Minitab command *corr c1 c2* yields r = 0.995.

$H_o: \rho = 0$
$H_1: \rho \neq 0$
$\alpha = 0.05$ and df = 38
C.V. $t = \pm t_{\alpha/2} = \pm 2.024$ [or $r = \pm 0.312$]
calculations:

$t_r = (r - \mu_r)/s_r$

$= (0.995 - 0)/\sqrt{[1 - (0.995)^2]/38}$

$= 0.995/0.0162 = 61.413$

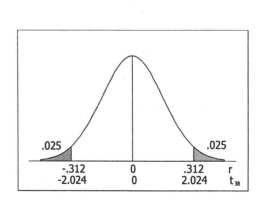

P-value = $2 \cdot \text{tcdf}(61.413,99,38) = 2(6.16 \cdot 10^{-40}) = 1.23 \cdot 10^{-39} \approx 0$
conclusion:
   Reject H_o; there is sufficient evidence to conclude that $\rho \neq 0$ (in fact, that $\rho > 0$).

31. a. For the n=29 sample observations, the Minitab command *corr c1 c2* yields r = 0.961.

H_o: $\rho = 0$
H_1: $\rho \neq 0$
$\alpha = 0.05$ and df = 27
C.V. $t = \pm t_{\alpha/2} = \pm 2.052$ [or r = ±0.361]
calculations:
   $t_r = (r - \mu_r)/s_r$

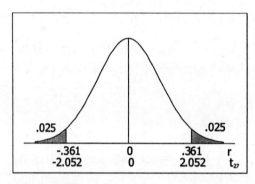

$= (0.961 - 0)/ \sqrt{[1 - (0.961)^2]/27}$
$= 0.961/0.0532 = 18.057$
   P-value = $2 \cdot \text{tcdf}(18.057,99,27) = 2(6.68 \cdot 10^{-17}) = 1.34 \cdot 10^{-16} \approx 0$
conclusion:
   Reject H_o; there is sufficient evidence to conclude that $\rho \neq 0$ (in fact, that $\rho > 0$).
   Yes; since the two variables are highly correlated, measuring both of them is unnecessary.

b. For the n=29 sample observations, the Minitab command *corr c3 c2* yields r = 0.863.

H_o: $\rho = 0$
H_1: $\rho \neq 0$
$\alpha = 0.05$ and df = 27
C.V. $t = \pm t_{\alpha/2} = \pm 2.052$ [or r = ±0.361]
calculations:
   $t_r = (r - \mu_r)/s_r$
   $= (0.863 - 0)/ \sqrt{[1 - (0.863)^2]/27}$
   $= 0.863/0.0972 = 8.876$
   P-value = $2 \cdot \text{tcdf}(8.876,99,27) = 2(8.59 \cdot 10^{-10}) = 1.72 \cdot 10^{-9} \approx 0$
conclusion:
   Reject H_o; there is sufficient evidence to conclude that $\rho \neq 0$ (in fact, that $\rho > 0$).
   Yes; since the two variables are highly correlated, measuring both of them is unnecessary.

c. Tar, because it has a higher correlation with nicotine.

33. A linear correlation coefficient very close to zero indicates <u>no</u> significant linear correlation and no tendencies can be inferred.

35. A linear correlation coefficient very close to zero indicates no significant <u>linear</u> correlation, but there may be some other type of relationship between the variables.

37. The following table gives the values for y, x, x^2, log x, $\sqrt{x}$ and 1/x. The rows at the bottom of the table give the sum of the values (Σv), the sum of the squares of the values (Σv^2), the sum of each value times the corresponding y value (Σvy), and the quantity $n\Sigma v^2 - (\Sigma v)^2$ needed in subsequent calculations.

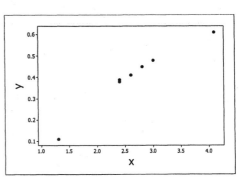

	y	x	x^2	log x	$\sqrt{x}$	1/x
	0.11	1.3	1.69	0.1139	1.1402	0.7692
	0.38	2.4	5.76	0.3802	1.5492	0.4167
	0.41	2.6	6.76	0.4150	1.6125	0.3846
	0.45	2.8	7.84	0.4472	1.6733	0.3571
	0.39	2.4	5.76	0.3802	1.5492	0.4167
	0.48	3.0	9.00	0.4771	1.7321	0.3333
	0.61	4.1	16.81	0.6128	2.0248	0.2439
Σv	2.83	18.6	53.62	2.8264	11.2814	2.9216
Σv^2	1.2817	53.62	539.95	1.2774	18.6000	1.3850
Σvy		8.258	25.495	1.2795	4.7989	1.0326
$n\Sigma v^2 - (\Sigma v)^2$	0.9630	29.38	904.55	0.9533	2.9300	1.1593

In general, $r = [n(\Sigma vy) - (\Sigma v)(\Sigma y)]/[\sqrt{n(\Sigma v^2) - (\Sigma v)^2} \sqrt{n(\Sigma y^2) - (\Sigma y)^2}]$

a. For v = x,
   $r = [7(8.258) - (18.6)(2.83)]/[\sqrt{29.38}\sqrt{0.9630}] = 0.9716$

b. For v = x^2,
   $r = [7(25.495) - (53.62)(2.83)]/[\sqrt{904.55}\sqrt{0.9630}] = 0.9053$

c. For v = log x,
   $r = [7(1.2795) - (2.8264)(2.83)]/[\sqrt{0.9533}\sqrt{0.9630}] = 0.9996$

d. For v = $\sqrt{x}$,
   $r = [7(4.7989) - (11.2814)(2.83)]/[\sqrt{2.9300}\sqrt{0.9630}] = 0.9918$

e. For v = 1/x,
   $r = [7(1.0326) - (2.9216)(2.83)]/[\sqrt{1.1593}\sqrt{0.9630}] = -0.9842$

In each case the critical values from Table A-6 for testing significance at the 0.05 level are ±0.754. While all the correlations are significant, the largest value for r occurs in part (c).

39. For r = 0.600, (1+r)/(1−r) = (1.6)/(0.4) = 4. Following the procedure in the text with $\alpha = 0.05$,
    Step a. $z_{\alpha/2} = 1.960$

    Step b. $W_L = \frac{1}{2} \cdot \ln(4) - 1.960/\sqrt{47} = 0.407$

    $W_R = \frac{1}{2} \cdot \ln(4) + 1.960/\sqrt{47} = 0.979$

    Step c. $(e^{.814} - 1)/(e^{.814} + 1) < \rho < (e^{1.958} - 1)/(e^{1.958} + 1)$

    $(1.258)/(3.258) < \rho < (6.086)/(8.086)$

    $0.386 < \rho < 0.753$

NOTE: While the distribution of r is not normal, the there is a transformation for which the transformed r's have a normal distribution. That transformation [which is actually the inverse hyperbolic tangent function] is $T(r) = \tanh^{-1}(r)$. Most mathematical/scientific calculators have the $\tanh(x)$ and $\tanh^{-1}$ functions.

$T(r)$ is normally distributed with $\mu_{T(r)} = T(\rho)$ and $\sigma_{T(r)} = \sqrt{1/(n-3)}$.

• The confidence interval for exercise #39 may be found by constructing a confidence interval for $T(\rho) = \tanh^{-1}(\rho)$ and then untransforming using the $\tanh(x)$ function.

$T(r) \pm z_{\alpha/2} \cdot \sigma_{T(r)}$

$\tanh^{-1}(0.600) \pm 1.96 \sqrt{1/47}$

0.693 ± 0.286

$0.407 < T(\rho) < 0.979$

$0.386 < \rho < 0.753$

• The value of the power for exercise #40 may be found in the same manner.
• The hyperbolic tangent function is defined as $y = (e^x - e^{-x})/(e^x + e^{-x})$. The inverse hyperbolic tangent function is found by solving for x in terms of y to get $x = (\frac{1}{2}) \cdot \ln[(1+y)/(1-y)]$, which is the transformation for r given in the text. The formulas in the text are the usual normal formulas with $(\frac{1}{2}) \cdot \ln[(1+r)/(1-r)]$ substituted for r and with $\sqrt{1/(n-3)}$ substituted for the standard deviation.

10-3 Regression

1. In the model $y = \beta_0 + \beta_1 x$, the independent variable x is called the predictor variable and the dependent variable y is called the response variable. Conceptually, a given x value is used to predict a value for y and the y value is considered the resulting response for a given x value.

3. It is not unusual in an applied problem for one regression line to apply within a certain range of x values, and for a different regression line to apply within a different range of x values. When a regression line is found using a given range of x values, therefore, it might not be the appropriate regression line to predict y for x values outside of that given range.

5. For n=20, C.V. = ±0.444. Since r = 0.870 > 0.444, use the regression line for prediction.
    $\hat{y} = -3.22 + 1.02x$
    $\hat{y}_{110} = -3.22 + 1.02(110) = 109.0$

7. For n=16, C.V. = ±0.497. Since r = 0.765 > 0.497, use the regression line for prediction.

    $\hat{y} = 3.46 + 1.01x$
    $\hat{y}_{.40} = 3.46 + 1.01(0.40) = 3.86$ calories/gram

NOTE: For exercises #9-#28, the exact summary statistics (i.e., without any rounding) are given on the right. While the intermediate calculations on the left are presented rounded to various degrees of accuracy, the entire unrounded values were preserved in the calculator until the end. When finding a predicted value, always verify that it is reasonable for the story problem and consistent with the given data points used to find the regression equation.

9. $\bar{x} = 2.4$

$\bar{y} = 2.4$

$b_1 = [n(\Sigma xy) - (\Sigma x)(\Sigma y)]/[n(\Sigma x^2) - (\Sigma x)^2]$

$\quad = 36/56 = 0.643$

$b_0 = \bar{y} - b_1\bar{x}$

$\quad = 2.4 - 0.643(2.4) = 0.857$

$\hat{y} = b_0 + b_1 x$

$\hat{y} = 0.857 + 0.643x$

$n = 5$
$\Sigma x = 12$
$\Sigma y = 12$
$\Sigma x^2 = 40$
$\Sigma y^2 = 42$
$\Sigma xy = 36$

11. a. $\bar{x} = 2.8$

$\bar{y} = 2.8$

$b_1 = [n(\Sigma xy) - (\Sigma x)(\Sigma y)]/[n(\Sigma x^2) - (\Sigma x)^2]$

$\quad = 576/636 = 0.906$

$b_0 = \bar{y} - b_1\bar{x}$

$\quad = 2.8 - 0.908(2.8) = 0.264$

$\hat{y} = b_0 + b_1 x$

$\hat{y} = 0.264 + 0.906x$

$n = 10$
$\Sigma x = 28$
$\Sigma y = 28$
$\Sigma x^2 = 142$
$\Sigma y^2 = 142$
$\Sigma xy = 136$

b. $\bar{x} = 2.0$

$\bar{y} = 2.0$

$b_1 = [n(\Sigma xy) - (\Sigma x)(\Sigma y)]/[n(\Sigma x^2) - (\Sigma x)^2]$

$\quad = 0/54 = 0$

$b_0 = \bar{y} - b_1\bar{x}$

$\quad = 2.0 - 0(2.0) = 2.0$

$\hat{y} = b_0 + b_1 x$

$\hat{y} = 2.0 + 0x$ [or simply $\hat{y} = 2.0$, for any x]

$n = 9$
$\Sigma x = 18$
$\Sigma y = 18$
$\Sigma x^2 = 42$
$\Sigma y^2 = 42$
$\Sigma xy = 36$

c. The results are very different – without the outlier, x has no predictive value for y. A single outlier can have a dramatic effect on the regression equation.

13. $\bar{x} = 128.75$

$\bar{y} = 86.00$

$b_1 = [n(\Sigma xy) - (\Sigma x)(\Sigma y)]/[n(\Sigma x^2) - (\Sigma x)^2]$

$\quad = 2440/9900 = 0.246$

$b_0 = \bar{y} - b_1\bar{x}$

$\quad = 86.00 - 0.246(128.75) = 54.27$

$\hat{y} = b_0 + b_1 x = 54.27 + 0.246x$

$\hat{y}_{100} = \bar{y} = 86.00$ min [no significant correlation]

$n = 8$
$\Sigma x = 1030$
$\Sigma y = 699$
$\Sigma x^2 = 133850$
$\Sigma y^2 = 60204$
$\Sigma xy = 8885$

Following an eruption with a height of 100 feet, the best predicted interval until the next eruption is 86 minutes.

15. $\bar{x} = 89.57$ $n = 7$
    $\bar{y} = 146.86$ $\Sigma x = 627$
    $b_1 = [n(\Sigma xy) - (\Sigma x)(\Sigma y)]/[n(\Sigma x^2) - (\Sigma x)^2]$ $\Sigma y = 1028$
    $\quad = 427949/123254 = 3.472$ $\Sigma x^2 = 73769$
    $b_0 = \bar{y} - b_1\bar{x}$ $\Sigma y^2 = 398600$
    $\quad = 146.86 - 3.472(89.57) = -164.14$ $\Sigma xy = 153215$
    $\hat{y} = b_0 + b_1 x = -164.14 + 3.472x$
    $\hat{y}_{40} = -164.14 + 3.472(40) = -25.26$ million dollars

For a movie with a 40 million dollar budget, the best predicted gross amount is a 25 million
dollar loss – assuming a linear relationship between the budget and the gross. Examination of
the data points, however, suggests that -25 is not the appropriate predicted value for 40. The
one movie with a large budget and large gross may be an outlier having a dramatic effect on
the linear regression equation, or it could be that a non-linear model might better fit the data.

17. $\bar{x} = 40.90$ $n = 10$
    $\bar{y} = 254.40$ $\Sigma x = 409$
    $b_1 = [n(\Sigma xy) - (\Sigma x)(\Sigma y)]/[n(\Sigma x^2) - (\Sigma x)^2]$ $\Sigma y = 2544$
    $\quad = 84544/6829 = 12.380$ $\Sigma x^2 = 17411$
    $b_0 = \bar{y} - b_1\bar{x}$ $\Sigma y^2 = 755448$
    $\quad = 254.40 - 12.380(40.90) = -251.95$ $\Sigma xy = 112504$
    $\hat{y} = b_0 + b_1 x = -251.95 + 12.380x$
    $\hat{y}_{50} = -251.95 + 12.380(50) = 367.06$ lbs

For a bear with a chest size of 50 inches, the best predicted weight is 367 pounds.

19. $\bar{x} = 133.93$ $n = 14$
    $\bar{y} = 88.64$ $\Sigma x = 1875$
    $b_1 = [n(\Sigma xy) - (\Sigma x)(\Sigma y)]/[n(\Sigma x^2) - (\Sigma x)^2]$ $\Sigma y = 1241$
    $\quad = 11447/14881 = 0.769$ $\Sigma x^2 = 252179$
    $b_0 = \bar{y} - b_1\bar{x}$ $\Sigma y^2 = 111459$
    $\quad = 88.64 - 0.769(133.93) = -14.38$ $\Sigma xy = 167023$
    $\hat{y} = b_0 + b_1 x = -14.38 + 0.769x$
    $\hat{y}_{140} = -14.38 + 0.769(140) = 93.31$ mmHg

For a person with a systolic blood pressure reading of 140, the best predicted diastolic reading
is 93 mmHg.

21. $\bar{x} = 23.525$ $n = 8$
    $\bar{y} = 6.50$ $\Sigma x = 188.2$
    $b_1 = [n(\Sigma xy) - (\Sigma x)(\Sigma y)]/[n(\Sigma x^2) - (\Sigma x)^2]$ $\Sigma y = 52.0$
    $\quad = -656.64/59381.08 = -0.0111$ $\Sigma x^2 = 11850.04$
    $b_0 = \bar{y} - b_1\bar{x}$ $\Sigma y^2 = 402.90$
    $\quad = 6.50 - (-0.0111)(23.525) = 6.76$ $\Sigma xy = 1141.33$
    $\hat{y} = b_0 + b_1 x = 6.76 - 0.0111x$
    $\hat{y}_2 = \bar{y} = 6.50$ million viewers [no significant correlation]

For a star with a salary of \$2,000,000, the best predicted number of viewers is 6.5 million.

23. $\bar{x} = 59.75$

$\bar{y} = 147.077$

$b_1 = [n(\Sigma xy) - (\Sigma x)(\Sigma y)]/[n(\Sigma x^2) - (\Sigma x)^2]$
$\quad = 128.994/4076 = 0.0316$

$b_0 = \bar{y} - b_1\bar{x}$
$\quad = 147.077 - 0.0316(59.75) = 145.186$

$\hat{y} = b_0 + b_1 x = 145.19 + 0.0316x$

$\hat{y}_{73} = \bar{y} = 147.077$ [no significant correlation]

$n = 8$
$\Sigma x = 478$
$\Sigma y = 1176.617$
$\Sigma x^2 = 29070$
$\Sigma y^2 = 173068.661845$
$\Sigma xy = 70318.990$

On a day when the temperature is 73°F, the predicted female winning time is 147.077 minutes. That compares reasonably well to the actual winning time of 150.750 minutes.

25. $\bar{x} = 1016.25$

$\bar{y} = 80.75$

$b_1 = [n(\Sigma xy) - (\Sigma x)(\Sigma y)]/[n(\Sigma x^2) - (\Sigma x)^2]$
$\quad = 53983.2/1032732 = 0.0523$

$b_0 = \bar{y} - b_1\bar{x}$
$\quad = 80.75 - 0.0523(1016.25) = 27.63$

$\hat{y} = b_0 + b_1 x = 27.63 + 0.0523x$

$\hat{y}_{1000} = 27.63 + 0.0523(1000) = 79.90$ °F

$n = 8$
$\Sigma x = 8130$
$\Sigma y = 646.0$
$\Sigma x^2 = 8391204$
$\Sigma y^2 = 52626.60$
$\Sigma xy = 663245.4$

When a cricket is producing 1000 chirps per minute, the best predicted temperature is 79.9°F.

27. $\bar{x} = 63.53$

$\bar{y} = 72.67$

$b_1 = [n(\Sigma xy) - (\Sigma x)(\Sigma y)]/[n(\Sigma x^2) - (\Sigma x)^2]$
$\quad = -104.0/918.8 = -0.113$

$b_0 = \bar{y} - b_1\bar{x}$
$\quad = 72.67 - (-0.113)(63.53) = 79.76$

$n = 12$
$\Sigma x = 762.4$
$\Sigma y = 872$
$\Sigma x^2 = 48514.38$
$\Sigma y^2 = 64032$
$\Sigma xy = 55392.4$

$\hat{y} = b_0 + b_1 x = 79.86 - 0.113x$

$\hat{y}_{66} = \bar{y} = 72.67$ beats/minute [no significant correlation]

For a woman who is 66 inches tall, the best predicted pulse rate is 73. Since all the pulse rates are multiples of 4 (as a result of counting for 15 seconds and multiplying by 4?), it appears that 73 might not be a possible rate in this situation – and one could argue that the best predicted rate in this context, following the practice of making predictions consistent with the accuracy of the original data, is 72.

29. For the n=40 sample observations, the Minitab command *regress c1 1 c2* yields

SELL = 99 + 0.979 LIST

$\quad \hat{y} = 99 + 0.979x$

$\hat{y}_{400,000} = 99 + 0.979(400000)$

$\quad\quad\quad = 391,699$ dollars

For a house that lists for $400,000, the best predicted selling price is $391,699.

31. a. For the n=29 sample observations, the Minitab command *regress c2 1 c1* yields

NICOTINE = 0.154 + 0.0651 TAR

$$\hat{y} = 0.154 + 0.0651x$$

$$\hat{y}_{15} = 0.154 + 0.0651(15) = 1.1305 \text{ mg}$$

For a cigarette with 15 mg of tar, the best predicted amount of nicotine is 1.1 mg.

b. For the n=29 sample observations, the Minitab command *regress c2 1 c3* yields

NICOTINE = 0.192 + 0.0606 CO

$$\hat{y} = 0.192 + 0.0606x$$

$$\hat{y}_{15} = 0.192 + 0.0606(15) = 1.101 \text{ mg}$$

For a cigarette with 15 mg of CO, the best predicted amount of nicotine is 1.1 mg.

33. If H_o:ρ=0 is true, there is no linear correlation between x and y and $\hat{y}=\overline{y}$ is the appropriate prediction for y for any x.

If H_o:β_1=0 is true, then the true regression line is $y = \beta_o + 0x = \beta_o$ and the best estimate for β_o is $b_o = \overline{y} - 0\overline{x} = \overline{y}$, producing the line $\hat{y}=\overline{y}$.

Since both hypotheses imply precisely the same result, they are equivalent.

35. The scatterplot of the original data is given below. While other scatterplots in this chapter appear to have a non-linear pattern, these data appear to lie almost exactly in a straight line and are not in need of being transformed.

The analyses that follow confirm that a linear regression using the original x values (r = 0.9968) is a better fit than a linear regression using the ln(x) values (r = 0.9631).

▶ original data

x	y
2.0	12.0
2.5	18.7
4.2	53.0
10.0	225.0

n = 4
$\Sigma x = 18.7$
$\Sigma y = 308.7$
$\Sigma x^2 = 127.89$
$\Sigma y^2 = 53927.69$
$\Sigma xy = 2543.35$

$b_1 = [n(\Sigma xy) - (\Sigma x)(\Sigma y)]/[n(\Sigma x^2) - (\Sigma x)^2]$
$= 4400.71/161.87 = 27.2$

$b_o = \overline{y} - b_1\overline{x}$
$= (308.7/4) - (27.2)(18.7/4) = -49.9$

$\hat{y} = b_o + b_1x = -49.9 + 27.2x$

$r = 4400.71/[\sqrt{161.87}\sqrt{120415.07}]$
$= 0.9968$

▶ using ln(x) for x   [i.e., "x" is really ln(x)]

x	y
0.693	12.0
0.916	18.7
1.435	53.0
2.393	225.0

n = 4
$\Sigma x = 5.347$
$\Sigma y = 308.7$
$\Sigma x^2 = 8.681$
$\Sigma y^2 = 53927.69$
$\Sigma xy = 619.594$

$b_1 = [n(\Sigma xy) - (\Sigma x)(\Sigma y)]/[n(\Sigma x^2) - (\Sigma x)^2]$
$= 827.7220/6.134071 = 134.9$

$b_o = \overline{y} - b_1\overline{x}$
$= (308.7/4) - (134.9)(5.347/4) = -103.2$

$\hat{y} = b_o + b_1\text{"x"} = -103.2 + 134.9 \cdot \ln(x)$

$r = 827.7720/[\sqrt{6.134071}\sqrt{120415.07}]$
$= 0.9631$

In each case, $r = [n(\Sigma xy) - (\Sigma x)(\Sigma y)]/[\sqrt{n(\Sigma x^2) - (\Sigma x)^2}\sqrt{n(\Sigma y^2) - (\Sigma y)^2}]$.

10-4 Variation and Prediction Intervals

1. A prediction interval is a confidence interval on a predicted value for y.

3. For each ordered pair (x,y), the explained deviation is the difference between the mean y value and the y value predicted by the regression equation. It has the original units of the problem and is equal to $\hat{y} - \bar{y}$. The explained variation is the sum of the squares of the explained deviations. It has the square of the units of the original problem and is equal to $\Sigma(\hat{y} - \bar{y})^2$.

5. The coefficient of determination is $r^2 = (0.3)^2 = 0.09$.
   The portion of the total variation in y explained by the regression is $r^2 = 0.09 = 9\%$.

7. The coefficient of determination is $r^2 = (-0.901)^2 = 0.812$.
   The portion of the total variation in y explained by the regression is $r^2 = 0.812 = 81.2\%$.

9. Since b_1 and r have the same sign, $b_1 = 0.0651$ implies that $r > 0$.
   Since $r^2 = 92.4\% = 0.924$ and $r > 0$, $r = +\sqrt{0.924} = 0.961$.
   From Table A-6 for n = 29, C.V. = ±0.361. Therefore r = 0.961 indicates a significant (positive) linear correlation.

11. $\hat{y}_{17} = 1.2599$, from the "fit" on the Minitab display. This also agrees with using the regression equation to obtain $\hat{y}_{17} = 0.154 + 0.0651(17) = 1.26$ mg.

NOTE: The following summary statistics apply to exercises #13-#16 and #17-#20. They are all that is necessary to use the chapter formulas to work the problems.

exercise #13	exercise #14	exercise #15	exercise #16
n = 7	n = 7	n = 14	n = 8
$\Sigma x = 21065$	$\Sigma x = 627$	$\Sigma x = 1875$	$\Sigma x = 8130$
$\Sigma y = 215$	$\Sigma y = 1028$	$\Sigma y = 1241$	$\Sigma y = 646.0$
$\Sigma x^2 = 65711765$	$\Sigma x^2 = 73769$	$\Sigma x^2 = 252179$	$\Sigma x^2 = 8391204$
$\Sigma y^2 = 6769$	$\Sigma y^2 = 398600$	$\Sigma y^2 = 111459$	$\Sigma y^2 = 52626.60$
$\Sigma xy = 628500$	$\Sigma xy = 153215$	$\Sigma xy = 167023$	$\Sigma xy = 663245.2$
see also 10.2-3 #16	see also 10.2-3 #15	see also 10.2-3 #19	see also 10.2-3 #25

13. The predicted values were calculated using the regression line $\hat{y} = 54.69504 - 0.007968918x$.

x	y	$\hat{y}$	$\bar{y}$	$\hat{y} - \bar{y}$	$(\hat{y} - \bar{y})^2$	$y - \hat{y}$	$(y - \hat{y})^2$	$y - \bar{y}$	$(y - \bar{y})^2$
3175	27	29.39	30.71	-1.32	1.733	-2.39	5.730	-3.71	13.764
3450	29	27.20	30.71	-3.51	12.304	1.80	3.232	-1.71	2.924
3225	27	29.00	30.71	-1.71	2.940	-2.00	3.981	-3.71	13.764
3985	24	22.94	30.71	-7.77	60.390	1.06	1.126	-6.71	45.024
2440	37	35.25	30.71	4.54	20.620	1.75	3.059	6.29	39.564
2500	34	34.77	30.71	4.06	16.506	-0.77	0.597	3.29	10.824
2290	37	36.45	30.71	5.74	32.904	0.55	0.307	6.29	39.564
21065	215	215.00	215.00	0.00	147.397	0.00	18.032	0.00	165.429

a. The explained variation is $\Sigma(\hat{y} - \bar{y})^2 = 147.397$
b. The unexplained variation is $\Sigma(y - \hat{y})^2 = 18.032$
c. The total variation is $\Sigma(y - \bar{y})^2 = 165.429$
d. $r^2 = \Sigma(\hat{y} - \bar{y})^2 / \Sigma(y - \bar{y})^2 = 147.397/165.429 = 0.8910$
e. $s_e^2 = \Sigma(y - \hat{y})^2/(n-2) = 18.032/5 = 3.6064$
   $s_e = \sqrt{3.6064} = 1.89905$

NOTE: A table such as the one in exercise #13 organizes the work and provides all the values needed to discuss variation. In such a table, the following must always be true (except for minor discrepancies due to rounding) and can be used as a check before proceeding.

* $\Sigma y = \Sigma \hat{y} = \Sigma \overline{y}$ * $\Sigma(\hat{y} - \overline{y}) = \Sigma(y - \hat{y}) = \Sigma(y - \overline{y}) = 0$ * $\Sigma(y - \hat{y})^2 + \Sigma(\hat{y} - \overline{y})^2 = \Sigma(y - \overline{y})^2$

15. The predicted values were calculated using the regression line $\hat{y} = -14.37981 + 0.7692359x$.

x	y	$\hat{y}$	$\overline{y}$	$\hat{y} - \overline{y}$	$(\hat{y} - \overline{y})^2$	$y - \hat{y}$	$(y - \hat{y})^2$	$y - \overline{y}$	$(y - \overline{y})^2$
138	82	91.77	88.64	3.13	9.809	-9.77	95.546	-6.64	44.127
130	91	85.62	88.64	-3.02	9.132	5.38	28.935	2.36	5.556
135	100	89.47	88.64	0.82	0.679	10.53	110.943	11.36	128.986
140	100	93.31	88.64	4.67	21.813	6.69	44.713	11.36	128.986
120	80	77.93	88.64	-10.71	114.796	2.07	4.291	-8.64	74.698
125	90	81.77	88.64	-6.87	47.171	8.23	67.656	1.36	1.842
120	80	77.93	88.64	-10.71	114.796	2.07	4.291	-8.64	74.698
130	80	85.62	88.64	-3.02	9.132	-5.62	31.594	-8.64	74.698
130	80	85.62	88.64	-3.02	9.132	-5.62	31.594	-8.64	74.698
144	98	96.39	88.64	7.75	60.022	1.61	2.592	9.36	87.557
143	105	95.62	88.64	6.98	48.694	9.38	87.967	16.36	267.558
140	85	93.31	88.64	4.67	21.813	-8.31	69.110	-3.64	13.270
130	70	85.62	88.64	-3.02	9.132	-15.62	244.011	18.64	347.554
150	100	101.01	88.64	12.36	152.838	-1.01	1.011	11.36	128.986
1875	1241	1241.00	1241.00	0.00	628.960	0.00	824.254	0.00	1453.214

a. The explained variation is $\Sigma(\hat{y} - \overline{y})^2 = 628.960$

b. The unexplained variation is $\Sigma(y - \hat{y})^2 = 828.254$

c. The total variation is $\Sigma(y - \overline{y})^2 = 1453.214$

d. $r^2 = \Sigma(\hat{y} - \overline{y})^2 / \Sigma(y - \overline{y})^2 = 628.960/1453.214 = 0.4328$

e. $s_e^2 = \Sigma(y - \hat{y})^2/(n-2) = 824.254/12 = 68.6878$

$s_e = \sqrt{68.6878} = 8.28781$

17. a. $\hat{y} = 54.69504 - 0.007968918x$

$\hat{y}_{3700} = 54.69504 - 0.007968918(3700)$

$= 25.210$, rounded to 25.2 mpg

b. preliminary calculations for n = 7

$\overline{x} = (\Sigma x)/n = 21065/7 = 3009.29$

$n\Sigma x^2 - (\Sigma x)^2 = 7(65711765) - (21065)^2 = 16248130$

$\alpha = 0.05$ and df = n–2 = 5

$$\hat{y} \pm t_{\alpha/2}s_e \sqrt{1 + 1/n + n(x_o - \overline{x})^2/[n\Sigma x^2 - (\Sigma x)^2]}$$

$\hat{y}_{3700} \pm t_{\alpha/2}(1.89905) \sqrt{1 + 1/7 + 7(3700-3009.29)^2/[16248130]}$

$25.210 \pm (2.571)(1.89905) \sqrt{1.34839}$

25.210 ± 5.670

$19.5 < y_{3700} < 30.9$ (mpg)

19. a. $\hat{y} = -14.37981 + 0.7692359x$

$\hat{y}_{120} = -14.37981 + 0.7692359(120)$

$= 77.92$, rounded to 78 mmHg

b. preliminary calculations for n = 14

$\bar{x} = (\Sigma x)/n = 1875/14 = 133.929$

$n\Sigma x^2 - (\Sigma x)^2 = 14(252179) - (1875)^2 = 14881$

$\alpha = 0.05$ and $df = n-2 = 12$

$$\hat{y} \pm t_{\alpha/2} s_e \sqrt{1 + 1/n + n(x_o - \bar{x})^2/[n\Sigma x^2 - (\Sigma x)^2]}$$

$$\hat{y}_{120} \pm t_{\alpha/2}(8.28781) \sqrt{1 + 1/14 + 14(120-133.929)^2/[14881]}$$

$77.92 \pm (2.179)(8.28781) \sqrt{1.25395}$

77.92 ± 20.22

$57.7 < y_{120} < 98.2$ (mmHg)

NOTE: Exercises #21–#24 refer to the chapter problem of Table 10-1. Use the following, which are calculated and/or discussed in the text,

$n = 8$ $\Sigma x = 1751$ $\Sigma x^2 = 399451$ $\hat{y} = 34.770 + 0.23406$ $s_e = 4.97392$

and the additional values

$\bar{x} = (\Sigma x)/n = 1751/8 = 218.875$

$n\Sigma x^2 - (\Sigma x)^2 = 8(399451) - (1751)^2 = 129607$

Using a slightly different regression equation for $\hat{y}$ or a slightly different value for s_e may result in slightly different values in exercises #21-#24.

21. $\hat{y}_{180} = 34.770 + 0.23406(180) = 76.901$

$\alpha = 0.01$ and $df = n-2 = 6$

$$\hat{y} \pm t_{\alpha/2} s_e \sqrt{1 + 1/n + n(x_o - \bar{x})^2/[n\Sigma x^2 - (\Sigma x)^2]}$$

$$\hat{y}_{180} \pm t_{\alpha/2}(4.97392) \sqrt{1 + 1/8 + 8(180-218.875)^2/[129607]}$$

$76.901 \pm (3.707)(4.97392) \sqrt{1.21828}$

76.901 ± 20.351

$56.5 < y_{180} < 97.3$ (min)

23. $\hat{y}_{200} = 34.770 + 0.23406(200) = 81.582$

$\alpha = 0.05$ and $df = n-2 = 6$

$$\hat{y} \pm t_{\alpha/2} s_e \sqrt{1 + 1/n + n(x_o - \bar{x})^2/[n\Sigma x^2 - (\Sigma x)^2]}$$

$$\hat{y}_{200} \pm t_{\alpha/2}(4.97392) \sqrt{1 + 1/8 + 8(200-218.875)^2/[129607]}$$

$81.582 \pm (2.447)(4.97392) \sqrt{1.14699}$

81.582 ± 13.035

$68.5 < y_{200} < 94.6$ (min)

25. Use the following, which are calculated and/or discussed in the text,

$n = 8$ $\Sigma x = 1751$ $\Sigma x^2 = 399451$ $\hat{y} = 34.770 + 0.23406$ $s_e = 4.97392$

and the additional values

$\bar{x} = (\Sigma x)/n = 1751/8 = 218.875$

$\Sigma x^2 - (\Sigma x)^2/n = 399451 - (1751)^2/8 = 16200.875$

a. $\alpha = 0.05$ and $df = n-2 = 6$

$$b_o \pm t_{\alpha/2}s_e \sqrt{1/n + (\overline{x})^2/[\Sigma x^2 - (\Sigma x)^2/n]}$$

$34.770 \pm t_{\alpha/2}(4.97392)\sqrt{1/8 + (218.875)^2/[16200.875]}$

$34.770 \pm (2.447)(4.97392)\sqrt{3.08202}$

34.770 ± 21.367

$13.4 < \beta_o < 56.1$ (minutes)

b. $\alpha = 0.05$ and $df = n-2 = 6$

$$b_1 \pm t_{\alpha/2}s_e / \sqrt{\Sigma x^2 - (\Sigma x)^2/n}$$

$0.23406 \pm t_{\alpha/2}(4.97392)/\sqrt{16200.875}$

$0.23406 \pm (2.447)(4.97392)/\sqrt{16200.875}$

0.23406 ± 0.09562

$0.1384 < \beta_1 < 0.3297$ (minutes/second)

NOTE: The confidence interval for $\beta_o = y_0$ may also be found as the confidence interval [as distinguished from the prediction interval, see exercise #26] for $x = 0$.

$\hat{y}_0 = 34.770 + 0.23406(0) = 34.770$

$\alpha = 0.05$ and $df = n-2 = 6$

$$\hat{y} \pm t_{\alpha/2}s_e \sqrt{1/n + n(x_o - \overline{x})^2/[n\Sigma x^2 - (\Sigma x)^2]}$$ modifies to become

$$\hat{y}_0 \pm t_{\alpha/2}s_e \sqrt{1/n + n(\overline{x})^2/[n\Sigma x^2 - (\Sigma x)^2]}$$

$34.770 \pm t_{\alpha/2}(4.97392)\sqrt{1/8 + 8(218.875)^2/[129607]}$

$34.770 \pm (2.447)(4.97392)\sqrt{3.08202}$

34.770 ± 21.367

$13.4 < \beta_o < 56.1$ (minutes)

10-5 Multiple Regression

1. Multiple regression is a method for exploring the relationship of one response variable to two or more predictor variables. It is a generalization of the regression equations in section 10-3, which used a single predictor variable.

3. No. In general, the requirements given at the beginning of the chapter for correlation and simple linear regression still apply – in particular, that the data are a random sample of independent quantitative measurements. Since eye color as typically measured is a categorical and not a quantitative variable, it is not a valid response variable in this context.

5. WEIGHT = -272 – 0.87(HEADLENGTH) + 0.55(LENGTH) + 12.2(CHEST)
   $\hat{y} = -271.71 - 0.870x_1 + 0.554x_2 + 12.153x_3$

NOTE: The more accurate values may be obtained from the "coef" [i.e., coefficient] column of the Minitab table. These more accurate values will be used in the remaining problems.

7. Yes, since the adjusted R^2 is 0.924. Although the P-value is less than 0.05, that alone in multiple regression problems does not necessarily indicate that the regression is of practical significance. Increasing the number of predictor variables, like increasing the sample size in the previous chapters, can create statistical significance that is not of practical value. The adjusted R^2 takes into account the number of x variables.

9. The best single predictor variable in a simple linear regression for weight is waist size, which explains about 79% of the variation in the men's weights. That regression has the highest R^2 (0.790) and adjusted R^2 (0.785), and the lowest P-value (0.000), among the three simple linear regressions.

11. Two of the given regression equations have the highest R^2 of 0.870 – but one of them achieves that level using only two variables. Since adding cholesterol level to height and waist size does not increase the adjusted R^2, the best and most efficient (statistically and financially) regression equation is $\hat{y}$ = -206 + 2.66(HEIGHT) + 2.15(WAIST).

13. a. The Minitab command *regress c2 1 c1* yields
      NICOTINE = 0.154 + 0.0651(TAR)
      adjusted R^2 = 0.921
   b. The Minitab command *regress c2 1 c3* yields
      NICOTINE = 0.192 + 0.0606(CO)
      adjusted R^2 = 0.736
   c. The Minitab command *regress c2 2 c1 c3* yields
      NICOTINE = 0.182 + 0.0818(TAR) – 0.0186(CO)
      adjusted R^2 = 0.928
   d. The best equation for predicting the amount of nicotine is the equation in (a). Although the adjusted R^2 is slightly higher in (c), the increase from adding another variable is negligible. In the interest of efficiency (both statistical and economic), the regression equation in (a) is the wisest choice.
   e. Yes, since the P-value of 0.000 in the "Regression" line of the Analysis of Variance Table indicates statistical significance.

15. To find the best variables for a regression to predict selling price SELL, consider the correlations. Arranged in one table, the correlations are as follows.

	SELL	LIST	AREA	ACRE	AGE	TAX	ROOM	BDRM	BATH
SELL	1.000								
LIST	0.995	1.000							
AREA	0.802	0.808	1.000						
ACRE	0.463	0.458	0.061	1.000					
AGE	-0.488	-0.493	-0.382	-0.126	1.000				
TAX	0.833	0.849	0.823	0.314	-0.572	1.000			
ROOM	0.635	0.654	0.821	-0.125	-0.484	0.700	1.000		
BDRM	0.552	0.555	0.610	-0.078	-0.333	0.479	0.726	1.000	
BATH	0.586	0.580	0.653	0.100	-0.406	0.647	0.639	0.486	1.000

The variable having the highest correlation [0.995] with SELL is LIST. LIST is the single best variable for predicting SELL. The next variable to add would be one with a moderately high correlation with SELL and a low correlation with LIST (so that it is not merely duplicating information provided by LIST). This suggests that ACRE or AGE would be appropriate variables to consider next. But since the seller presumably considered the other variables when determining LIST, there is a sense in which their information is already contained in LIST. The best multiple regression equation with SELL as the dependent variable is probably
      SELL = 99 + 0.979(LIST)
      adjusted R^2 = 0.990
The adjusted R^2 and overall P-value [0.000] indicate that this is a suitable regression equation for predicting selling price. Since the adjusted R^2 is so close to 1.00, there is virtually no room for improvement and no additional variable could be of any real value.

17. a. original claim: $\beta_1 = 0$

H_o: $\beta_1 = 0$ minutes/second

H_1: $\beta_1 \neq 0$ minutes/second

$\alpha = 0.05$ [assumed] and df = 5

C.V. $t = \pm t_{\alpha/2} = \pm 2.571$

calculations:

$\quad t_{b_1} = (b_1 - \mu_{b_1})/s_{b_1}$

$\qquad = (0.24464 - 0)/0.04486$

$\qquad = 5.45$ [Minitab]

$\quad$ P-value $= 2 \cdot P(t_5 > 5.45) = 0.003$ [Minitab]

conclusion:

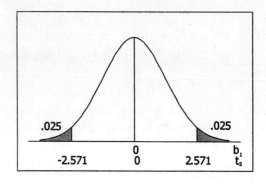

$\quad$ Reject H_o; there is sufficient evidence to reject the claim that $\beta_1 = 0$ and conclude that $\beta_1 \neq 0$ (in fact, that $\beta_1 > 0$).

The result in (a) implies that β_1 is significantly different from 0 and is appropriate for inclusion in the regression equation. The result below in (b), however, implies that β_2 is not significantly different from 0 and should be dropped from the regression equation.

b. original claim: $\beta_2 = 0$

H_o: $\beta_2 = 0$

H_1: $\beta_2 \neq 0$

$\alpha = 0.05$ [assumed] and df = 5

C.V. $t = \pm t_{\alpha/2} = \pm 2.571$

calculations:

$\quad t_{b_2} = (b_2 - \mu_{b_2})/s_{b_2}$

$\qquad = (0.-0.0983 - 0)/0.1623$

$\qquad = -0.61$ [Minitab]

$\quad$ P-value $= 2 \cdot P(t_5 < -0.61) = 0.571$ [Minitab]

conclusion:

$\quad$ Do not eject H_o; there is not sufficient evidence to reject the claim that $\beta_2 = 0$.

19. The Minitab commands *let c3 = 2 – c3*

$\qquad\qquad\qquad\qquad$ *regress c9 2 c1 c3*

yield the multiple regression equation

$\quad$ WEIGHT = 3.1 + 2.91(AGE) + 82.4(SEX).

$\quad \hat{y} = 3.1 + 2.91x_1 + 82.4x_2$

Yes, but not merely because the coefficient 82.4 is so large. Minitab indicates that for the test H_o: $\beta_2 = 0$, the sample value $b_2 = 82.4$ results in the test statistic $t_{51} = 3.96$ and P-value =.000. As suggested by (a) and (b) below, sex does have a significant effect on the weight of a bear.

a. $\hat{y} = 3.1 + 2.91x_1 + 82.4x_2$

$\quad \hat{y}_{20,0} = 3.1 + 2.91(20) + 82.4(0)$

$\qquad = 61.3$ lbs

b. $\hat{y} = 3.1 + 2.91x_1 + 82.4x_2$

$\quad \hat{y}_{20,1} = 3.1 + 2.91(20) + 82.4(1)$

$\qquad = 143.7$ lbs

10-6 Modeling

1. A mathematical model is an equation, typically expressing one variable as a function of one or more other variables, that fits real world data.

3. The year 3000 is too far beyond the scope of the data used in the model to assert that the same conditions and assumptions that applied to the data used to build the model would still be relevant.

5. The graph appears to be that of a straight line function.
   • Try a linear regression of the form $y = ax + b$.
   The Minitab command *regress c2 1 c1* yields
   $$y = 3 + 2x$$
   The adjusted $R^2 = 100.0\%$ indicates a perfect fit.
   • Choose the linear model $y = 3 + 2x$.

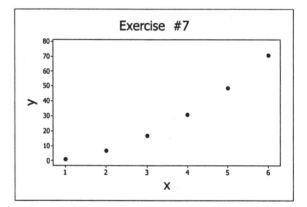

7. The graph appears to be that of a quadratic function.
   • Try a quadratic regression of the form $y = ax^2 + bx + c$.
   The Minitab commands *let c3= c1*c1* and *regress c2 2 c1 c3* yield
   $$y = 2x^2 + 0x - 1$$
   The adjusted $R^2 = 100.0\%$ indicates a perfect fit.
   • Choose the quadratic model $y = 2x^2 - 1$.

9. The graph could be that of any of several functions.
   • Try a linear regression of the form $y = ax + b$.
   The Minitab command *regress c2 1 c1* yields
   $$y = 21.7 + 2.74x$$
   The adjusted $R^2 = 76.5\%$ indicates a reasonable fit.
   • Try a quadratic regression of the form $y = ax^2 + bx + c$.
   The Minitab commands *let c3= c1*c1* and *regress c2 2 c1 c3* yield
   $$y = 0.0256x^2 + 2.15x + 24.1$$
   The adjusted $R^2 = 75.5\%$ indicates a reasonable fit.
   • Try a power function regression of the form $y = a \cdot x^b$
   $$\ln y = \ln (a \cdot x^b)$$
   $$\ln y = \ln a + b \cdot (\ln x)$$

The Minitab commands *let c3 = loge(c1)*, *let c4 = loge(c2)* and *regress c4 1 c3* yield
   $\ln y = 2.8946 + 0.455(\ln x)$
   The adjusted $R^2 = 77.9\%$ indicates a reasonable fit.
   solving for the original parameter: $\ln a = 2.8946$ $\qquad a = e^{2.8946}$ $\qquad a = 18.07$
   $y = a \cdot x^b = 18.07x^{0.455}$
•Even though the power function model has a slightly high adjusted R^2, the difference is negligible and not worth the extra mathematical complication. Unless there are actuarial or other reasons to suspect that this phenomenon should follow a power function, choose the linear model $y = 21.7 + 2.74x$.
   NOTE: This is a judgment call – the quadratic and power function are also acceptable.

11. The graph appears to be that of a quadratic function.
    •Try a quadratic regression of the form $y = ax^2 + bx + c$.
    The Minitab commands *let c3= c1*c1* and *regress c2 2 c1 c3* yield
       $y = 4.90x^2 - 0.0286x + 0.0048$
    The adjusted $R^2 = 100.0\%$ indicates a perfect fit.
    •Choose the quadratic model $y = 4.90x^2 - 0.0286x + 0.0048$.
       $\hat{y}_{12} = 4.90(12)^2 - 0.0286(12) + 0.0048 = 705.3$ meters.

    But if the building from which the ball is dropped is only 50 meters tall, the ball will hit the ground and stop falling long before 12 seconds elapse.

13. NOTE: The following analysis codes the years so that 1971 is x=0 and determines a regression equation. Coding the years so that 1971 is x=1 gives the different equation $y = 1.382 \cdot (1.424)^x$, which considers 1970 as the starting year x=0 but gives the same numerical predictions for each year. In general, the recoding of the years is arbitrary – and while a different equation may result, the individual predictions and other key characteristics will be identical.
    Consider the pattern in the box at the right.
    For a variable that starts with value $y = a$ at year 0 and doubles every 18 months,
       $y = a \cdot 2^{x/2} = a \cdot 2^{(1/2)x} = a \cdot (2^{1/2})^x = a \cdot (\sqrt{2})^x.$

year	doubles every 12 months	doubles every 18 months
0	$a = a \cdot 2^0$	$a = a \cdot 2^0$
1	$2a = a \cdot 2^1$	
2	$4a = a \cdot 2^2$	$2a = a \cdot 2^1$
3	$8a = a \cdot 2^3$	
4	$16a = a \cdot 2^4$	$4a = a \cdot 2^2$
...	...	...
x	$a \cdot 2^x$	$a \cdot 2^{x/2}$

    a. If Moore's law applies as indicated, and the years are coded with 1971 = 0, the data should be a good fit to the exponential model
       $y = 2.3 \cdot (1.414)^x.$

b. Try an exponential regression of the form $y = a \cdot b^x$.
$$\ln y = \ln (a \cdot b^x)$$
$$\ln y = \ln a + x \cdot (\ln b)$$

The Minitab commands *let c3 = loge(c2)* and *regress c3 1 c1* yield
$\ln y = 0.6767 + 0.35333x$

The adjusted $R^2 = 98.2\%$ indicates an excellent fit.

solving for the original parameters: $\ln a = 0.6767$ $\ln b = 0.35333$
$$a = e^{0.6767} \qquad\qquad b = e^{0.35333}$$

$y = a \cdot b^x = 1.967 \cdot (1.424)^x$ $a = 1.967$ $b = 1.424$

Choose the exponential model $y = 1.967 \cdot (1.424)^x$

c. Yes. The $1.424 \approx 1.414$ indicates that the y value is doubling approximately every 18 months. In addition, the starting value for 1971 (x=0) of 2.0 (1.967, rounded to one decimal) is close to the actual value of 2.3.

15. The table below was obtained using $\hat{y}_{lin} = -61.93 + 27.20x$ and $\hat{y}_{quad} = 2.77x^2 -6.00 x + 10.01$

year	pop	$\hat{y}_{lin}$	$y - \hat{y}_{lin}$	$(y - \hat{y}_{lin})^2$	$\hat{y}_{quad}$	$y - \hat{y}_{quad}$	$(y - \hat{y}_{quad})^2$
1	5	-34.727	39.7273	1578.26	6.776	-1.77622	3.1550
2	10	-7.527	17.5273	307.21	9.074	0.92587	0.8572
3	17	19.673	-2.6727	7.14	16.906	0.09417	0.0089
4	31	46.873	-15.8727	251.94	30.271	0.72867	0.5310
5	50	74.073	-24.0727	579.50	49.171	0.82937	0.6879
6	76	101.273	-25.2727	638.71	73.604	2.39627	5.7421
7	106	128.473	-22.4727	505.02	103.571	2.42937	5.9018
8	132	155.673	-23.6727	560.40	139.071	-7.07133	50.0037
9	179	182.873	-3.8727	15.00	180.106	-1.10583	1.2229
10	227	210.073	16.9273	286.53	226.674	0.32587	0.1062
11	281	237.273	43.7273	1912.07	278.776	2.22378	4.9452
68	1114	1114.000	0.0000	6641.78	1114.000	0.00000	73.1618

a. $\Sigma(y - \hat{y})^2 = 6641.78$ for the linear model

b. $\Sigma(y - \hat{y})^2 = 73.16$ for the quadratic model

c. Since $73.16 < 6641.78$, the quadratic model is better – using the sum of squares criterion.

Statistical Literacy and Critical Thinking

1. Correlation is a measure of the degree of association between two variables. Regression is a procedure to identify the formula that describes that association.

2. If the linear correlation coefficient is $r = 0$, then there is no <u>linear</u> relationship between the variables – but there might be a relationship defined by something other than a straight line graph.

3. The conclusion of cause-and-effect cannot be justified solely on the basis of the existence of a correlation. If there is a medical condition that happens to be associated with low cholesterol levels, for example, then people taking a drug to combat that condition would have low cholesterol – but it would not be because of the drug. If there is a carefully designed study with patients randomly assigned to a treatment group and a control group in such a way that the only thing different between the two groups is the drug, then such a conclusion could be justified.

4. Not necessarily. The accuracy of the prediction depends on s_e, the spread around the regression line. There can be a definite correlation, but little confidence in the prediction. For children between birth and 12 years, for example, there is an undeniable correlation between age and weight – you won't find any 12 year olds weighing 7 pounds, 8 ounces! But just knowing that a child is twelve years old gives you no information about his weight within the very large weight range of all 12 year olds.

Review Exercises

1. preliminary calculations

$n = 11$

$\Sigma x = 750$

$\Sigma y = 517$

$\Sigma xy = 35669$

$\Sigma x^2 = 53654$

$\Sigma y^2 = 32827$

$n(\Sigma xy) - (\Sigma x)(\Sigma y) = 11(35669) - (750)(517)$
$= 4609$

$n(\Sigma x^2) - (\Sigma x)^2 = 11(53654) - (750)^2$
$= 27694$

$n(\Sigma y^2) - (\Sigma y)^2 = 11(32827) - (517)^2$
$= 93775$

a. From Table A-6 with $n = 11$ and $\alpha = 0.05$, C.V. $= \pm 0.602$.

$r = [n(\Sigma xy) - (\Sigma x)(\Sigma y)]/[\sqrt{n(\Sigma x^2) - (\Sigma x)^2} \sqrt{n(\Sigma y^2) - (\Sigma y)^2}]$

$= 4609/[\sqrt{27694}\sqrt{93775}] = 0.0904$

There is no significant linear correlation.

b. $\bar{x} = (\Sigma x)/n = 750/11 = 68.182$

$\bar{y} = (\Sigma y)/n = 517/11 = 47.000$

$b_1 = [n(\Sigma xy) - (\Sigma x)(\Sigma y)]/[n(\Sigma x^2) - (\Sigma x)^2]$
$= 4609/27694 = 0.1664$

$b_0 = \bar{y} - b_1\bar{x}$
$= 47.000 - 0.1664(68.182) = 35.652$

$\hat{y} = b_0 + b_1x$
$= 35.652 + 0.166x$

c. $\hat{y}_{50} = \bar{y} = 47.00$ deaths [no significant correlation]

For a year with 50 deaths from encounters with boats, the best predicted number of natural deaths is 47.

2. preliminary calculations

$n = 8$

$\Sigma x = 1751$

$\Sigma y = 1030$

$\Sigma xy = 227185$

$\Sigma x^2 = 399451$

$\Sigma y^2 = 133850$

$n(\Sigma xy) - (\Sigma x)(\Sigma y) = 8(227185) - (1751)(1030)$
$= 13950$

$n(\Sigma x^2) - (\Sigma x)^2 = 8(399451) - (1751)^2$
$= 129607$

$n(\Sigma y^2) - (\Sigma y)^2 = 8(133850) - (1030)^2$
$= 9900$

a. From Table A-6 with $n = 8$ and $\alpha = 0.05$, C.V. $= \pm 0.707$.

$r = [n(\Sigma xy) - (\Sigma x)(\Sigma y)]/[\sqrt{n(\Sigma x^2) - (\Sigma x)^2} \sqrt{n(\Sigma y^2) - (\Sigma y)^2}]$

$= 13950/[\sqrt{129607}\sqrt{9900}] = 0.3894$

There is no significant linear correlation.

b. $\bar{x} = (\Sigma x)/n = 1751/8 = 218.875$
$\bar{y} = (\Sigma y)/n = 1030/8 = 128.750$
$b_1 = [n(\Sigma xy) - (\Sigma x)(\Sigma y)]/[n(\Sigma x^2) - (\Sigma x)^2]$
$= 13950/129607 = 0.1076$
$b_0 = \bar{y} - b_1\bar{x}$
$= 128.750 - 0.1076(218.875) = 105.192$
$\hat{y} = b_0 + b_1x$
$= 105.192 + 0.1076x$

c. $\hat{y}_{180} = \bar{y} = 128.750$ feet [no significant correlation]
For an eruption that has a duration of 180 seconds, the best predicted height is 128.75 feet.

3. preliminary calculations

$n = 10$ $n(\Sigma xy) - (\Sigma x)(\Sigma y) = 10(6716083.75 - (26481)(2473.02)$
$\Sigma x = 26481$ $= 1672794.88$
$\Sigma y = 2473.02$ $n(\Sigma x^2) - (\Sigma x)^2 = 10(71605869) - (26481)^2$
$\Sigma xy = 6716083.75$ $= 14815329$
$\Sigma x^2 = 71605869$ $n(\Sigma y^2) - (\Sigma y)^2 = 10(634142.1730) - (2473.02)^2$
$\Sigma y^2 = 634142.1730$ $= 225593.8096$

a. $r = [n(\Sigma xy) - (\Sigma x)(\Sigma y)]/[\sqrt{n(\Sigma x^2) - (\Sigma x)^2}\sqrt{n(\Sigma y^2) - (\Sigma y)^2}]$
$= 1672794.88/[\sqrt{14815329}\sqrt{225593.8096}] = 0.9150$

$H_0: \rho = 0$
$H_1: \rho \ne 0$
$\alpha = 0.05$ and df $= 8$
C.V. $t = \pm t_{\alpha/2} = \pm 2.306$ [or $r = \pm.632$]
calculations:
$t_r = (r - \mu_r)/s_r$
$= (0.9150 - 0)/\sqrt{[1 - (0.9150)^2]/8}$
$= 0.9150/0.1426$
$= 6.415$

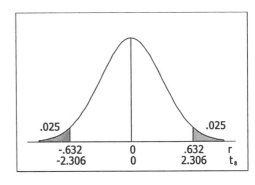

P-value $= 2\cdot$tcdf$(6.415,99,8) = 2(0.0001) = 0.0002$
conclusion:
Reject H_0; there is sufficient evidence to conclude that $\rho \ne 0$ (in fact, that $\rho > 0$).

b. $r^2 = (0.9150)^2 = 0.837 = 83.7\%$

c. $\bar{x} = (\Sigma x)/n = 26481/10 = 2648.1$
$\bar{y} = (\Sigma y)/n = 2473.02/10 = 247.302$
$b_1 = [n(\Sigma xy) - (\Sigma x)(\Sigma y)]/[n(\Sigma x^2) - (\Sigma x)^2]$
$= 1672794.88/14815329 = 0.1129$
$b_0 = \bar{y} - b_1\bar{x}$
$= 247.302 - 0.1129(2648.1) = -51.694$
$\hat{y} = b_0 + b_1x$
$= -51.694 + 0.1136x$

d. $\hat{y}_{3000} = -51.694 + 0.1129(3000) = 287.03$ dollars
For a time when 3000 kwh of electricity is used, the best predicted cost is $287.03.

4. preliminary calculations

$n = 10$

$n(\Sigma xy) - (\Sigma x)(\Sigma y) = 10(127784.85) - (514)(2473.02)$

$\Sigma x = 514$

$= 6716.22$

$\Sigma y = 2473.02$

$n(\Sigma x^2) - (\Sigma x)^2 = 10(29180) - (514)^2$

$\Sigma xy = 127784.85$

$= 27604$

$\Sigma x^2 = 29180$

$n(\Sigma y^2) - (\Sigma y)^2 = 10(634142.1730) - (2473.02)^2$

$\Sigma y^2 = 634142.1730$

$= 225593.8096$

a. $r = [n(\Sigma xy) - (\Sigma x)(\Sigma y)]/[\sqrt{n(\Sigma x^2) - (\Sigma x)^2} \sqrt{n(\Sigma y^2) - (\Sigma y)^2}]$

$= 6716.22/[\sqrt{27604}\sqrt{225593.8096}] = 0.0851$

$H_o: \rho = 0$

$H_1: \rho \neq 0$

$\alpha = 0.05$ and df = 8

C.V. $t = \pm t_{\alpha/2} = \pm 2.306$ [or $r = \pm.632$]

calculations:

$t_r = (r - \mu_r)/s_r$

$= (0.0851 - 0)/\sqrt{[1 - (0.0851)^2]/8}$

$= 0.0851/0.3522$

$= 0.242$

P-value = $2 \cdot tcdf(0.242, 99, 8) = 2(0.4076) = 0.8152$

conclusion:

Do not reject H_o; there is not sufficient evidence to conclude that $\rho \neq 0$.

b. $r^2 = (0.0851)^2 = 0.007 = 0.7\%$

c. $\overline{x} = (\Sigma x)/n = 514/10 = 51.4$

$\overline{y} = (\Sigma y)/n = 2473.02/10 = 247.302$

$b_1 = [n(\Sigma xy) - (\Sigma x)(\Sigma y)]/[n(\Sigma x^2) - (\Sigma x)^2]$

$= 6716.22/27604 = 0.2433$

$b_o = \overline{y} - b_1\overline{x}$

$= 247.302 - 0.2433(51.4) = 234.796$

$\hat{y} = b_o + b_1x$

$= 234.796 + 0.2433x$

d. $\hat{y}_{40} = \overline{y} = 247.302$ (dollars) [no significant correlation]

For a time when the average daily temperature is 40°F, the best predicted cost is $247.30.

5. The Minitab command *regress c3 2 c1 c2* yields

$\hat{y} = -128 + 0.123x_1 + 0.955x_2$

$R^2 = 0.942$

adjusted $R^2 = 0.925$

overall P-value = 0.000

Yes; considering the adjusted R^2 and the overall P-value, this is an excellent regression for predicting cost. No; since the adjusted R^2 here is so much higher that the one in exercise #3 (0.925 > 0.817), this is the best regression for predicting cost.

Cumulative Review Exercises

The following summary statistics apply to the exercises in the cumulative review.

$n = 8$	$n(\Sigma xy) - (\Sigma x)(\Sigma y) = 8(4028535) - (411)(79602)$
$\Sigma x = 411$	$= -488142$
$\Sigma y = 79602$	$n(\Sigma x^2) - (\Sigma x)^2 = 8(22023) - (411)^2$
$\Sigma xy = 4028535$	$= 7263$
$\Sigma x^2 = 22023$	$n(\Sigma y^2) - (\Sigma y)^2 = 8(814142944 - (79602)^2$
$\Sigma y^2 = 814142944$	$= 176665148$

1. $r = [n(\Sigma xy) - (\Sigma x)(\Sigma y)]/[\sqrt{n(\Sigma x^2) - (\Sigma x)^2} \sqrt{n(\Sigma y^2) - (\Sigma y)^2}]$

$= -488142/[\sqrt{7263}\sqrt{176665148}] = -0.4309$

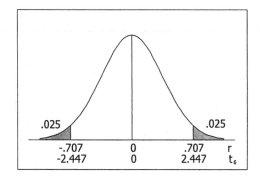

$H_0: \rho = 0$

$H_1: \rho \neq 0$

$\alpha = 0.05$ and $df = 6$

C.V. $t = \pm t_{\alpha/2} = \pm 2.447$ [or $r = \pm.707$]

calculations:

$t_r = (r - \mu_r)/s_r$

$= (-0.4309 - 0)/\sqrt{[1 - (-0.4309)^2]/6}$

$= -0.4309/0.3684$

$= -1.1696$

P-value $= 2 \cdot tcdf(-99,-1.1696,6) = 2(0.1433) = 0.2865$

conclusion:

Do not reject H_0; there is not sufficient evidence to conclude that $\rho \neq 0$.
There is no significant linear correlation between Super Bowl points and the DJIA. Yes, this is the expected result.

2. $\bar{x} = (\Sigma x)/n = 411/8 = 51.375$

$\bar{y} = (\Sigma y)/n = 79602/8 = 9950.250$

$b_1 = [n(\Sigma xy) - (\Sigma x)(\Sigma y)]/[n(\Sigma x^2) - (\Sigma x)^2]$

$= -488142/7263 = -67.209$

$b_0 = \bar{y} - b_1\bar{x}$

$= 9950.250 - (-67.209)(51.375) = 13403.134$

$\hat{y} = b_0 + b_1 x$

$= 13403.134 - 67.209x$

$\hat{y}_{50} = \bar{y} = 9950.25$ [no significant correlation]

For a year in which there are 50 points scored in the Super Bowl, the best predicted high value for DJIA is 9950.

3. Yes. It is possible to carry out the mathematical calculations to make the test, but it would have no meaning. Because the two quantities have different units, differences would not be meaningful.

4. $\bar{x} = (\Sigma x)/n = 411/8 = 51.375$
   $s^2 = [n(\Sigma x^2) - (\Sigma x)^2]/[n(n-1)]$
   $\quad = [7263]/[8(7)] = 129.6964$
   $s = 11.388$
   $\alpha = 0.05$ and $df = 7$
   $\qquad \bar{x} \pm t_{\alpha/2}{\cdot}s/\sqrt{n}$
   $51.4 \pm 2.365{\cdot}(11.388)/\sqrt{8}$
   51.4 ± 9.5
   $41.9 < \mu < 60.9$ (points)

5. A confidence interval is based on all the data and applies to the population from which the data were selected. When there is a trend over time, the population is changing. Data collected from populations changing over time are relevant for making predictions for specific times only when the time factor is built into the model – as when "year" is one of the variables in a regression. Since ordinary confidence intervals use the data only from the variable of interest and do not consider the time factor, they are not appropriate for making predictions about populations that are changing over time.

6. The corresponding z scores in the table below were determined as follows.
   (1) Arrange the n=8 scores in order and place them in the x column.
   (2) For each x_i, calculate the cumulative probability using $cp_i = (2i-1)/2n$ for $i = 1,2,\ldots,n$.
   (3) For each cp_i, find the z_i for which $P(z<z_i) = cp_i$ for $i = 1,2,\ldots,n$.
   The resulting normal quantile plot is relatively close to a straight line with no obvious patterns, indicating that the data appear to come from a population with a distribution that is approximately normal.

i	x	cp	z
1	37	0.0625	-1.53
2	39	0.1875	-0.89
3	41	0.3125	-0.49
4	53	0.4375	-0.16
5	55	0.5625	0.16
6	56	0.6875	0.49
7	61	0.8125	0.89
8	69	0.9375	1.53

7. $\bar{x} = (\Sigma x)/n = 411/8 = 51.375$, rounded to 51.4
   $s^2 = [n(\Sigma x^2) - (\Sigma x)^2]/[n(n-1)]$
   $\quad = [7263]/[8(7)] = 129.6964$
   $s = 11.388$, rounded to 11.4

8. normal distribution
   $\mu = 51.375$
   $\sigma = 11.388$
   $P(x < 40)$
   $\quad = P(z<-1.00)$
   $\quad = 0.1587$

Chapter 11

Multinomial Experiments and Contingency Tables

11-2 Multinomial Experiments: Goodness of Fit

1. A test for "goodness-of-fit" examines the differences between values observed in the sample and values predicted by a specific mathematical model. If those differences are too large, then there is not a good fit between the observed data and the hypothesized model.

3. An observed frequency is the number of sample observations falling into a particular category. An expected frequency is the number of observations that the hypothesized model predicts will fall into a particular category.

5. a. H_o: $p_1 = p_2 = p_3$
   b. $\Sigma O = 30$. Under the hypothesis of equally likely categories, the expected frequency for each category is $\Sigma O/n = 30/3 = 10$.
   c. $\chi^2 = \Sigma[(O - E)^2/E] = [(5-10)^2/10] + [(5-10)^2/10] + [(20-10)^2/10]$
   $$= 25/10 + 25/10 + 100/10$$
   $$= 2.5 + 2.5 + 10.0 = 15.0$$
   d. For $\alpha = 0.05$ and df = 2, the critical value is $\chi^2_\alpha = 5.991$.
   e. Reject H_o; there is sufficient evidence to conclude that at least one proportion is not as claimed – i.e., that the categories are not equally likely.

7. a. For df=37 and α=0.10, the critical value is $\chi^2_\alpha = 51.805$.
   b. For a χ^2 distribution with df=37, 90% of the probability falls above 29.051 and 10% falls above 51.805. Since $29.051 < 38.232 < 51.805$, $0.10 < \text{P-value} < 0.90$.
   c. There is not sufficient evidence to reject the claim that the 38 results are equally likely.

9. H_o: $p_1 = p_2 = p_3 = \ldots = p_6 = 1/6$
   H_1: at least $p_i \neq 1/6$
   $\alpha = 0.05$ and df = 5
   C.V. $\chi^2 = \chi^2_\alpha = 11.071$

   calculations:

outcome	O	E	$(O-E)^2/E$
1	27	33.33	1.2033
2	31	33.33	0.1633
3	42	33.33	2.2533
4	40	33.33	1.3333
5	28	33.33	0.8533
6	32	33.33	0.0533
	200	200.00	5.8600

   $\chi^2 = \Sigma[(O - E)^2/E] = 5.860$
   P-value = χ^2cdf $(5.860,99,5) = 0.3201$

   conclusion:
   Do not reject Ho; there is not sufficient evidence to conclude that at least one the proportions is different from 1/6.
   No. This particular loaded die did not behave noticeably different from a fair die.

NOTES FOR THE REMAINING EXERCISES IN THIS SECTION:
(1) In multinomial problems, always verify that $\Sigma E = \Sigma O$ before proceeding. If these sums are not equal, then an error has been made and further calculations have no meaning.
(2) As in the previous uses of the χ^2 distribution, the accompanying illustrations follow the "usual" shape – even though that shape is not correct for df = 1 and df = 2.
(3) As in the previous sections, a subscript (the df) may be used to identify which χ^2 distribution to use in the tables.
(4) Exact P-values are found as in previous sections using the χ^2cdf command on the TI-83/84$^+$ calculator.

11. H_o: $p_{Sun} = p_{Mon} = p_{Tue} = \ldots = p_{Sat} = 1/7$
H_1: at least one $p_i \neq 1/7$
$\alpha = 0.05$ and df = 6
C.V. $\chi^2 = \chi^2_\alpha = 12.592$

calculations:

day	O	E	$(O-E)^2/E$
Sun	132	117	1.9231
Mon	98	117	3.0855
Tue	95	117	4.1368
Wed	98	117	3.0855
Thu	105	117	1.2308
Fri	133	117	2.1880
Sat	158	117	14.3675
	819	819	30.0171

$\chi^2 = \Sigma[(O - E)^2/E] = 30.017$
P-value = χ^2cdf (30.017,99,6) = 0.00004
conclusion:
Reject Ho; there is sufficient evidence to conclude that at least one the proportions is different from 1/7 – i.e., that car crash fatalities occur with different frequencies on different days of the week.
Drinking and/or increased recreational travel on the weekends could explain the differences – and the larger number on Saturday may be due to early morning crashes as a result of Friday night parties that last past midnight.

13. H_o: $p_{Jan} = p_{Feb} = p_{Mar} = \ldots = p_{Dec} = 1/12$
H_1: at least one $p_i \neq 1/12$
$\alpha = 0.05$ and df = 11
C.V. $\chi^2 = \chi^2_\alpha = 19.675$

calculations:

month	O	E	$(O-E)^2/E$
Jan	6	16.67	6.82667
Feb	8	16.67	4.50667
Mar	10	16.67	2.66667
Apr	16	16.67	0.02667
May	22	16.67	1.70667
Jun	28	16.67	7.70667
Jul	24	16.67	3.22667
Aug	28	16.67	7.70667
Sep	26	16.67	5.22667
Oct	14	16.67	0.42667
Nov	10	16.67	2.66667
Dec	8	16.67	4.50667
	200	200.00	47.20000

$\chi^2 = \Sigma[(O-E)^2/E] = 47.200$

P-value $= \chi^2\text{cdf}\,(47.200,99,11) = 0.000002$

conclusion:

Reject Ho; there is sufficient evidence to reject the claim that motorcycle deaths occur with equal frequency in the different months.

There are probably fewer motorcycle deaths during the winter months because fewer persons ride their motorcycles when the weather is cold or the roads are icy.

15. H_o: $p_{Jan} = p_{Feb} = p_{Mar} = \ldots = p_{Dec} = 1/12$

H_1: at least one $p_i \neq 1/12$

$\alpha = 0.05$ and df $= 11$

C.V. $\chi^2 = \chi_\alpha^2 = 19.675$

calculations:

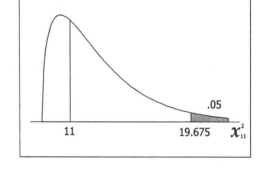

month	O	E	$(O-E)^2/E$
Jan	7	6.33	0.07018
Feb	3	6.33	1.75439
Mar	7	6.33	0.07018
Apr	7	6.33	0.07018
May	8	6.33	0.43860
Jun	7	6.33	0.07018
Jul	6	6.33	0.01754
Aug	6	6.33	0.01754
Sep	5	6.33	0.28070
Oct	6	6.33	0.01754
Nov	9	6.33	1.12281
Dec	5	6.33	0.28070
	76	76.00	4.21053

$\chi^2 = \Sigma[(O-E)^2/E] = 4.211$

P-value $= \chi^2\text{cdf}\,(4.211,99,11) = 0.9633$

conclusion:

Do not reject Ho; there is not sufficient evidence to reject the claim that Oscar-winning actresses are born in the different months with the same frequency.

No; there is no reason why more Oscar-winning actresses would be born in some months more than others.

NOTE: This exercise has remarkable uniformity in the data – the chance a random sample of births has such little monthly variation is $1-0.9633 = 0.0367$. While the critical region in goodness-of-fit tests is entirely in the upper tail to identify too much deviation from the null hypothesis, this calculated $\chi^2 = 4.211$ is significant in the lower tail. In some rare instances (e.g., to catch people suspected of fabricating data) a lower tail chi-square test is used to detect a fit that is "too good."

17. H_0: $p_{Jan} = p_{Feb} = p_{Mar} = \ldots = p_{Dec} = 1/12$
    H_1: at least one $p_i \neq 1/12$
    $\alpha = 0.05$ and df = 11
    C.V. $\chi^2 = \chi^2_\alpha = 19.675$
    calculations:

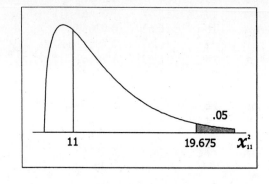

month	O	E	$(O-E)^2/E$
Jan	5	9.17	1.89394
Feb	8	9.17	0.14848
Mar	6	9.17	1.09394
Apr	8	9.17	0.14848
May	11	9.17	0.36667
Jun	14	9.17	2.54848
Jul	10	9.17	0.07576
Aug	9	9.17	0.00303
Sep	10	9.17	0.07576
Oct	12	9.17	0.87576
Nov	8	9.17	0.14848
Dec	9	9.17	0.00303
	110	110.00	7.38182

$\chi^2 = \Sigma[(O - E)^2/E] = 7.382$
P-value = χ^2cdf $(7.382,99,11) = 0.7674$

conclusion:

   Do not reject Ho; there is not sufficient evidence to reject the claim that weddings are held in different months with the same frequency.

The results refute the claim that "most" marriages occur in June – or even that June is significantly different from other months.

19. H_0: $p_4 = 2/16$, $p_5 = 4/16$, $p_6 = 5/16$, $p_7 = 5/16$
    H_1: at least one p_i is not as claimed
    $\alpha = 0.05$ and df = 3
    C.V. $\chi^2 = \chi^2_\alpha = 7.815$
    calculations:

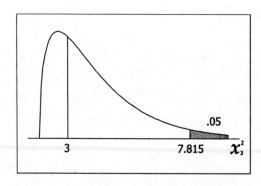

games	O	E	$(O-E)^2/E$
4	18	12.1250	2.84665
5	20	24.2500	0.74485
6	22	30.3125	2.27951
7	37	30.3125	1.47539
	97	97.0000	7.34639

$\chi^2 = \Sigma[(O - E)^2/E] = 7.346$
P-value = χ^2cdf $(7.346,99,3) = 0.0616$

conclusion:

   Do not reject Ho; there is not sufficient evidence to reject that claim that the observed frequencies agree with the theoretical proportions.

The evidence does not support the claim that 7-game series occur more often than expected.

21. H_o: $p_{gr} = 0.16$, $p_{or} = 0.20$, $p_{ye} = 0.14$, $p_{bl} = 0.24$, $p_{re} = .13$, $p_{br} = 0.13$
   H_1: at least one p_i is not as claimed
   $\alpha = 0.05$ and $df = 5$
   C.V. $\chi^2 = \chi^2_\alpha = 11.071$
   calculations:

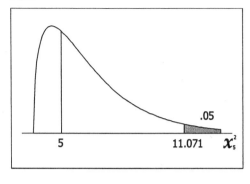

color	O	E	$(O-E)^2/E$
green	19	16	0.56250
orange	25	20	1.25000
yellow	8	14	2.57143
blue	27	24	0.37500
red	13	13	0.00000
brown	8	13	1.92308
	100	100	6.68210

   $\chi^2 = \Sigma[(O - E)^2/E] = 6.682$
   P-value $= \chi^2 cdf\ (6.682,99,5) = 0.2454$

   conclusion:

   Do not reject Ho; there is not sufficient evidence to reject the claim that the color distribution is as claimed by Mars, Inc.

23. H_o: $p_w = 0.757$, $p_H = 0.091$, $p_b = 0.108$, $p_A = 0.038$, $p_N = 0.007$
   H_1: at least one p_i is not as claimed
   $\alpha = 0.05$ and $df = 4$
   C.V. $\chi^2 = \chi^2_\alpha = 9.488$
   calculations:

race	O	E	$(O-E)^2/E$
white	644	569.264	9.8117
Hisp	23	68.432	30.1623
black	69	81.216	1.8375
Asian	14	28.576	7.4349
Native	2	5.264	2.0239
	752	752.000	51.2703

   $\chi^2 = \Sigma[(O - E)^2/E] = 51.270$
   P-value $= \chi^2 cdf\ (51.270,99,4) = 1.96 \times 10^{-10} \approx 0$

   conclusion:

   Reject Ho; there is sufficient evidence to reject the claim that the participants fit the same distribution as the US population.

   If the participants are not racially representative of the US, and if there are racial differences in cancer occurrence and/or responses to treatments, then the unadjusted sample mean would not describe the US population. A correct national mean could still be obtained by keeping sample data by race and calculating a weighted sample mean – but if there are racial differences, then each race should be examined separately anyway and a national mean has no real value. The underlying concern here may be for representative inclusion in clinical trials more for appearances than for medical or statistical reasons.

25. H_o: the leading digit proportions conform to Benford's law

H_1: at least one p_i is not as claimed

$\alpha = 0.05$ and df $= 8$

C.V. $\chi^2 = \chi^2_\alpha = 15.507$

calculations:

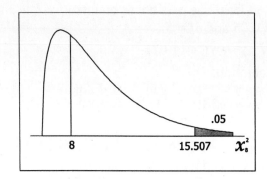

digit	O	p	E	$(O-E)^2/E$
1	72	0.301	60.2	2.3130
2	23	0.176	35.2	4.2284
3	26	0.125	25.0	0.0400
4	20	0.097	19.4	0.0186
5	21	0.079	15.8	1.7114
6	18	0.067	13.4	1.5791
7	8	0.058	11.6	1.1172
8	8	0.051	10.2	0.4745
9	4	0.046	9.2	2.9391
	200	1.000	200.0	14.4213

$\chi^2 = \Sigma[(O - E)^2/E] = 14.421$

P-value $= \chi^2$cdf $(14.421,99,8) = 0.0714$

conclusion:

Do not reject Ho; there is not sufficient evidence to reject the claim that the author's leading digits conform to Benford's law.

27. Use the lower tail of the χ^2 distribution to determine the critical region. If the null hypothesis is true and only random fluctuation exists, the expected value of $\chi^2 = \Sigma[(O - E)^2/E]$ is df – i.e., the degrees of freedom. If $\chi^2 = \Sigma[(O - E)^2/E]$ is significantly greater than df, then there is more variation than can be expected by chance and the data do not fit the null hypothesis. If $\chi^2 = \Sigma[(O - E)^2/E]$ is significantly less than df, then there is less deviation than can be expected by chance and the data fit the null hypothesis too well.

29. a. Use the binomial formula $P(x) = [n!/(n-x)!x!] \cdot p^x \cdot (1-p)^{n-x}$ with n=3 and p=1/3.

$P(x=0) = 1(1/3)^0(2/3)^3 =$ 8/27 $= 0.296$

$P(x=1) = 3(1/3)^1(2/3)^2 = 12/27 = 0.444$

$P(x=2) = 3(1/3)^2(2/3)^1 =$ 6/27 $= 0.222$

$P(x=3) = 1(1/3)^3(2/3)^0 =$ 1/27 $= 0.037$

b. Included with the expected frequencies are the usual χ^2 calculations needed for part (c).

x	O	E	$(O-E)^2/E$
0	89	88.89	0.000
1	133	133.33	0.001
2	52	66.67	3.227
3	26	11.11	19.951
	300	300.00	23.179

c. H_o: there is goodness of fit to the binomial distribution with n=3 and p = 1/3

H_1: there is not goodness of fit

$\alpha = 0.05$ and df $= 3$

C.V. $\chi^2 = \chi^2_\alpha = 7.815$

calculations:

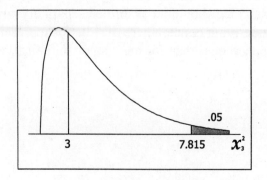

$\chi^2 = \Sigma[(O - E)^2/E] = 23.179$

P-value $= \chi^2$cdf $(23.179,99,3) = 0.00004$

conclusion:
  Reject Ho; there is sufficient evidence to reject the claim that the observed
  frequencies fit a binomial distribution with n=3 and p = 1/3.
NOTE: While the phrase "goodness of fit" is the accepted statistical terminology in this
context, the test could also be expressed as
H_o: $p_0 = 8/27$, $p_1 = 12/27$, $p_2 = 6/27$, $p_3 = 1/27$
H_1: at least one p_i is not as claimed

11-3 Contingency Tables: Independence and Homogeneity

1. In this section, the chi-square statistic measures the amount of disagreement between the
   observed frequencies and the frequencies expected if the null hypothesis were true.

3. In general, a contingency is a related event or the relationship between two events. In this
   section, a contingency table is an array of frequencies of related events. The methods of this
   section seek to determine whether the relationship between two variables is one of
   independence, or whether the value of one of them is contingent upon the value of the other.

5. H_o: ethnicity and being stopped are independent*
   H_1: ethnicity and being stopped are related*
   $\alpha = 0.05$ and df = 1
   C.V. $\chi^2 = \chi_\alpha^2 = 3.841$
   calculations:

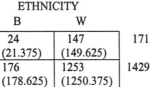

		ETHNICITY			
		B	W		
STOPPED?	Y	24	147	171	
		(21.375)	(149.625)		
	N	176	1253	1429	
		(178.625)	(1250.375)		
		200	1400	1600	

   $\chi^2 = \Sigma[(O - E)^2/E]$
      $= 0.3224 + 0.0461$
        $0.0386 + 0.0055 = 0.413$ [Minitab]
   P-value = χ^2cdf (0.413,99,1) = 0.521 [Minitab]
conclusion:
  Do not reject Ho; there is not sufficient evidence to reject the claim that ethnicity and being
  stopped are independent.
No. We cannot conclude that racial profiling is being used.
*NOTE: While the table claims to summarize "results for randomly selected drivers stopped by
police," it appears that fixed numbers (200 black, 1400 white) of the ethnic groups were asked
whether or not they had been stopped by the police. If so, the test should use "H_o: the
proportion of drivers stopped is the same for both ethnic groups" to test homogeneity and not
independence. The tests are equivalent mathematically, but there is a subtle difference in
interpretation. This comment could apply to several other of the exercises in this section.

NOTES FOR THE REMANINING EXERCISES:
  (1) For each row and each column it must be true that $\Sigma O = \Sigma E$. After the marginal row and
    column totals are calculated, both the row totals and column totals must sum to produce the
    same grand total. If either of the preceding is not true, then an error has been made and
    further calculations have no meaning.

(2) All contingency table analyses in this manual use the following conventions.
   • The E values for each cell are given in parentheses below the O values.
   • The addends used to calculate the χ^2 test statistic follow the arrangement of the cells in the contingency table – making it easier to monitor the large number of intermediate steps involved and helping to prevent errors caused by missing or double-counting cells.
   • As before, P-values are obtained using the χ^2cdf command on the TI-83/84[+] calculator.
   • The accompanying illustrations follow the "usual" chi-square shape, even though that shape is not correct for df=1 or df=2.

7. H_o: developing the flu and treatment used are independent
   H_1: developing the flu and treatment used are related
   $\alpha = 0.05$ and df $= 1$
   C.V. $\chi^2 = \chi^2_\alpha = 3.841$
   calculations:

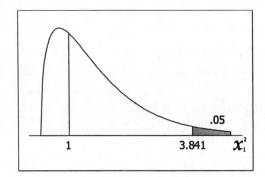

		FLU?		
		YES	NO	
TRT	VACCINE	14	1056	1070
		(72.8)	(997.2)	
	PLACEBO	95	437	532
		(36.2)	(495.8)	
		109	1493	1602

$\chi^2 = \Sigma[(O - E)^2/E]$
   $= 47.495 + 3.467$
   $95.526 + 6.974 = 153.462$
P-value $= \chi^2$cdf $(153.462,999,1) \approx 0$

conclusion:
   Reject Ho; there is sufficient evidence to reject the claim that developing the flu is independent of the treatment used.
Yes. The vaccine appears to be effective.

9. H_o: gender and left-handedness are independent
   H_1: gender and left-handedness are related
   $\alpha = 0.05$ and df $= 1$
   C.V. $\chi^2 = \chi^2_\alpha = 3.841$
   calculations:

		LEFTY?		
		Y	N	
GENDER	M	83	17	100
		(89)	(11)	
	F	184	16	200
		(178)	(22)	
		267	33	300

$\chi^2 = \Sigma[(O - E)^2/E]$
   $= 0.4045 + 3.2727$
   $0.2022 + 1.6364 = 5.516$
P-value $= \chi^2$cdf $(5.516,99,1) = 0.0188$

conclusion:
   Reject Ho; there is sufficient evidence to reject the claim that gender and left-handedness are independent. It appears that the proportion of left-handers is higher for males than it is for females.

11. H_o: subject truthfulness and polygraph reading are independent
H_1: subject truthfulness and polygraph reading are related
$\alpha = 0.05$ and df = 1
C.V. $\chi^2 = \chi_\alpha^2 = 3.841$

calculations:

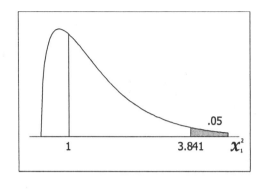

POLYGRAPH

		T	L	
SUBJECT	T	65 (54.40)	15 (25.60)	80
	L	3 (13.60)	17 (6.40)	20
		68	32	100

$\chi^2 = \Sigma[(O - E)^2/E]$
$= 2.0654 + 4.3891$
$8.2618 + 17.5562 = 32.273$
P-value $= \chi^2 cdf\,(32.273,99,1) = 0.00000001$

conclusion:

Reject Ho; there is sufficient evidence to reject the claim that a subject's truthfulness and polygraph reading are independent – and conclude that a person's truthfulness and the polygraph reading are related, so that the polygraph reading does indicate something about a person's truthfulness.

The results suggest that the two variables are related in that a person's truthfulness can be predicted from the polygraph reading – but the results do not suggest how accurate that prediction might be. Mathematically the chi square test addresses only the existence of a relationship and not the strength of a relationship – as an increase in sample size will increase the test statistic, even if all the proportions remain the same. Despite the extremely low P-value above, for example, the polygraph was incorrect $(3+15)/100 = 18\%$ of the time.

13. H_o: plea and sentence are independent
H_1: plea and sentence are related
$\alpha = 0.05$ and df = 1
C.V. $\chi^2 = \chi_\alpha^2 = 3.841$

calculations:

PLEA

		G	NG	
SENTENCE	P	392 (418.48)	58 (31.52)	450
	NP	564 (537.52)	14 (40.48)	578
		956	72	1028

$\chi^2 = \Sigma[(O - E)^2/E]$
$= 1.6759 + 22.2518$
$1.3047 + 17.3241 = 42.557$
P-value $= \chi^2 cdf\,(42.557,99,1) \approx 0$

conclusion:

Reject Ho; there is sufficient evidence to reject the claim that plea and sentence are independent. There appears to be a relationship between one's plea and whether or not he is sentenced to prison – and those who plead guilty are less likely to be sentenced to prison.
Yes. Assuming those who are really guilty will be convicted with a trial, these results suggest that a guilty plea should be encouraged. But since the study included only those with guilty pleas who were convicted in a trial, perhaps a guilty person who has a chance for acquittal

should plead not guilty. If 50 guilty people with not guilty pleas were acquitted, as shown below, including them in the table in the no prison category changes the above conclusion.

		PLEA			
		G	NG		
SENTENCE	P	392	58	450	
		(418.48)	(31.52)		
	NP	564	64	628	
		(537.52)	(40.48)		
		956	112	1078	

$$\chi^2 = \Sigma[(O - E)^2/E]$$
$$= 0.125 + 0.982$$
$$0.090 + 0.704 = 1.901$$

15. H_o: coin results and method used are independent
H_1: coin results and method used are related
$\alpha = 0.05$ and $df = 1$
C.V. $\chi^2 = \chi_\alpha^2 = 3.841$

calculations:

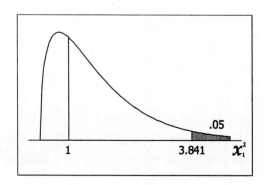

		RESULT			
		H	T		
METHOD	FLIP	14709	14306	29015	
		(14030.6)	(14984.4)		
	SPIN	9197	11225	20422	
		(9875.4)	(10546.6)		
		23906	25531	49437	

$$\chi^2 = \Sigma[(O - E)^2/E]$$
$$= 32.798 + 30.710$$
$$46.598 + 43.633 = 153.739$$

P-value $= \chi^2cdf\,(153.739,999,1) \approx 0$

conclusion:

Reject Ho; there is sufficient evidence to reject the claim that coin results are independent of the method used – it appears that for a penny more heads result from flipping than from spinning.

NOTE: In general, the data in the exercises in the text are real – i.e., not fabricated. As a protection against gross errors, if not for intellectual satisfaction, one should always stop to consider whether the conclusion in a test of hypothesis agrees with common sense. At first glance, the above coin results may seem surprising. But coins have different designs for their heads and tails, and they are not necessarily designed to be balanced. It appears that spinning, more than flipping, allows the coin enough time for the laws of physics and the unbalanced nature of the coin to affect the outcome.

17. The following table is used for the calculations in exercise #17.

		OCCUPATION				
		POLICE	CASHIER	TAXI.D	GUARD	
DEATH BY HOMICIDE?	Y	82 (112.92)	107 (75.28)	70 (64.25)	59 (65.55)	318
	N	92 (61.08)	9 (40.72)	29 (34.75)	42 (35.45)	172
		174	116	99	101	490

H_o: occupation and manner of death are independent
H_1: occupation and manner of death are related
$\alpha = 0.05$ [assumed] and df $= 3$
C.V. $\chi^2 = \chi_\alpha^2 = 7.815$

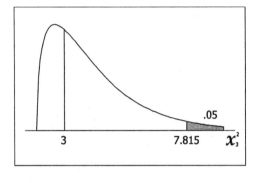

calculations:
$$\chi^2 = \Sigma[(O - E)^2/E]$$
$$= 8.468 + 13.364 + 0.515 + 0.654$$
$$15.655 + 24.708 + 0.952 + 1.209$$
$$= 65.524$$
P-value $= \chi^2$cdf $(65.524, 99, 3) \approx 0$

conclusion:

Reject Ho; there is sufficient evidence to reject the claim that occupation and manner of death are independent.

Yes. The occupation cashier appears to be significantly more prone to death by homicide.

19. The following table is used for the calculations in exercise #19.

		NUMBER OF CIGARETTES				
		0	1-14	15-34	35+	
USE BELT?	Y	175 (171.53)	20 (19.59)	42 (43.94)	6 (7.94)	243
	N	149 (152.47)	17 (17.41)	41 (39.06)	9 (7.06)	216
		324	37	83	15	459

H_o: amount of smoking and seat belt use are independent
H_1: amount of smoking and seat belt use are related
$\alpha = 0.05$ [assumed] and df $= 3$
C.V. $\chi^2 = \chi_\alpha^2 = 7.815$

calculations:
$$\chi^2 = \Sigma[(O - E)^2/E]$$
$$= 0.0702 + 0.0087 + 0.0858 + 0.4745$$
$$0.0790 + 0.0097 + 0.0965 + 0.5338$$
$$= 1.358$$
P-value $= \chi^2$cdf $(1.358, 99, 3) = 0.7154$
conclusion:

Do not reject Ho; there is not sufficient
evidence to reject the claim that amount of smoking and seat belt use are independent.
No. The theory is not supported by the sample data. While the deviation from the null hypothesis is in the direction of the theory, it is even less deviation from the null hypothesis than could be expected by chance alone.

21. The following table is used for the calculations in exercise #21.

		COLOR					
		black	white	yellow	red	blue	
RESULT	no injury	491 (509.5)	377 (353.9)	31 (28.2)	170 (173.7)	55 (58.6)	1124
	injury	213 (194.5)	112 (135.1)	8 (10.8)	70 (66.3)	26 (22.4)	429
		704	489	39	240	81	1553

H_o: injuries and helmet color are independent
H_1: injuries and helmet color are related
$\alpha = 0.05$ [assumed] and df = 4
C.V. $\chi^2 = \chi_\alpha^2 = 9.488$

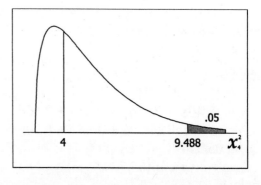

calculations:

$\chi^2 = \Sigma[(O - E)^2/E]$
$= 0.6737 + 1.5053 + 0.2725 + 0.0789 + 0.2241$
$\quad 1.7651 + 3.9438 + 0.7139 + 0.2068 + 0.5871$
$= 9.971$

P-value $= \chi^2$cdf (9.971,99,4) $= 0.0409$

conclusion:

Reject Ho; there is sufficient evidence to reject the claim that injuries are independent of helmet color.

Yes. These data lead to the same conclusion reached in the example of this section.

23. H_o: ethnicity and being stopped are independent
H_1: ethnicity and being stopped are related
$\alpha = 0.05$ and df = 1
C.V. $\chi^2 = \chi_\alpha^2 = 3.841$

calculations:

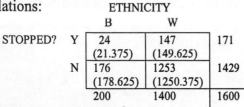

		ETHNICITY		
		B	W	
STOPPED?	Y	24 (21.375)	147 (149.625)	171
	N	176 (178.625)	1253 (1250.375)	1429
		200	1400	1600

$\chi^2 = \Sigma[(|O - E| - 0.5)^2/E]$
$= 0.2113 + 0.0302$
$\quad 0.0253 + 0.0036 = 0.270$

P-value $= \chi^2$cdf (0.270,99,1) $= 0.6033$

conclusion:

Do not reject Ho; there is not sufficient evidence to reject the claim that ethnicity and being stopped are independent.

No. We cannot conclude that racial profiling is being used.

Without the correction for continuity [see the solution for exercise #5 for the details] the calculated test statistic is 0.413. Since $(|O-E|-0.5)^2 < (O-E)^2$ whenever $|O-E| > 0.25$, Yates' correction for continuity generally lowers the calculated test statistic.

11-4 McNemar's Test for Matched Pairs

1. In section 11-3 the row and column variables/responses are different (but possibly related) and apply to the same subject. In this section the row and column variables/responses are the same and apply to two different (but matched) subjects. McNemar's test is appropriate for data in a 2x2 table when
   a. the variables/responses for the rows and columns are the same.
   b. the data consist of matched pairs.
   c. the row variable/responses apply to one of the pair, and the column variable/responses apply to its match.
   d. the frequency in each cell is the number of matched pairs that are described by that cell.

3. Discordant pairs of results are pairs which do not share the same result for the binomial variable being measured.

Exercises #5-#12 refer to the following 2x2 table, given both in general and with the specific frequencies observed in the experiment.

		BEFORE	
		smoker	non-smoker
AFTER	smoker	a	b
	non-smoker	c	d

		BEFORE	
		smoker	non-smoker
AFTER	smoker	50	6
	non-smoker	8	80

5. The total number of subjects in the experiment is $50 + 6 + 8 + 80 = 144$.

7. The number of subjects who appear unaffected by the experiment is $50 + 80 = 130$.

9. Discordant pairs are pairs for which the BEFORE and AFTER states do not agree.
   a. No. The data from the a=50 subjects who were smokers BEFORE and smokers AFTER do not represent discordant pairs of results.
   b. Yes. The data from the b=8 subjects who were smokers BEFORE and non-smokers AFTER represent discordant pairs of results.
   c. Yes. The data from the c=6 subjects who were non-smokers BEFORE and smokers AFTER represent discordant pairs of results.
   d. No. The data from the d=80 subjects who were non-smokers BEFORE and non-smokers AFTER do not represent discordant pairs of results.

11. For $\alpha = 0.01$ and $df = 1$, the critical value is $\chi^2 = 6.635$.

13. H_o: there is no difference between the treatment and the placebo
    H_1: there is a difference between the treatment and the placebo
    $\alpha = 0.05$ and $df = 1$
    C.V. $\chi^2 = \chi^2_\alpha = 3.841$
    calculations:
    $\quad$ b = 12, c = 22
    $\quad \chi^2 = (|b-c| - 1)^2/(b+c)$
    $\quad\quad = (10-1)^2/(12+22)$
    $\quad\quad = 81/34 = 2.382$
    $\quad$ P-value = χ^2cdf $(2.382,99,1) = 0.1227$
    conclusion: Do not reject Ho; there is not
    $\quad\quad$ sufficient evidence to reject the claim that
    $\quad\quad$ there is no difference between the treatment and the placebo.

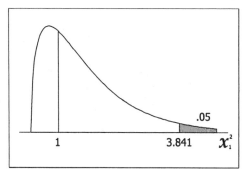

15. H_o: there is no difference between the PET-CT and MRI technologies
   H_1: there is a difference between the PET-CT and MRI technologies
   $\alpha = 0.05$ [assumed] and df = 1
   C.V. $\chi^2 = \chi_\alpha^2 = 3.841$
   calculations:

   $b = 1, c = 11$
   $\chi^2 = (|b\text{-}c| - 1)^2/(b+c)$
   $\quad = (10\text{-}1)^2/(1+11)$
   $\quad = 81/12 = 6.750$
   P-value $= \chi^2\text{cdf}\,(6.750,99,1) = 0.0094$

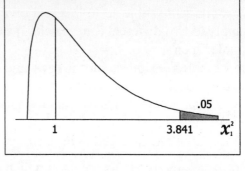

conclusion: Reject Ho; there is sufficient evidence
   to reject the claim that there is no difference
   between the PET-CT and MRI technologies. Since the majority of the discordant pairs are
   within the PET-CT=correct/MRI=incorrect cell, conclude that the PET-CT technology is
   more accurate.

17. $b=8, c=40, \chi^2 = (b\text{-}c)^2/(b+c) = (8\text{-}40)^2/(8+40) = 1024/48 = 21.333$
   The uncorrected test statistic of 21.333 is slightly larger than the corrected for continuity test
   statistic of 20.021. Since $(|b\text{-}c|\text{-}1)^2 < (b\text{-}c)^2$ whenever $|b\text{-}c| > 0.5$, the correction for continuity
   generally lowers the calculated test statistic.

19. This exercise follows the usual format for testing hypotheses with two modifications.
   • In order to compare two different techniques, the decision is based on comparing the α
     level and the P-value. This approach is valid for any hypothesis test and any test statistic,
     and it makes comparisons easier.
   • There are two "calculations" sections – one for each of the two techniques being
     compared.
   H_o: there is no difference between the two treatments
   H_1: there is a difference between the two treatments
   $\alpha = 0.05$
   calculations using the exact binomial:

   $x = 6 =$ largest frequency in cells
   $\qquad$ representing discordant pairs
   P-value $= 2 \cdot P(x \geq 6)$
   $\qquad = 2 \cdot [P(x\text{=}6) + P(x\text{=}7) + P(x\text{=}8)]$
   $\qquad = 2 \cdot [28(0.5)^6(0.5)^2 + 8(0.5)^7(0.5)^1 + 1(0.5)^8(0.5)^0]$
   $\qquad = 2 \cdot [0.109375 + 0.03125 + 0.00390625]$
   $\qquad = 2 \cdot [0.14453125]$
   $\qquad = 0.2891$

calculations using the χ^2 approximation:
   $b = 2, c = 6$
   $\chi^2 = (|b\text{-}c| - 1)^2/(b+c)$
   $\quad = (4\text{-}1)^2/(2+6)$
   $\quad = 9/8 = 1.125$
   P-value $= \chi^2\text{cdf}\,(1.125,99,1) = 0.2888$

conclusion: Do not reject Ho; there is not sufficient
   evidence to reject the claim that there is no difference between the two treatments.

Statistical Literacy and Critical Thinking

1. Categorical data are responses that can be placed in groups identified according to a non-numeric characteristic.

2. Nothing is wrong with the project – unless she tries to generalize the results to an inappropriate population. Without knowing how the sample was selected and how she worded her conclusion, it's difficult to make a precise critique. If she selected the "classmates" at random from a large university, for example, the results apply to the students of that university – but not necessarily to another school, and certainly not to the US population in general. If "classmates" means those students who are in one of her classes, however, the sample is a convenience sample and the results apply only to the students selected.

3. a. True. Chi-square formulas involve squaring a difference, and squares are never negative.
   b. False. The chi-square distribution is positively skewed – there is a lower bound at 0, but there is no upper bound.
   c. True. Each chi-square distribution "bunches up" around its degrees of freedom.
   d. False. In contingency tables, for example, the degrees of freedom is (r-1)(c-1) – regardless of the number of subjects that were surveyed to construct the table.
   e. False. Failure to have a random sample, in any situation, can create a bias that cannot be overcome merely by taking a larger sample.

4. Yes. The requirement is that all of the <u>expected</u> frequencies be at least 5, not that all of the <u>observed</u> frequencies be at least 5. No requirement is attached to the numbers of observed frequencies.

Review Exercises

1. H_0: $p_{Sun} = p_{Mon} = p_{Tue} = \ldots = p_{Sat} = 1/7$
   H_1: at least one $p_i \neq 1/7$
   $\alpha = 0.05$ and df = 6
   C.V. $\chi^2 = \chi^2_\alpha = 12.592$

   calculations:

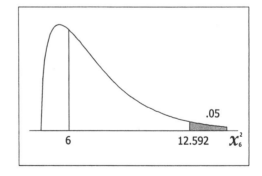

day	O	E	$(O-E)^2/E$
Sun	40	30.857	2.709
Mon	24	30.857	1.524
Tue	25	30.857	1.112
Wed	28	30.857	0.265
Thu	29	30.857	0.112
Fri	32	30.857	0.042
Sat	38	30.857	1.653
	216	216.000	7.417

   $\chi^2 = \Sigma[(O-E)^2/E] = 7.417$
   P-value = χ^2cdf (7.417,99,6) = 0.2840

   conclusion:
   Do not Reject Ho; there is not sufficient evidence to reject the claim that fatal crashes occur on the different days with equal frequency.

There were more crashes on Friday, Saturday and Sunday – which supports the Friday night and Saturday night (extending into Sunday morning) binge theory. But the increases were not large enough to be 95% sure that they were not random fluctuations from an equal frequency for each day. While the daily drinker (i.e., equal frequency) theory cannot be rejected, the results do not provide confirming support for either theory.

2. H_o: response and company status are independent
   H_1: response and company status are related
   $\alpha = 0.05$ and df = 1
   C.V. $\chi^2 = \chi_\alpha^2 = 3.841$
   calculations:

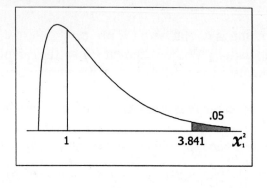

	OPINION		
	Y	N	
STATUS W	192 (181.60)	244 (254.40)	436
B	40 (50.40)	81 (70.60)	121
	232	325	557

$\chi^2 = \Sigma[(O-E)^2/E]$
$\quad = 0.5954 + 0.4250$
$\qquad 2.1454 + 1.5316 = 4.698$
P-value = χ^2cdf (4.697,99,1) = 0.0302
conclusion:
   Reject Ho; there is sufficient evidence to reject the claim that response and company status are independent. It appears that workers are more likely than bosses to think that it is unethical to monitor employee e-mail.
Yes; the decision changes if the 0.01 level is used – since P-value = 0.0302 > 0.01.
No; workers and bosses do not appear to agree on this issue.

3. H_o: type of crime and criminal/victim connection are independent
   H_1: type of crime and criminal/victim connection are related
   $\alpha = 0.05$ and df = 2
   C.V. $\chi^2 = \chi_\alpha^2 = 5.991$
   calculations:

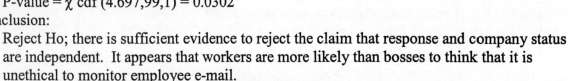

	CRIME			
	homicide	robbery	assault	
C/V S	12 (29.93)	379 (284.64)	727 (803.43)	1118
A	39 (21.07)	106 (200.36)	642 (565.57)	787
	51	485	1369	1905

$\chi^2 = \Sigma[(O-E)^2/E]$
$\quad = 10.7418 + 31.2847 + \ 7.2715$
$\qquad 15.2600 + 44.4425 + 10.3298 = 119.330$
P-value = χ^2cdf (119.330,999,2) ≈ 0
conclusion:
   Reject Ho; there is sufficient evidence to reject the claim that the type of crime and the criminal/victim connection are independent and conclude that the variables are related.
Police should not overlook acquaintances and relatives when investigating homicides.

4. H_o: there is no difference between the treatment and the placebo
   H_1: there is a difference between the treatment and the placebo
   $\alpha = 0.05$ and df $= 1$
   C.V. $\chi^2 = \chi^2_\alpha = 3.841$
   calculations:
       b = 8, c = 32
       $\chi^2 = (|b-c| - 1)^2/(b+c)$
          $= (24-1)^2/(8+32)$
          $= 528/40 = 13.225$
       P-value $= \chi^2\text{cdf} (13.225,99,1) = 0.0003$

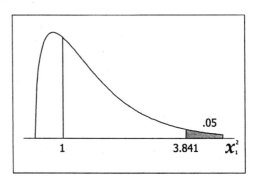

    conclusion: Reject Ho; there is sufficient evidence to reject the claim that there is no difference between the treatment and the placebo. Since the majority of the discordant pairs are within the treatment=relief/placebo=norelief cell, conclude that the treatment is effective.

Cumulative Review Exercises

1. scores in order: 70 73 75 80 80 84 85 90; n = 8 $\Sigma x = 637$ $\Sigma x^2 = 51035$
   $\bar{x} = 637/8 = 79.6$
   $\tilde{x} = (80 + 80)/2 = 80.0$
   $R = 90 - 70 = 20$
   $s^2 = [8(51035) - (637)^2]/[8(7)] = 44.8$
   $s = 6.7$
   five number summary:
       $x_1 = 70$
       $P_{25} = (73+75)/2 = 74.0$ $[(0.25)8 = 2$, a whole number, use $x_{2.5}]$
       $P_{50} = \tilde{x} = 80.0$
       $P_{75} = (84+85)/2 = 84.5$ $[(0.75)8 = 6$, a whole number, use $x_{6.5}]$
       $x_8 = 90$

2. Consider the table at the right.
   a. $P(C) = 153/637 = 0.240$

   b. $P(M) = 330/637 = 0.518$

   c. $P(C$ or $M) = P(C) + P(M) - P(C$ and $M)$
           $= 153/637 + 330/637 - 80/637$
           $= 403/637 = 0.633$

   d. $P(F_1$ and $F_2) = P(F_1) \cdot P(F_2|F_1)$
            $= (307/637)(306/636) = 0.232$

   e. $P(B|F) = 84/307 = 0.274$

	A	B	C	D	
M	85	90	80	75	330
F	80	84	73	70	307
	165	174	153	145	637

3. The following table is used for the calculations in exercise #3.

		SELECTION				
		A	B	C	D	
GENDER	M	85	90	80	75	330
		(85.48)	(90.14)	(79.26)	(75.12)	
	F	80	84	73	70	307
		(79.52)	(83.86)	(73.74)	(69.88)	
		165	174	153	145	637

This test follows the pattern established in section 11-3.

H_o: gender and selection are independent

H_1: gender and selection are related

$\alpha = 0.05$ [assumed] and df = 3

C.V. $\chi^2 = \chi^2_\alpha = 7.815$

calculations:

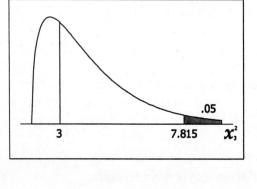

$\chi^2 = \Sigma[(O - E)^2/E]$

$= 0.00268 + 0.00022 + 0.00687 + 0.00018$

$\quad 0.00288 + 0.00024 + 0.00738 + 0.00020$

$= 0.0207$

P-value $= \chi^2$cdf $(0.0207, 99, 3) = 0.9992$

conclusion:

Do not reject Ho; there is not sufficient evidence to reject the claim that men and women choose the different answers in the same proportions.

4. This test follows the pattern established in section 10-2.

$n = 4 \qquad \Sigma x = 330 \qquad \Sigma x^2 = 27350 \qquad n(\Sigma x^2) - (\Sigma x)^2 = 4(27350) - (330)^2 = 500$

$\qquad\qquad \Sigma y = 307 \qquad \Sigma y^2 = 23685 \qquad n(\Sigma y^2) - (\Sigma y)^2 = 4(23685) - (307)^2 = 491$

$\qquad\qquad\qquad\qquad \Sigma xy = 25450 \qquad n(\Sigma xy) - (\Sigma x)(\Sigma y) = 4(25450) - (330)(307) = 490$

$r = [n(\Sigma xy) - (\Sigma x)(\Sigma y)]/[\sqrt{n(\Sigma x^2) - (\Sigma x)^2} \sqrt{n(\Sigma y^2) - (\Sigma y)^2}]$

$\quad = 490/[\sqrt{500}\sqrt{491}] = 0.9889$

H_o: $\rho = 0$

H_1: $\rho \neq 0$

$\alpha = 0.05$ and df = 2

C.V. $t = \pm t_{\alpha/2} = \pm 4.303$ [or $r = \pm 0.950$]

calculations:

$t_r = (r - \mu_r)/s_r$

$\quad = (0.9889 - 0)/\sqrt{[1 - (0.9889)^2]/2}$

$\quad = 0.9889/0.1049$

$\quad = 9.43$

P-value $= 2 \cdot$tcdf$(9.43, 99, 2) = 2(0.0055) = 0.0110$

conclusion:

Reject H_o; there is sufficient evidence to conclude that $\rho \neq 0$ (in fact, that $\rho > 0$).

5. This test follows the pattern established in section 9-4.

$d = x - y =$ pre-training $-$ post-training: 5 6 7 5

$n = 4$ $\Sigma d = 23$ $\Sigma d^2 = 135$ $\bar{d} = 5.75$ $s_d = 0.9574$

original claim: $\mu_d = 0$

H_0: $\mu_d = 0$

H_1: $\mu_d \neq 0$

$\alpha = 0.05$ [assumed] and df $= 3$

C.V. $t = -t_{\alpha/2} = \pm 3.182$

calculations

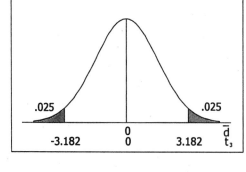

$$t_{\bar{d}} = (\bar{d} - \mu_{\bar{d}})/s_{\bar{d}}$$

$$= (5.75 - 0)/(0.9574/\sqrt{4})$$

$$= 5.75/0.4787$$

$$= 12.011$$

P-value $= 2 \cdot \text{tcdf}(12.011,99,3) = 2(0.0006) = 0.0012$

conclusion:

Reject H_0; there is sufficient evidence to reject the claim that $\mu_d = 0$ and conclude that $\mu_d \neq 0$ (in fact, that $\mu_d > 0$ – i.e., the pre-training scores are significantly higher than the post-training scores).

6. Let the men be group 1. This test follows the pattern established in section 9-3.

group 1: men (n=4)	group 2: women (n=4)
$\Sigma x = 330$	$\Sigma x = 307$
$\Sigma x^2 = 27350$	$\Sigma x^2 = 23685$
$\bar{x} = 82.50$	$\bar{x} = 76.75$
$s^2 = 41.667$ (s=6.45)	$s^2 = 40.917$ (s=6.40)

$\bar{x}_1 - \bar{x}_2 = 82.50 - 76.75 = 5.75$

original claim: $\mu_1 - \mu_2 = 0$

H_0: $\mu_1 - \mu_2 = 0$

H_1: $\mu_1 - \mu_2 \neq 0$

$\alpha = 0.05$ [assumed] and df $= 3$

C.V. $t = \pm t_{\alpha/2} = \pm 3.182$

calculations:

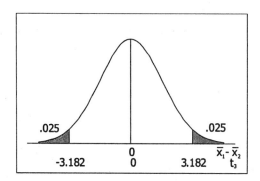

$$t_{\bar{x}_1 - \bar{x}_2} = (\bar{x}_1 - \bar{x}_2 - \mu_{\bar{x}_1 - \bar{x}_2})/s_{\bar{x}_1 - \bar{x}_2}$$

$$= (5.75 - 0)/\sqrt{41.667/4 + 40.917/4}$$

$$= 5.75/4.544 = 1.265$$

P-value $= 2 \cdot \text{tcdf}(1.265,99,3) = 2(0.1475) = 0.2951$

conclusion:

Do not reject H_0; there is not sufficient evidence to reject the claim that $\mu_1 - \mu_2 = 0$.

Chapter 12

Analysis of Variance

12-2 One-Way ANOVA

1. One-way analysis of variance tests the equality of two or more population means by analyzing sample variances. It finds a difference in the population means if the variance between the sample means is larger than can expected from the variance within the samples.

3. The variance between samples is an adjusted variance of the sample means – obtained by treating the sample means as individual scores. The variance within samples is a weighted average of the individual sample variances.

5. a. The null hypothesis is that $\mu_1 = \mu_2 = \mu_3 = \mu_4$, where the subscripts identify the four treatments.
   b. The alternative hypothesis is that at least one μ_i is different – i.e., that at least one of the treatments has a population mean that is different from the others.
   c. The test statistic is F = 8.448.
   d. From Table A-5, the critical value for $\alpha = 0.05$ and $df_{num} = 3$, $df_{den} = 15$ is $F_\alpha = 3.2874$.
   e. The P-value is $P(F_{15}^3 > 8.448) = 0.0016$.
   f. The conclusion is as follows. Reject Ho; there is sufficient evidence to reject the claim that $\mu_1 = \mu_2 = \mu_3 = \mu_4$ and to conclude that at least one μ_i is different.
   g. Removing the outlier changes the df for the variance within samples from 16 to 15, the calculated F statistic from 5.731 to 8.448, and the P-value from 0.0073 to 0.0016. The conclusion does not change. Removing the outlier made the lack of equality among the means slightly easier to identify, but it did not have much of an effect on the results.

7. H_o: $\mu_1 = \mu_2 = \mu_3 = \mu_4$
   H_1: at least one μ_i is different
   $\alpha = 0.05$ and $df_{num} = 3$, $df_{den} = 156$
   C.V. $F = F_\alpha = 2.6626$ [Excel]
   calculations:

   $F = ns_{\bar{x}}^2/s_p^2$

   $= 11.99995/34.08497$

   $= 0.35206$ [Excel]

   P-value = $P(F_{156}^3 > 0.3521) = 0.7877$ [Excel]

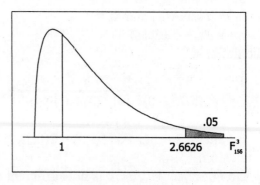

conclusion:

   Do not reject Ho; there is not sufficient evidence to reject the claim that $\mu_1 = \mu_2 = \mu_3 = \mu_4$. Given the four mean losses that range from -2.1 lbs to 3.3 lbs, it appears that one year of following the diet does not result in a weight loss worth the effort. The effort may be justified, however, when other potential benefits are considered – viz., maintaining one's weight (i.e., not gaining weight), enjoying a healthy lifestyle, etc.

NOTE: This section is calculation-oriented. Do not get so involved with the formulas that you miss the concepts. This manual arranges the calculations to promote both computational efficiency and understanding of the underlying principles. The following notation is used.

k = the number of groups

n_i = the number of scores in group i (where i = 1,2,…,k)

$\overline{x}_i$ = the mean of group i

s_i^2 = the variance of group i

$\overline{\overline{x}}$ = the overall mean of all the scores in all the groups

$\quad = (\Sigma n_i \overline{x}_i)/\Sigma n_i$ = the (weighted) mean of the group means

$\quad = \Sigma \overline{x}_i /k$ = simplified form when each group has equal size n

s_B^2 = the variance between the groups

$\quad = \Sigma n_i(\overline{x}_i - \overline{\overline{x}})^2/(k-1)$

$\quad = n \Sigma(\overline{x}_i - \overline{\overline{x}})^2/(k-1) = ns_{\overline{x}}^2$ = simplified form when each group has equal size n

s_p^2 = the variance within the groups

$\quad = (\Sigma df_i s_i^2)/\Sigma df_i$

$\quad = \Sigma s_i^2 /k$ = simplified form when each group has equal size n

numerator df = k-1

denominator df = Σdf_i

$\qquad\qquad\quad = k(n-1)$ = simplified form when each group has equal size n

$F = s_B^2/s_p^2$ = (variance between groups)/(variance within groups)

P-value = Fcdf(F, 99, numerator df, denominator df) from the TI-83/84$^+$ calculator

As a crude check against errors, and to help get a feeling for the problem, always verify that the following "overall" values are realistic in that

$\overline{\overline{x}}$ is a value between the lowest and the highest of the $\overline{x}_i$ values.

s_p^2 is a value between the lowest and the highest of the s_i^2 values.

As in previous chapters, the superscripts and subscripts (the numerator df and denominator df) may be used to identify which F distribution to look up in the tables.

9. H_0: $\mu_1 = \mu_2 = \mu_3 = \mu_4 = \mu_5 = \mu_6$
   H_1: at least one μ_i is different
   $\alpha = 0.05$ and $df_{num} = 5$, $df_{den} = 94$
   C.V. F = $F_\alpha = 2.3113$ [Statdisk]
   calculations:

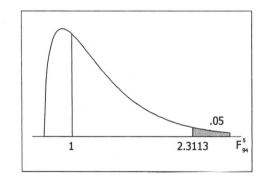

$\quad F = ns_{\overline{x}}^2/s_p^2$

$\qquad = 0.001/0.003$

$\qquad = 0.443$ [Statdisk]

$\quad$ P-value = $P(F_{94}^5 > 0.443) = 0.8173$ [Statdisk]

conclusion:

$\quad$ Do not reject Ho; there is not sufficient evidence to reject the claim that

$\quad \mu_1 = \mu_2 = \mu_3 = \mu_4 = \mu_5 = \mu_6$.

No. The results do not suggest that there is any problem requiring corrective action.

11. Since each group has equal size n=5, use the simplified form of the calculations.
The following preliminary values are identified.

	subcompact	compact	midsize	fullsize
n	5	5	5	5
Σx	3344	2779	2434	2689
Σx^2	2470638	1577659	1297312	1541765
$\overline{x}$	668.8	555.8	486.8	537.8
s^2	58542.7	8272.8	28110.2	23905.2

k = 4 $\overline{\overline{x}} = \Sigma \overline{x}_i/k$

n = 5   = 562.3

$s_{\overline{x}}^2 = \Sigma(\overline{x}_i - \overline{\overline{x}})^2/(k-1)$ $s_p^2 = \Sigma s_i^2/k$

   = 5895   = 29707.725

$H_o: \mu_1 = \mu_2 = \mu_3 = \mu_4$
$H_1:$ at least one μ_i is different
$\alpha = 0.05$ and $df_{num} = 3$, $df_{den} = 16$
C.V. $F = F_\alpha = 3.2389$
calculations:

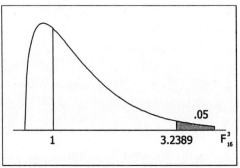

   $F = ns_{\overline{x}}^2/s_p^2$

      = 5(5895)/29707.725

      = 0.9922

   P-value = Fcdf(0.9922,99,3,16) = 0.4216

conclusion:

   Do not reject Ho; there is not sufficient evidence to reject the claim that $\mu_1 = \mu_2 = \mu_3 = \mu_4$.
No. The head injury data do not suggest that larger cars are safer.

13. Since each group has equal size n=6, use the simplified form of the calculations.
The following preliminary values are identified.

	female/black	male/black	female/white	male/white
n	6	6	6	6
Σx	632.34	732.67	671.66	692.66
Σx^2	67822.3156	89653.7045	75344.7156	80798.5756
$\overline{x}$	105.390	122.112	111.943	115.443
s^2	236.001	37.230	31.371	167.119

k = 4 $\overline{\overline{x}} = \Sigma \overline{x}_i/k$

n = 6   = 113.722

$s_{\overline{x}}^2 = \Sigma(\overline{x}_i - \overline{\overline{x}})^2/(k-1)$ $s_p^2 = \Sigma s_i^2/k$

   = 48.647   = 117.930

$H_o: \mu_1 = \mu_2 = \mu_3 = \mu_4$
$H_1:$ at least one μ_i is different
$\alpha = 0.05$ and $df_{num} = 3$, $df_{den} = 20$
C.V. $F = F_\alpha = 3.0984$
calculations:

   $F = ns_{\overline{x}}^2/s_p^2$

      = 6(48.647)/117.930

      = 2.4750

   P-value = Fcdf(2.4750,99,3,20) = 0.0911

conclusion:

   Do not reject Ho; there is not sufficient evidence to reject the claim that $\mu_1 = \mu_2 = \mu_3 = \mu_4$.
Yes. By this measure, these groups can be considered to be samples from the same population.

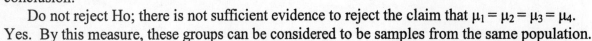

15. This exercise requires the following summary statistics from Data Set 14.

	Indian	wheat	pre-1983	post-1983
n	5	35	35	37
$\overline{x}$	3.02754	3.08307	3.07478	2.49910
s	0.0849148	0.0492909	0.0391000	0.0164800

$\overline{\overline{x}} = (\Sigma n_i \overline{x}_i)/\Sigma n_i$

$\quad = [5(3.02754) + 35(3.08307) + 35(3.07478) + 37(2.44910)]/[5 + 35 + 35 + 37] = 323.12915/112 = 2.88508$

$\Sigma n_i(\overline{x}_i - \overline{\overline{x}})^2 = 5(3.02754\text{-}2.88508)^2 + 35(3.08307\text{-}2.88508)^2 + 35(3.07478\text{-}2.88508)^2 + 37(2.49910\text{-}2.88508)^2$

$\quad\quad = 8.245270$

$\Sigma df_i s_i^2 = 4(0.0849148)^2 + 34(0.0492909)^2 + 34(0.0391000)^2 + 36(0.0164800)^2 = 0.173205$

$s_B^2 = \Sigma n_i(\overline{x}_i - \overline{\overline{x}})^2/(k-1) = 8.245270/3 = 2.74842$

$s_p^2 = (\Sigma df_i s_i^2)/\Sigma df_i = 0.173205/108 = 0.001604$

$H_0: \mu_1 = \mu_2 = \mu_3 = \mu_4$
$H_1:$ at least one μ_i is different
$\alpha = 0.05$ and $df_{num} = 3$, $df_{den} = 108$
C.V. $F = F_\alpha = 2.6802$
calculations:

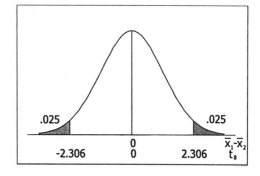

$\quad F = s_B^2/s_p^2$

$\quad\quad = 2.74842/0.001604$

$\quad\quad = 1713.747$

$\quad\quad$ P-value $= Fcdf(1713.747, 9999, 3, 108) \approx 0$
conclusion:

> Reject Ho; there is sufficient evidence to reject the claim that $\mu_1 = \mu_2 = \mu_3 = \mu_4$ and conclude that at least one μ_i is different.

No. It appears that machines cannot treat the weights of all pennies the same way.

17. Let the "none" treatment be group 1, and use the summary statistics given in the text.

a. Testing $H_0: \mu_1 = \mu_2$ using the pooled variance t test of section 9-3.

$\overline{x}_1\text{-}\overline{x}_2 = 0.184 - 0.332 = -0.148$ $\quad s_p^2 = [(df_1)s_1^2 + (df_2)s_2^2]/[df_1 + df_2]$

$df = df_1 + df_2 = 4 + 4 = 8$ $\quad\quad = [(4)(0.127)^2 + (4)(0.565)^2]/[4+4]$

$\quad\quad\quad\quad\quad = 1.341416/8 = 0.167677$

$H_0: \mu_1 - \mu_2 = 0$
$H_1: \mu_1 - \mu_2 \neq 0$
$\alpha = 0.05$ and $df = 8$
C.V. $t = \pm t_{\alpha/2} = \pm 2.306$
calculations:

$\quad t_{\overline{x}_1\text{-}\overline{x}_2} = (\overline{x}_1\text{-}\overline{x}_2 - \mu_{\overline{x}_1\text{-}\overline{x}_2})/s_{\overline{x}_1\text{-}\overline{x}_2}$

$\quad\quad = (-0.148 - 0)/\sqrt{0.167677/5 + 0.167677/5}$

$\quad\quad = -0.148/0.25898$

$\quad\quad = -0.57147$

$\quad\quad$ P-value $= 2 \cdot tcdf(-99, -0.57147, 8) = 2(0.2917) = 0.5834$
conclusion:

> Do not reject H_0; there is not sufficient evidence to reject the claim that $\mu_1 - \mu_2 = 0$.

b. Testing H_o: $\mu_1 = \mu_2$ using the analysis of variance test of this section.
Since each of the $k=2$ groups has equal size $n=5$, use the simplified form of the calculations.

$\overline{\overline{x}} = \Sigma\overline{x}_i/k = (0.184 + 0.332)/2 = 0.258$

$s_{\overline{x}}^2 = \Sigma(\overline{x}_i - \overline{\overline{x}})^2/(k-1) = [(0.184-0.258)^2 + (0.332-0.258)^2]/1 = 0.010952/1 = 0.010952$

$s_p^2 = \Sigma s_i^2/k = [(0.127)^2 + (0.565)^2]/2 = 0.335354/2 = 0.167677$

H_o: $\mu_1 = \mu_2$
H_1: at least one μ_i is different
$\alpha = 0.05$ and $df_{num} = 1$, $df_{den} = 8$
C.V. $F = F_\alpha = 5.3177$
calculations:

$F = ns_{\overline{x}}^2/s_p^2$

$= 5(0.010952)/0.167677$

$= 0.3266$

P-value $= Fcdf(0.3266,99,1,8) = 0.5834$

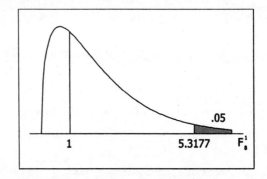

conclusion:

Do not reject H_o; there is not sufficient evidence to reject the claim that $\mu_1 = \mu_2$.

c. The tests are equivalent, as demonstrated by the P-values. In general $(t_{df})^2 = F_{df}^1$.
In this particular case, for $\alpha = 0.05$,

critical values: $(\pm t_{8,\alpha/2})^2 = (\pm 2.306)^2 = 5.3176 \approx 5.3177 = F_{8,\alpha}^1$ [except for round-off]

calculated statistic: $t^2 = (-0.57147)^2 = 0.3266 = F$

19. One convenient way to summarize and display differences between means is to list the means
in order, and place letters after each mean in such a way that means which are significantly
different do not share the same letter. Applying this technique to the Bonferroni results in the
text and the Tukey results in this exercise yields the following comparison.

Bonferroni Test				Tukey Test		
trt	mean			trt	mean	
3	0.164	a		3	0.164	a
1	0.184	a		1	0.184	a
2	0.332	a		2	0.332	a
4	1.334	b		4	1.334	b

The Bonferroni and Tukey tests completely agree in that both declare that treatment 4
(fertilizer and irrigation) is significantly different from the others, and there are no other
significant differences. But the two tests will not always completely agree. The tables in the
text indicate that the Bonferroni test is slightly more conservative than the Tukey test in that it
gives slightly higher P-values for declaring differences to be significant and its confidence
intervals on the treatment differences are slightly wider than those given with the Tukey test.

12-3 Two-Way ANOVA

1. Analysis of variance is a test to discover differences in population means by examining
population variances. The test is two-way when the experimental units can be classified
according to levels of two different factors so that the data can be arranged in a two-
dimensional table – with the rows and columns representing the two different factors.

3. When there is interaction, it is not possible to make an overall statement about the results for each factor – because the results for each factor vary according to the various levels of the other factor.

NOTE: The formulas and principles in this section are logical extensions of those in the previous section. In particular,

$$SS_{Row} = \Sigma n_i (\overline{x}_i - \overline{\overline{x}})^2 \text{ for } i = 1,2,3\ldots \text{ [for each row]}$$

$$SS_{Col} = \Sigma n_j (\overline{x}_j - \overline{\overline{x}})^2 \text{ for } j = 1,2,3\ldots \text{ [for each column]}$$

$$SS_{Tot} = \Sigma (x - \overline{\overline{x}})^2 \text{ for all the x's}$$

When there is only one observation per cell, the unexplained variation is

$$SS_{Err} = SS_{Tot} - SS_{Row} - SS_{Col}, \text{ and there is not enough data to measure interaction.}$$

When there is more than one observation per cell, the unexplained variation (i.e., the failure of experimental units in the same cell to respond the same) is

$$SS_{Err} = \Sigma (x - \overline{x}_{ij})^2 = \Sigma df_{ij} s_{ij}^2 \text{ [for each cell – i.e., for each i,j (row,col) combination]}$$

and the interaction sum of squares is

$$SS_{Int} = SS_{Tot} - SS_{Row} - SS_{Col} - SS_{Err}.$$

Since the data will be analyzed with statistical software, however, the above formulas need not be used by hand.

5. H_o: there is no site-treatment interaction
H_1: there is site-treatment interaction
$\alpha = 0.05$ [assumed] and $df_{num} = 3$, $df_{den} = 32$
C.V. $F = F_\alpha = 2.9233$
calculations:
   $F = MS_{Int}/MS_{Err}$
   $= 0.635336/0.227360$
   $= 2.79$ [Minitab]
   P-value $= P(F_{32}^3 > 2.79) = 0.056$ [Minitab]

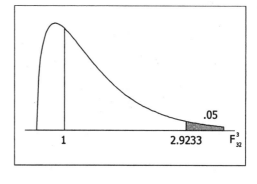

conclusion:
   Do not reject H_o; there is not sufficient evidence to reject the claim that there is no interaction between site and treatment.

7. H_o: $\mu_1 = \mu_2 = \mu_3 = \mu_4$ [there is no treatment effect]
H_1: at least one treatment μ_i is different
$\alpha = 0.05$ [assumed] and $df_{num} = 3$, $df_{den} = 32$
C.V. $F = F_\alpha = 2.9233$
calculations:
   $F = MS_{Trt}/MS_{Err}$
   $= 0.583949/0.227360$
   $= 2.567$ [Minitab]
   P-value $= P(F_{32}^3 > 2.57) = 0.072$ [Minitab]

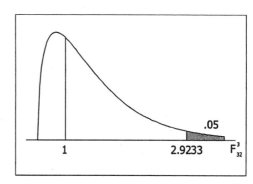

conclusion:
   Do not reject H_o; there is not sufficient evidence to support the claim that treatment has an effect on weight.

9. H_0: $\mu_1 = \mu_2$ [there is no gender effect]
   H_1: at least one gender μ_i is different
   $\alpha = 0.05$ [assumed] and $df_{num} = 1$, $df_{den} = 36$
   C.V. $F = F_\alpha = 4.0847$
   calculations:
   $\quad F = MS_{Gen}/MS_{Err}$
   $\quad\quad = 52635.0/10465.2$
   $\quad\quad = 5.03$ [Minitab]
   $\quad$ P-value $= P(F_{36}^1 > 5.03) = 0.031$ [Minitab]

conclusion:
   Reject H_0; there is sufficient evidence to support the claim that gender has an effect on SAT scores.

11. H_0: $\mu_1 = \mu_2 \ldots = \mu_{24}$ [there is no subject effect]
    H_1: at least one subject μ_i is different
    $\alpha = 0.05$ [assumed] and $df_{num} = 23$, $df_{den} = 69$
    C.V. $F = F_\alpha = 1.7001$
    calculations:
    $\quad F = MS_{Subj}/MS_{Err}$
    $\quad\quad = 140.5/36.3$
    $\quad\quad = 3.87$ [Minitab]
    $\quad$ P-value $= P(F_{69}^{23} > 3.87) = 0.000$ [Minitab]

conclusion:
    Reject H_0; there is sufficient evidence to support the claim that the choice of subject has an effect on the hearing test score.
This makes practical sense, because all people do not have the same level of hearing.

13. While it is assumed the ANOVA table will be obtained using statistical software, the actual calculations necessary to construct the table by hand are given below. There are n=4 observations in each of the indicated age-gender cells.

GENDER	under 20	20 – 40	over 40	
	$\Sigma x = 288$	$\Sigma x = 288$	$\Sigma x = 288$	
	$\Sigma x^2 = 21536$	$\Sigma x^2 = 21120$	$\Sigma x^2 = 21152$	
male	$\overline{x} = 72$	$\overline{x} = 72$	$\overline{x} = 72$	$\overline{x} = 72$
	$s^2 = 266.667$	$s^2 = 128$	$s^2 = 138.667$	
	$\Sigma x = 284$	$\Sigma x = 300$	$\Sigma x = 264$	
	$\Sigma x^2 = 20272$	$\Sigma x^2 = 22736$	$\Sigma x^2 = 17504$	
female	$\overline{x} = 71$	$\overline{x} = 75$	$\overline{x} = 66$	$\overline{x} = 70.667$
	$s^2 = 36$	$s^2 = 78.667$	$s^2 = 26.667$	
	$\overline{x} = 71.5$	$\overline{x} = 73.5$	$\overline{x} = 69$	$\Sigma x = 1712$ $\overline{\overline{x}} = 71.333$
				$\Sigma x^2 = 124320$ $s^2 = 95.536$

$SS_{Gen} = \Sigma n_i (\overline{x}_i - \overline{\overline{x}})^2 = 12(72 - 71.333)^2 + 12(70.667 - 71.7333)^2 = 10.667$

$SS_{Age} = \Sigma n_j (\overline{x}_j - \overline{\overline{x}})^2 = 8(71.5 - 71.333)^2 + 8(73.5 - 71.333)^2 + 8(69 - 71.333)^2 = 81.333$

$SS_{Tot} = \Sigma (\overline{x} - \overline{\overline{x}})^2 = df \cdot s^2 = 23 \cdot (95.536) = 2197.333$

$SS_{Err} = \Sigma (x - \overline{x}_{ij})^2 = \Sigma df_{ij} s_{ij}^2 = 3(266.667) + 3(128) \quad + 3(138.667)$
$\quad\quad\quad\quad\quad\quad\quad\quad\quad 3(36) \quad\quad + 3(78.667) + 3(26.667) \quad = 2024$

$SS_{Int} = SS_{Tot} - SS_{Gen} - SS_{Age} - SS_{Err} = 2107.333 - 10.667 - 81.333 - 2024 = 81.333$

This is the ANOVA table for the tests of hypotheses in this exercise.

Source	df	SS	MS	F
Gender	1	10.667	10.667	0.0949
Age	2	81.333	40.667	0.3617
Interaction	2	81.333	40.667	0.3617
Error	18	2024.000	112.444	
Total	23	2197.333		

H_0: there is no gender-age interaction
H_1: there is gender-age interaction
$\alpha = 0.05$ [assumed] and $df_{num} = 2$, $df_{den} = 18$
C.V. $F = F_\alpha = 3.5546$
calculations:

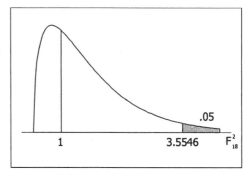

$\quad F = MS_{Int}/MS_{Err}$
$\quad\quad = 40.667/112.444$
$\quad\quad = 0.3617$
$\quad$ P-value = Fcdf(0.3617,99,2,18) = 0.7015
conclusion:

Do not reject H_0; there is not sufficient evidence to reject the claim that there is no interaction between gender and age.

H_0: $\mu_1 = \mu_2$ [there is no gender effect]
H_1: at least one gender μ_i is different
$\alpha = 0.05$ [assumed] and $df_{num} = 1$, $df_{den} = 18$
C.V. $F = F_\alpha = 4.4139$
calculations:

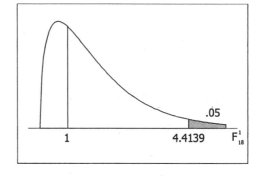

$\quad F = MS_{Gen}/MS_{Err}$
$\quad\quad = 10.667/112.444$
$\quad\quad = 0.0949$
$\quad$ P-value = Fcdf(0.0949,99,1,18) = 0.7616
conclusion:

Do not reject H_0; there is not sufficient evidence to support the claim that gender has an effect on pulse rate.

H_0: $\mu_1 = \mu_2 = \mu_3$ [there is no age effect]
H_1: at least one age μ_i is different
$\alpha = 0.05$ [assumed] and $df_{num} = 2$, $df_{den} = 18$
C.V. $F = F_\alpha = 3.5546$
calculations:

$\quad F = MS_{Age}/MS_{Err}$
$\quad\quad = 40.667/112.444$
$\quad\quad = 0.3617$
$\quad$ P-value = Fcdf(0.3617,99,2,18) = 0.7015
conclusion:

Do not reject H_0; there is not sufficient evidence to support the claim that age has an effect on pulse rate.

15. a. No change. The ANOVA calculated F statistics are ratios of variances. Since adding the same constant to each score does not affect the variances, the ANOVA table will not change.

   b. No change. The ANOVA calculated F statistics are ratios of variances. Since multiplying each score by the same nonzero constant will multiply all the variances by the square of that constant, the numerators and denominators of each ANOVA F ratio will be multiplied by the same constant – and the ratios will not change.

   c. No change. The same values will appear in different positions, but each value will refer to the same factor it did before the transposition.

   d. Depends. The change will affect the mean for row 1, the mean for column 1, and the variability within the first cell. In general, all calculated ANOVA statistics will change.

Statistical Literacy and Critical Thinking

1. The analysis of variance procedure tests the equality of two or more population means by analyzing sample variances. It finds a difference in the population means if the variance between the sample means is larger than can expected from the variance within the samples. And so even though the procedure is used to make decisions about means, it is called "analysis of variance."

2. The clinical trial described should be analyzed using a one-way ANOVA of section 12-2.

3. The clinical trial as modified should be analyzed using the t-test of section 9-3 for two independent means. The one-way ANOVA test of section 12-2 could be used with k=2, but the test of section 9-3 is preferred because
   (1) a one-tailed test is possible.
   (2) there is a possibility of using the test (and constructing straightforward confidence intervals) with or without the assumption of equal variances.

4. No. Eye color as commonly reported is not a quantitative variable. One of the analysis of variance requirements is that the response variable be continuous and quantitative – in particular, that the sample values in each cell come from a population that is approximately normal.

Review Exercises

1. H_o: $\mu_A = \mu_B = \mu_C$
   H_1: at least one μ_i is different
   $\alpha = 0.05$ and $df_{num} = 2$, $df_{den} = 14$
   C.V. $F = F_\alpha = 3.7389$
   calculations:

   $$F = s_B^2/s_p^2$$
   $$= 0.003826/0.0000816$$
   $$= 46.90 \text{ [Minitab]}$$

   P-value $= P(F_{14}^2 > 46.90) = 0.000$ [Minitab]

   conclusion:
   Reject Ho; there is sufficient evidence to reject the claim that the three groups have the same mean blood alcohol level.

2. Since each group has equal size n=12, use the simplified form of the calculations.
The following preliminary values are identified.

	Clancy	Rowling	Tolstoy	total	k = 3
n	12	12	12	36	n = 12
Σx	848.8	969.0	793.8	2611.6	
Σx^2	61449.54	78487.82	53189.10		
$\overline{x}$	70.733	80.750	66.150	72.544	
s^2	128.281	21.915	61.748		

$$s_{\overline{x}}^2 = \Sigma(\overline{x}_i - \overline{\overline{x}})^2/(k-1) = 111.500/2 = 55.750 \qquad s_p^2 = \Sigma s_i^2/k = 211.944/3 = 70.648$$

H_o: $\mu_1 = \mu_2 = \mu_3$
H_1: at least one μ_i is different
$\alpha = 0.05$ and $df_{num} = 2$, $df_{den} = 33$
C.V. $F = F_\alpha = 3.3158$
calculations:

$$F = ns_{\overline{x}}^2/s_p^2$$
$$= 12(55.750)/70.648$$
$$= 9.4695$$

P-value = Fcdf(9.4695,99,2,33) = 0.0006

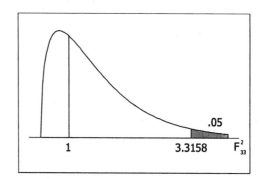

conclusion:

Reject Ho; there is sufficient evidence to reject the claim that $\mu_1 = \mu_2 = \mu_3$ and to conclude that at least one μ_i is different.

No. Based on these results, the books do not appear to have the same reading level.

3. H_o: there is no gender-major interaction
H_1: there is gender-major interaction
$\alpha = 0.05$ [assumed] and $df_{num} = 2$, $df_{den} = 12$
C.V. $F = F_\alpha = 3.8853$
calculations:

$$F = MS_{Int}/MS_{Err}$$
$$= 7.0556/37.7778$$
$$= 0.19 \text{ [Minitab]}$$

P-value = $P(F_{12}^2 > 0.19) = 0.832$ [Minitab]

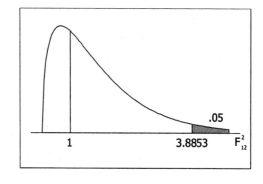

conclusion:

Do not reject H_o; there is not sufficient evidence to reject the claim that there is no interaction between gender and major.

4. H_o: $\mu_1 = \mu_2$ [there is no gender effect]
H_1: at least one gender μ_i is different
$\alpha = 0.05$ [assumed] and $df_{num} = 1$, $df_{den} = 12$
C.V. $F = F_\alpha = 4.7472$ [Excel 4.747221]
calculations:

$$F = MS_{Gen}/MS_{Err}$$
$$= 29.3889/37.7778$$
$$= 0.78 \text{ [Minitab]}$$

P-value = $P(F_{12}^1 > 0.78) = 0.395$ [Minitab]

conclusion:

Do not reject H_o; there is not sufficient evidence to support the claim that estimated length is affected by gender.

5. H_o: $\mu_1 = \mu_2 = \mu_3$ [there is no major effect]
   H_1: at least one major μ_i is different
   $\alpha = 0.05$ [assumed] and $df_{num} = 2$, $df_{den} = 12$
   C.V. $F = F_\alpha = 3.8853$
   calculations:

   $F = MS_{Maj}/MS_{Err}$
   $= 5.0556/37.7778$
   $= 0.13$ [Minitab]

   P-value $= P(F_{12}^2 > 0.13) = 0.876$ [Minitab]

   conclusion:

   Do not reject H_o; there is not sufficient evidence to support the claim that estimated length is affected by major.

6. The calculations to construct the ANOVA table are given below, starting with the mean and variance of the n=4 observations in each gender-smoking cell.

	SMOKING?		
GENDER	Yes	No	
male	$\overline{X} = 98.7000$	$\overline{X} = 97.9500$	$\overline{X} = 98.3250$
	$s^2 = 0.226667$	$s^2 = 0.543333$	
female	$\overline{X} = 98.4700$	$\overline{X} = 98.2500$	$\overline{X} = 98.3625$
	$s^2 = 0.129167$	$s^2 = 0.363333$	
	$\overline{X} = 98.5875$	$\overline{X} = 98.1000$	$\overline{X} = 98.34375$ [$s^2 = 0.334625$]

$SS_{Gen} = \Sigma n_i (\overline{X}_i - \overline{\overline{X}})^2 = 8(98.3250 - 98.34375)^2 + 8(98.3625 - 98.34375)^2 = 0.005625$

$SS_{Smo} = \Sigma n_j (\overline{X}_j - \overline{\overline{X}})^2 = 8(98.5875 - 98.34375)^2 + 8(98.1000 - 98.34375)^2 = 0.950625$

$SS_{Tot} = \Sigma (\overline{X} - \overline{\overline{X}})^2 = df \cdot s^2 = 15 \cdot (0.334625) = 5.019375$

$SS_{Err} = \Sigma (x - \overline{X}_{ij})^2 = \Sigma df_{ij} s_{ij}^2 = 3(0.226667) + 3(0.543333) + 3(0.129167) + 3(0.363333) = 3.787500$

$SS_{Int} = SS_{Tot} - SS_{Gen} - SS_{Age} - SS_{Err} = 5.019375 - 0.005625 - 0.950625 - 3.787500 = 0.275625$

This is the ANOVA table for the tests of hypotheses.

Source	df	SS	MS	F
Gender	1	0.005625	0.005625	0.0178
Smoking	1	0.950625	0.950625	3.0119
Interaction	1	0.275625	0.275625	0.8733
Error	12	3.787500	0.315625	
Total	15	5.019375		

H_o: there is no gender-smoking interaction
H_1: there is gender-smoking interaction
$\alpha = 0.05$ and $df_{num} = 1$, $df_{den} = 12$
C.V. $F = F_\alpha = 4.7472$
calculations:

   $F = MS_{Int}/MS_{Err}$
   $= 0.275625/0.315625 = 0.8733$

   P-value $= Fcdf(0.8733, 99, 1, 12) = 0.3685$

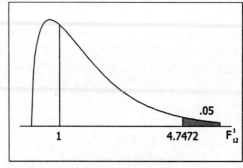

conclusion:

   Do not reject H_o; there is not sufficient evidence to reject the claim that there is no body temperature interaction between gender and whether a person smokes.

H_o: $\mu_1 = \mu_2$ [there is no gender effect]
H_1: at least one gender μ_i is different
$\alpha = 0.05$ and $df_{num} = 1$, $df_{den} = 12$
C.V. $F = F_\alpha = 4.7472$
calculations:
　$F = MS_{Gen}/MS_{Err}$
　　$= 0.005625/0.315625 = 0.0178$

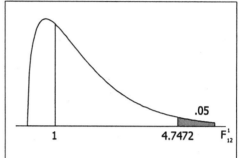

　P-value = Fcdf(0.0178,99,1,12) = 0.8960
conclusion:

　Do not reject H_o; there is not sufficient evidence to support the claim that gender has an effect on body temperature.

H_o: $\mu_1 = \mu_2$ [there is no smoking effect]
H_1: at least one smoking μ_i is different
$\alpha = 0.05$ and $df_{num} = 1$, $df_{den} = 12$
C.V. $F = F_\alpha = 4.7472$
calculations:
　$F = MS_{Smo}/MS_{Err}$
　　$= 0.950625/0.315625 = 3.0119$
　P-value = Fcdf(3.0119,99,1,12) = 0.1082
conclusion:

　Do not reject H_o; there is not sufficient evidence to support the claim that whether a person smokes has an effect on body temperature.

7. For all 6 scores, $\overline{\overline{x}} = 11.667$ and $s^2 = 2.2667$. The row and column means necessary for the following formulas may be obtained by inspection.

$SS_{Tra} = \Sigma n_i (\overline{x}_i - \overline{\overline{x}})^2 = 3(12.000 - 11.667)^2 + 3(11.333 - 11.667)^2 = 0.66667$

$SS_{Cyl} = \Sigma n_j (\overline{x}_j - \overline{\overline{x}})^2 = 2(10.000 - 11.667)^2 + 2(12.000 - 11.667)^2 + 2(13.000 - 11.667)^2 = 9.33333$

$SS_{Tot} = \Sigma (\overline{x} - \overline{\overline{x}})^2 = df \cdot s^2 = 5 \cdot (2.2667) = 11.33333$
$SS_{Err} = SS_{Tot} - SS_{Tra} - SS_{Cyl} = 11.33333 - 0.66667 - 9.33333 = 1.33333$
This is the ANOVA table for the tests of hypotheses.

Source	df	SS	MS	F
Transmission	1	0.66667	0.66667	1.0000
Cylinder	2	9.33333	4.66667	7.0000
Error	2	1.33333	0.66667	
Total	5	11.33333		

a. H_o: $\mu_1 = \mu_2$ [there is no transmission effect]
　H_1: at least one transmission μ_i is different
　$\alpha = 0.05$ [assumed] and $df_{num} = 1$, $df_{den} = 2$
　C.V. $F = F_\alpha = 18.513$
　calculations:
　　$F = MS_{Tra}/MS_{Err}$
　　　$= 0.66667/0.66667$
　　　$= 1.0000$

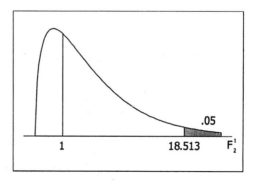

　　P-value = Fcdf(1.0000,9999,1,2) = 0.4225
　conclusion:

　　Do not reject H_o; there is not sufficient evidence to support the claim that the amounts of emitted greenhouse gases are affected by the type of transmission.

b. H_o: $\mu_1 = \mu_2 = \mu_3$ [there is no cylinder effect]
H_1: at least one cylinder μ_i is different
$\alpha = 0.05$ [assumed] and $df_{num} = 2$, $df_{den} = 2$
C.V. $F = F_\alpha = 19.000$
calculations:

$F = MS_{Smo}/MS_{Err}$
$= 4.66667/0.66667$
$= 7.0000$

P-value = Fcdf(7.0000,9999,2,2) = 0.1250

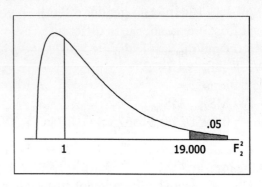

conclusion:

Do not reject H_o; there is not sufficient evidence to support the claim that the amounts of emitted greenhouse gases are affected by the numbers of cylinders.

c. No. We cannot conclude there is no effect, but only that there is not enough evidence to be 95% certain that there is an effect. The very small sample sizes make it unlikely that the data would provide enough evidence to reject any fairly reasonable null hypothesis.

8. The following preliminary values are identified.

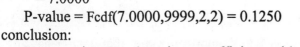

	presidents	popes	monarchs
n	37	24	14
Σx	557	315	318
Σx^2	11603	5981	11722
$\overline{x}$	15.054	13.125	22.714
s^2	89.3859	80.2880	346.0659

$\overline{\overline{x}} = (\Sigma n_i \overline{x}_i)/\Sigma n_i = [37(15.054) + 24(13.125) + 14(22.714)]/[37+24+14] = 1190/75 = 15.867$

$\Sigma n_i(\overline{x}_i - \overline{\overline{x}})^2 = 37(15.054-15.867)^2 + 24(13.125-15.867)^2 + 14(22.714-15.867)^2 = 861.241$

$\Sigma df_i s_i^2 = 36(89.3859) + 23(80.2880) + 13(346.0659) = 9562.405$

$s_B^2 = \Sigma n_i(\overline{x}_i - \overline{\overline{x}})^2/(k-1) = 861.241/2 = 430.621$

$s_p^2 = (\Sigma df_i s_i^2)/\Sigma df_i = 9562.405/72 = 132.811$

H_o: $\mu_1 = \mu_2 = \mu_3$
H_1: at least one μ_i is different
$\alpha = 0.05$ [assumed] and $df_{num} = 2$, $df_{den} = 72$
C.V. $F = F_\alpha = 3.1504$
calculations:

$F = s_B^2/s_p^2$

$= 430.621/132.811$
$= 3.2424$

P-value = Fcdf(3.2424,99,2,72) = 0.0448

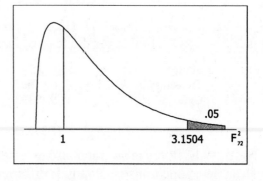

conclusion:

Reject Ho; there is sufficient evidence to reject the claim that $\mu_1 = \mu_2 = \mu_3$ and to conclude that the survival times for the three groups differ.

Cumulative Review Exercises

1. The following preliminary values are identified.

	presidents	popes	monarchs
n	37	24	14
Σx	557	315	318
Σx^2	11603	5981	11722
$\bar{x}$	15.054	13.125	22.714
s^2	89.3859	80.2880	346.0659

a. $\bar{x}_{pres} = 15.1$ yrs

   $\bar{x}_{pope} = 13.1$ yrs

   $\bar{x}_{mon} = 22.7$ yrs

   Since there are essentially no minimum age requirements for monarchs, one expects that their lengths of life after assuming office could be longer.

b. Taking the square roots of the variances in the table of preliminary values,
   $s_{pres} = 9.5$ yrs
   $s_{popes} = 9.0$ yrs
   $s_{mon} = 18.6$ yrs

c. Let the presidents be group 1.
   $\bar{x}_1 - \bar{x}_2 = 15.054 - 22.714 = -7.660$
   original claim: $\mu_1 - \mu_2 \neq 0$ yrs
   H_o: $\mu_1 - \mu_2 = 0$ yrs
   H_1: $\mu_1 - \mu_2 \neq 0$ yrs
   $\alpha = 0.05$ [assumed] and df = 13
   C.V. $t = \pm t_{\alpha/2} = \pm 2.160$
   calculations:

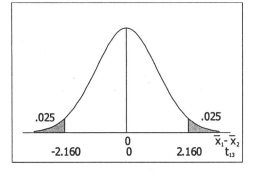

   $$t_{\bar{x}_1-\bar{x}_2} = (\bar{x}_1 - \bar{x}_2 - \mu_{\bar{x}_1-\bar{x}_2})/s_{\bar{x}_1-\bar{x}_2}$$
   $$= (-7.660 - 0)/\sqrt{89.3859/37 + 346.0659/14}$$
   $$= -7.660/5.209 = -1.471$$
   P-value = $2 \cdot tcdf(-99,-1.471,13) = 2(0.0826) = 0.1652$
   conclusion:
   Do not reject H_o; there is not sufficient evidence to conclude that $\mu_1 - \mu_2 \neq 0$.

d. Answers will vary. According to the frequency distribution at the right, the data appear not to be normally distributed – although grouping the data differently may give a different impression. The data appear to be uniformly distributed – or positively skewed, with the lowest class having the highest frequency.

time (years)	frequency
0 – 4	8
5 – 9	3
10 – 14	6
15 – 19	7
20 – 24	6
25 – 29	6
30 – 34	0
35 – 39	1
	37

e. $\alpha = 0.05$ and df = 36
   $$\bar{x} \pm t_{\alpha/2} \cdot s/\sqrt{n}$$
   $15.054 \pm 2.028(9.454)/\sqrt{37}$
   15.054 ± 3.152
   $11.9 < \mu < 18.2$ (years)

2. For each color, the scores are listed in order at the right.
The following preliminary values are identified.

	Red	Green	Blue
n	20	20	20
Σx	19210	19600	20899
Σx^2	19030374	20298724	22792193
$\overline{x}$	960.50	980.00	1044.95
s^2	30482.579	57406.526	50199.103

#	Red	Green	Blue
01	621	499	706
02	699	583	793
03	743	630	821
04	813	780	848
05	855	793	866
06	858	828	892
07	896	907	915
08	896	916	939
09	898	993	996
10	908	996	1004
11	921	1025	1013
12	996	1071	1039
13	1030	1111	1068
14	1092	1121	1097
15	1095	1147	1131
16	1130	1153	1159
17	1133	1180	1244
18	1179	1188	1370
19	1190	1229	1408
20	1257	1450	1590

a. $\overline{x}_R = 960.5$, $\overline{x}_G = 980.0$, $\overline{x}_B = 1044.95$

No. While the means are fairly close, the one for the blue
M&M's appears to be significantly higher – a statistical test
is needed to decide for sure.

b. $\tilde{x}_R = (908+921)/2 = 914.5$

$\tilde{x}_G = (996+1025)/2 = 1010.5$

$\tilde{x}_B = (1004+1013)/2 = 1008.5$

No. While the medians are fairly close, the one for the red
M&M's appears to be significantly lower – a statistical test
is needed to decide for sure.

c. $s_R = 174.6$, $s_G = 239.6$, $s_B = 224.1$

No. While the standard deviations are fairly close, the one for the red M&M's appears to be
significantly lower – a statistical test is needed to decide for sure.

d. Let the red M&M's be group 1.
$\overline{x}_1 - \overline{x}_2 = 960.5 - 980.0 = -19.5$
original claim: $\mu_1 - \mu_2 = 0$
H_0: $\mu_1 - \mu_2 = 0$
H_1: $\mu_1 - \mu_2 \neq 0$
$\alpha = 0.05$ [assumed] and df = 19
C.V. $t = \pm t_{\alpha/2} = \pm 2.093$
calculations:

$$t_{\overline{x}_1 - \overline{x}_2} = (\overline{x}_1 - \overline{x}_2 - \mu_{\overline{x}_1 - \overline{x}_2})/s_{\overline{x}_1 - \overline{x}_2}$$

$$= (-19.5 - 0)/\sqrt{30483/20 + 57407/20}$$

$$= -19.5/66.291 = -0.294$$

P-value = $2 \cdot$ tcdf$(-99, -0.294, 19) = 2(0.3859) = 0.7718$
conclusion:

Do not reject H_0; there is not sufficient evidence to reject the claim that $\mu_1 - \mu_2 = 0$.

e. $\alpha = 0.05$ and df = 19

$$\overline{x} \pm t_{\alpha/2} \cdot s/\sqrt{n}$$

$960.5 \pm 2.093(174.593)/\sqrt{20}$

960.5 ± 81.7

$878.8 < \mu < 1042.2$

f. Since each group has equal size n=20, use the simplified form of the calculations. In addition to the summary statistics above, the following preliminary values are noted.

$k = 3$

$\bar{\bar{x}} = \Sigma\bar{x}_i/k = 995.15$

$s_{\bar{x}}^2 = \Sigma(\bar{x}_i - \bar{\bar{x}})^2/(k-1) = 3910.182/2 = 1955.0925$

$s_p^2 = \Sigma s_i^2/k = 138088.208/3 = 46029.4027$

$H_0: \mu_1 = \mu_2 = \mu_3$
$H_1:$ at least one μ_i is different
$\alpha = 0.05$ and $df_{num} = 2$, $df_{den} = 57$
C.V. $F = F_\alpha = 3.1504$
calculations:

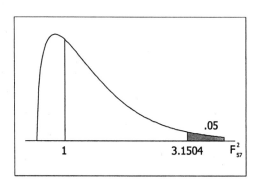

$F = ns_{\bar{x}}^2/s_p^2$

$= 20(1955.0925)/46029.4027$

$= 0.8495$

P-value = Fcdf(0.8495,99,2,57) = 0.4330

conclusion:

Do not reject Ho; there is not sufficient evidence to reject the claim that $\mu_1 = \mu_2 = \mu_3$.

3. Let x = the number of offspring with blue eyes.
binomial: n=100 and p=0.25
normal approximation appropriate since

$np = 100(0.25) = 25 \geq 5$

$nq = 100(0.75) = 75 \geq 5$

$\mu = np = 100(0.25) = 25$

$\sigma = \sqrt{npq} = \sqrt{100(0.25)(0.75)} = 4.330$

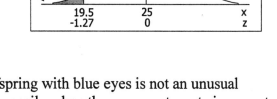

$P(x \leq 19)$

$= P_C(x < 19.5)$

$= P(z < -1.27)$

$= 0.1020$

No. Since 0.1020 > 0.05, getting 19 or fewer offspring with blue eyes is not an unusual occurrence – but rather one that could occur fairly easily when the one-quarter rate is correct.
NOTE: The exact answer can be found using the Minitab commands

   cdf;

   binomial 19 0.25

to produce the entire cumulative distribution and the row P(x≤19) = 0.09953.

4. a. normal distribution

$\mu = 7.54$

$\sigma = 1.09$

$P(x>8.00)$

$= P(z>0.42)$

$= 1 - P(z<0.42)$

$= 1 - 0.6628$

$= 0.3372$

b. normal distribution, since the original distribution is so

$\mu_{\bar{x}} = \mu = 7.54$

$\sigma_{\bar{x}} = \sigma/\sqrt{n} = 1.09/\sqrt{16} = 0.2725$

$P(\bar{x}>8.00)$

$= P(z>1.69)$

$= 1 - P(z<1.69)$

$= 1 - 0.9545$

$= 0.0455$

c. Let B = a baby weighs more than 7.54 lbs.

$P(B) = 0.5000$, for each birth

$P(B_1 \text{ and } B_2 \text{ and } B_3) = P(B_1) \cdot P(B_2) \cdot P(B_3)$

$= (0.5000)(0.5000)(0.5000)$

$= 0.125$

Chapter 13

Nonparametric Statistics

13-2 Sign Test

1. The sign test is "nonparametric" or "distribution-free" because it does not require the data to come from a particular distribution defined in terms of certain parameters.

3. The conclusion must be that there is not enough evidence to support the claim. In order for the data to support the claim the number of girl births would have to be significantly greater than $(\frac{1}{2}) \cdot (80) = 40$.

5. 15 +'s and 4 −'s
   $n = 19 \leq 25$; use C.V. = 4 from Table A-7
   Since $\min(15,4) = 4 \leq 4 =$ C.V., reject the hypothesis of no difference.
   Since there were more +'s, conclude that the first variable has the larger scores.

7. 30 +'s and 35 −'s
   $n = 65 > 25$; use C.V. = -1.96 from the z table
   $z = [(x+0.5) - n/2]/[\sqrt{n}/2] = [30.5 - 32.5]/[\sqrt{65}/2] = -2.0/4.031 = -0.496$
   Since -0.496 > -1.96, do not reject the hypothesis of no difference.

NOTES FOR THE REMAINING EXERCISES IN THIS SECTION.
(1) FOR $n \leq 25$. Table A-7 gives only x_L, the <u>lower</u> critical value for the sign test. And so the text lets x be the <u>smaller</u> of the number of +'s or the number of −'s, and warns the reader to use common sense to avoid concluding the reverse of what the data indicates. But the problem's symmetry means that the upper critical value is $x_U = n - x_L$ and that $\mu_x = n/2$, the natural expected value of x when H_o is true. For completeness, this manual indicates those values whenever using the sign test – and uses a normal curve for illustration, even though the distribution of x is discrete.

Letting x always be the number of +'s is an alternative approach that maintains the natural agreement between the alternative hypothesis and the critical region – and is consistent with the logic and notation of parametric tests.
(2) FOR $n > 25$. The correction for continuity is a conservative adjustment (intending to make less likely a false rejection of H_o) by shifting the x value 0.5 units toward the middle. When x is the smaller of the number of +'s or the number of −'s, this always involves replacing x with $(x + 0.5)$. The alternate approach suggested above (of letting x always be the number of +'s) maintains the natural agreement between the alternative hypothesis and the critical region – and is consistent with the logic and notation of parametric tests. In this approach, x is replaced with either $(x + 0.5)$ or $(x - 0.5)$ according to which one shifts the value toward the middle – i.e., with $(x + 0.5)$ when $x < \mu_x = n/2$, and with $(x - 0.5)$ when $x > \mu_x = n/2$.
The formula given is the usual one for converting a score into its standard score, using the

correction for continuity: $\quad z_x = \dfrac{x - \mu_x}{\sigma_x} = \dfrac{(x \pm 0.5) - (n/2)}{\sqrt{n}/2}$.

9. Let the Friday the 6th values be group 1.

month	1	2	3	4	5	6
6th–13th	−	−	−	+	−	−

n = 6: 1 +'s and 5 −'s

H_0: median difference = 0 admissions

H_1: median difference ≠ 0 admissions

α = 0.05

C.V. $x = x_{L,\alpha/2} = 0$

    $x = x_{U,\alpha/2} = 6 - 0 = 6$

calculations:

    x = 1

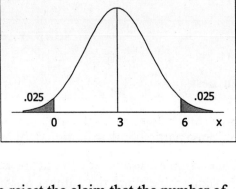

conclusion:

Do not reject H_0; there is not sufficient evidence to reject the claim that the number of admissions are not affected by it being Friday the 13th.

NOTE: Remember that the critical values are included in the critical region – so that we reject H_0 in exercise #9, for example, if x≤0 or x≥6. Some exercises in this section also appeared previously in a parametric context in chapters 8 and 9. Because comparison of the parametric and non-parametric results are usually informative and provide statistical insights, we list those exercises.

this section	#9	#10	#11	#12	#15	#16	#17	#18
previously	9-4 #12	9-rev #8	9-4 #16	9-4 #14	8-3 #10	8-3 #11	8-3 #12	8-3 #18

11. Let the reported heights be group 1.

subject#	1	2	3	4	5	6	7	8	9	10	11	12
rep-mea	+	+	−	+	+	−	+	−	−	−	+	+

n = 12: 7 +'s and 5 −'s

H_0: median difference = 0 in

H_1: median difference ≠ 0 in

α = 0.05

C.V. $x = x_{L,\alpha/2} = 2$

    $x = x_{U,\alpha/2} = 12 - 2 = 10$

calculations:

    x = 5 (using less frequent count)

    x = 7 (using + count)

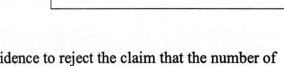

conclusion:

Do not reject H_0; there is not sufficient evidence to support the claim that there is a difference between the self-reported heights and the measured heights.

13. claim: median < 98.6

temp	1	2	3	4	5	6	7	8	9	10	11	12
t−98.6	−	−	0	−	−	+	−	−	−	−	−	

n = 11: 1 + and 10 −'s

H_0: median = 98.6 °F

H_1: median < 98.6 °F

α = 0.05

C.V. $x = x_{L,\alpha} = 2$

calculations:

    x = 1

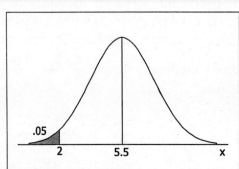

conclusion:
> Reject H_o; there is sufficient evidence to support the claim that these body temperatures come from a population with a median that is less than 98.6°F.

15. Let x = the number of boy births.
claim: $p > 0.5$
n = 51: 39 +'s and 12 –'s [the data are in the proper direction to proceed]
Since n>25, use z with
$\mu_x = n/2 = 51/2 = 25.5$
$\sigma_x = \sqrt{n}/2 = \sqrt{51}/2 = 3.571$
H_o: $p = 0.5$
H_1: $p > 0.5$
$\alpha = 0.01$
C.V. $z = -z_\alpha = -2.326$
calculations:

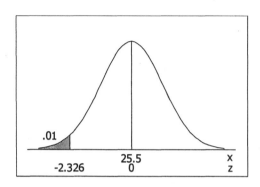

x = 12
$z = [(x+0.5) - n/2]/[\sqrt{n}/2]$
$= [12.5 - 25.5]/3.571 = -13.0/3.571 = -3.64$
P-value = P(z<-3.64) = 0.0001
conclusion:
> Reject H_o; there is sufficient evidence to support the claim that with this method the probability of a boy is greater than 0.5.

Yes. The method appears to work

NOTE: Working the exercise using the alternate method preserves the natural agreement between the alternative hypothesis and the critical region as follows.
H_o: $p = 0.5$
H_1: $p > 0.5$
$\alpha = 0.01$
C.V. $z = z_\alpha = 2.326$
calculations:

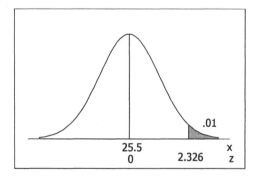

x = 39
$z = [(x - 0.5) - n/2]/[\sqrt{n}/2]$
$= [38.5 - 25.5]/3.571 = 3.64$
conclusion:
> Reject H_o; conclude the same as above.

17. Let x = the number of Internet users who use the Internet for making travel plans.
claim: $p < 0.5$
n = 734: 360 +'s and 374 –'s [the data are in the proper direction to proceed]
Since n>25, use z with
$\mu_x = n/2 = 734/2 = 367$
$\sigma_x = \sqrt{n}/2 = \sqrt{734}/2 = 13.546$
H_o: $p = 0.5$
H_1: $p < 0.5$
$\alpha = 0.01$
C.V. $z = -z_\alpha = -2.326$

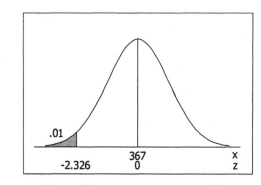

calculations:
x = 360

$z = [(x+0.5) - n/2]/[\sqrt{n}/2] = [360.5 - 367]/13.546 = -6.5/13.546 = -0.48$

P-value = P(z<-0.48) = 0.3156

conclusion:

Do not reject H_o; there is not sufficient evidence to support the claim that less than 50% of Internet users use the Internet for making travel plans.

Yes. Since each person who makes a reservation on the Internet is one less person who needs the services of a travel agent, it is important for travel agents to keep abreast of the trends in that area.

19. claim: median < 100

results: 40 +'s and 60 –'s and 21 0's

The three requested tests are given on the following page. As seen by the increasing P-values, and as suggested by common sense, the tests are increasingly conservative. In the third method of part (c), counting all the zeros to favor the null hypothesis actually renders the test unnecessary – and so the alternative method is used so that a comparison can be made.

a. Use the usual method, discarding the 0's

n = 100: 40 +'s and 60 –'s, so that x = 40

Since n>25, use z with

$\mu_x = n/2 = 100/2 = 50$

$\sigma_x = \sqrt{n}/2 = \sqrt{100}/2 = 5.000$

H_o: median = 100

H_1: median < 100

α = 0.05

C.V. $z = -z_{\alpha/2} = -1.645$

calculations:

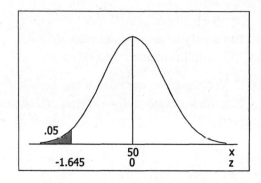

$z = [(x+0.5) - n/2]/[\sqrt{n}/2]$

$= [40.5 - 50]/5.000 = -9.5/5.000 = -1.900$

P-value = P(z<-1.90) = 0.0287

conclusion:

Reject H_o; there is sufficient evidence to support the claim that median < 100.

b. Use the second method, counting half the 0's each way (and discarding the odd zero).

n = 120: 50 +'s and 70 –'s, so that x = 50

Since n>25, use z with

$\mu_x = n/2 = 120/2 = 60$

$\sigma_x = \sqrt{n}/2 = \sqrt{120}/2 = 5.477$

H_o: median = 100

H_1: median < 100

α = 0.05

C.V. $z = -z_{\alpha} = -1.645$

calculations:

$z = [(x+0.5) - n/2]/[\sqrt{n}/2]$

$= [50.5 - 60]/5.477 = -9.5/5.477 = -1.734$

P-value = P(z<-1.73) = 0.0418

conclusion:
   Reject H_o; there is sufficient evidence to support the claim that median < 100.

c. Use the third method, counting all the 0's to favor H_o (i.e., against H_1).
   n = 121: 61 +'s and 60 −'s, so that x=61 (alternate method)
   Since n>25, use z with

   μ_x = n/2 = 121/2 = 60.5
   σ_x = $\sqrt{n}$ /2 = $\sqrt{121}$ /2 = 5.500
   H_o: median = 100
   H_1: median < 100
   α = 0.05
   C.V. z = $-z_\alpha$ = -1.645
   calculations:

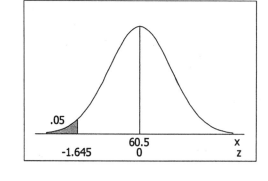

   $z = [(x-.5) - n/2]/[\sqrt{n}/2]$
      $= [60.5 - 60.5]/5.500 = 0/5.500 = 0$
      P-value = P(z<0.00) = 0.5000
   conclusion:
      Do not reject H_o; there is not sufficient evidence to support the claim that median < 100.

21. Let p = the proportion of those hired that are female.
    claim: p < 0.5
    a. Use the sign test.
       n = 54: 18 +'s and 36 −'s
       Since n>25, use z with

       μ_x = n/2 = 54/2 = 27
       σ_x = $\sqrt{n}$ /2 = $\sqrt{54}$ /2 = 3.674
       H_o: p = 0.5
       H_1: p < 0.5
       α = 0.01
       C.V. z = $-z_\alpha$ = -2.326
       calculations:

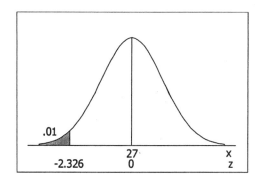

          x = 18
          $z = [(x+0.5) - n/2]/[\sqrt{n}/2]$
             $= [18.5 - 27]/3.674 = -8.5/3.674 = -2.313$
          P-value = P(z<-2.31) = 0.0104
       conclusion:
          Do not reject H_o; there is not sufficient evidence to support the claim that p < 0.5.
       NOTE: The exercise describes the formula $z = [(x+0.5) - n/2]/[\sqrt{n}/2]$ as the normal
       approximation to the binomial because it follows directly from the normal approximation of
       the sampling distribution of the $\hat{p}$'s as used to test H_o: p = 0.5 given in section 8-3.

       $$z_{\hat{p}} = \frac{\hat{p} - 0.5}{\sqrt{(0.5)(0.5)/n}} = \frac{x/n - 0.5}{0.5/\sqrt{n}} \times \frac{n}{n} = \frac{x - 0.5n}{0.5\sqrt{n}} = \frac{x - (n/2)}{\sqrt{n}/2}$$

b. Use the exact binomial distribution.

H_o: p = 0.5
H_1: p < 0.5
$\alpha = 0.01$
calculations:

P-value = P(x≤18|n=54 and p = 0.5)

= $\Sigma[54!/(54\text{-}x)!x!]\cdot(0.5)^x\cdot(0.5)^{54\text{-}x}$ for x = 0,1,2,3,...,18

Using the Minitab commands *cdf; binomial 54 0.5.* yields P-value = 0.00922 in the
P(x≤18) row of the resulting table.

conclusion:

Reject Ho; there is sufficient evidence to support the claim that p < 0.5

13-3 Wilcoxon Signed-Ranks Test for Matched Pairs

1. The parametric methods of section 9-4 require that the differences obtained from the matched pairs come from a population that is approximately normal. The nonparametric Wilcoxon signed-ranks test requires only that the differences obtained from the matched pairs come from a population that is approximately symmetric. When the differences come from a population that is approximately normal, the parametric methods of section 9-4 are to be preferred.

3. When the differences come from a population that is approximately symmetric, the Wilcoxon signed-ranks test is more efficient than the sign test – i.e., it can reach the same conclusion with a smaller sample size. Table 13-2 indicates that when working with normal populations the Wilcoxon signed-ranks test and sign test have efficiency ratings of 0.95 and 0.63 respectively.

5. Let the x values be group 1.

x-y	25	28	42	48	39	36	42	33
R	1	2	6.5	8	5	4	6.5	3

$\Sigma R\text{--} = 0$ n = 8 non-zero ranks
$\Sigma R+ = 36$ check: $\Sigma R = n(n+1)/2 = 8(9)/2 = 36$
$\Sigma R = 36$

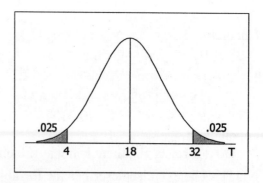

H_o: median difference = 0
H_1: median difference ≠ 0
$\alpha = 0.05$
C.V. $T = T_{L,\alpha/2} = 4$
       $T = T_{U,\alpha/2} = 36 - 4 = 32$
calculations:

T = 0 (using smaller ranks)
T = 36 (using positive ranks)

conclusion:

Reject H_o; there is sufficient evidence to conclude that median difference ≠ 0 (in fact, that the x scores are greater).

NOTE FOR THE REMAINING EXERCISES IN THIS SECTION:
Table A-8 gives only T_L, the <u>lower</u> critical value for the signed-ranks test. And so the text lets T be the <u>smaller</u> of the sum of the positive ranks or the sum of the negative ranks, and warns the reader to use common sense to avoid concluding the reverse of what the data indicates. But the problem's symmetry means that the upper critical value is $T_U = \Sigma R - T_L$ and that $\mu_T = \Sigma R/2$, the natural expected value of T when H_o is true. For completeness, this manual indicates those values

whenever using the signed-ranks test – and uses a normal curve for illustration, even though the distribution of T is discrete. And remember that the critical values are included in the critical region – so that we reject H_o in exercise #5, for example, if $T \le 4$ or $T \ge 32$.

Letting T always be the sum of the positive ranks is an alternative approach that maintains the natural agreement between the alternative hypothesis and the critical region – and is consistent with the logic and notation of parametric tests.

Finally, this manual uses a minus sign preceding ranks associated with negative differences. The ranks themselves are not negative, but use of the minus sign helps to organize the information.

NOTE: The following exercises in this section were worked in chapter 9 using the parametric test for the difference between two means and in section 13-2 using the sign test. Because comparison of the results are usually informative and provide statistical insights, we list those exercises.

this section	#7	#8	#9	#10
chapter 9	9-4 #12	review #8	9-4 #16	9-4 #14
section 13-2	#9	#10	#11	#12

7. Let the Friday the 6th values be group 1.

6th–13th	-4	-6	-3	1	-1	-7
R	-4	-5	-3	1.5	-1.5	-6

ΣR– = 19.5   n = 6 non-zero ranks
ΣR+ = 1.5   check: $\Sigma R = n(n+1)/2 = 6(7)/2 = 21$
ΣR = 21
H_o: median difference = 0 admissions
H_1: median difference $\ne$ 0 admissions
α = 0.05
C.V. $T = T_{L,\alpha/2} = 1$
      $T = T_{U,\alpha/2} = 21 - 1 = 20$
calculations:
   T = 1.5
conclusion:

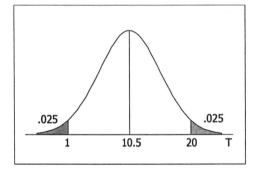

   Do not reject H_o; there is not sufficient evidence to conclude that median difference $\ne$ 0.

9. Let the reported values be group 1.

rep-mea	.1	1.1	-1.9	1.7	.7	-0.6	.5	-3.0	-1.6	-11.2	1.0	1.2
R	1	6	-10	9	4	-3	2	-11	-8	-12	5	7

ΣR– = 44   n = 12 non-zero ranks
ΣR+ = 34   check: $\Sigma R = n(n+1)/2 = 12(13)/2 = 78$
ΣR = 78
H_o: median difference = 0 in
H_1: median difference $\ne$ 0 in
α = 0.05
C.V. $T = T_{L,\alpha/2} = 14$
      $T = T_{U,\alpha/2} = 78 - 14 = 64$
calculations:
   T = 34
conclusion:
   Do not reject H_o; there is not sufficient evidence to conclude that median difference $\ne$ 0.

11. The 35 "actual – predicted" differences and their signed ranks are as follows.

35 (actual – predicted) values

1	-9	2	2	-7	-3	0	-7	7	2	-3	3	6	-2	0	2	-1	-10
-12	2	0	-1	3	1	0	4	0	-8	-2	-10	-1	-5	0	3	-4	

35 R_{a-p} values

3.0	-26.0	9.0	9.0	-23.0	-15.0	*	-23.0	23.0	9.0	-15.0	15.0	21.0	-9.0	*	9.0	-3.0	-27.5
-29.0	9.0	*	-3.0	15.0	3.0	*	18.5	*	-25.0	-9.0	-27.5	-3.0	-20.0	*	15.0	-18.5	

$\Sigma R- = 276.5$ $n = 29$ non-zero ranks

$\Sigma R+ = 158.5$ check: $\Sigma R = n(n+1)/2 = 29(30)/2 = 435$

$\Sigma R = 435$

H_0: median difference $= 0$ °F

H_1: median difference $\neq 0$ °F

$\alpha = 0.05$

C.V. $T = T_{L,\alpha/2} = 127$

$T = T_{U,\alpha/2} = 435 - 127 = 308$

calculations:

$T = 158.5$

conclusion:

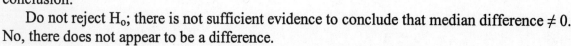

Do not reject H_0; there is not sufficient evidence to conclude that median difference $\neq 0$. No, there does not appear to be a difference.

13. The 106 differences obtained by subtracting 98.6 from each temperature, and the signed ranks of those differences, are as follows.

106 (x – 98.6) values

0.0	0.0	-0.6	-0.6	0.4	-0.2	-0.2	-0.2	-0.2	0.0	0.0
0.2	0.0	-1.6	-1.6	0.2	-1.0	-0.9	0.2	-0.6	-0.6	-0.3
-0.1	-1.3	0.1	-1.2	0.3	0.0	0.9	-1.1	-1.3	-1.0	-0.4
1.0	0.1	0.8	-0.4	-0.6	0.0	0.0	-1.4	-0.2	0.0	-0.4
-0.6	-0.8	-0.6	-0.2	0.0	0.0	-0.8	0.4	-2.1	-1.0	-0.6
-1.7	-1.0	-1.5	-0.7	-0.2	-1.3	-0.6	-1.1	-1.0	-0.4	-0.1
0.2	0.1	-0.8	-0.6	-1.5	-1.2	0.8	-0.2	0.0	-0.2	-0.1
0.0	-0.3	0.1	0.2	0.5	0.0	-0.7	0.2	-0.6	0.1	-0.1
0.3	-0.2	0.0	-1.5	-0.7	0.2	0.1	-1.0	-0.4	0.6	-0.8
-0.6	-0.2	-0.8	-0.2	-1.2	-0.6	-1.6				

106 $R_{x-98.6}$ values

*	*	-48.5	-48.5	37.0	-20.0	-20.0	-20.0	-20.0	*	*
20.0	*	-88.0	-88.0	20.0	-71.0	-66.5	20.0	-48.5	-48.5	-31.5
-5.5	-81.0	5.5	-78.0	31.5	*	66.5	-75.5	-81.0	-71.0	-37.0
71.0	5.5	62.0	-37.0	-48.5	*	*	-83.0	-20.0	*	-37.0
-48.5	-62.0	-48.5	-20.0	*	*	-62.0	37.0	-91.0	-71.0	-48.5
-90.0	-71.0	-85.0	-57.0	-20.0	-81.0	-48.5	-75.5	-71.0	-37.0	-5.5
20.0	5.5	-62.0	-48.5	-85.0	-78.0	62.0	-20.0	*	-20.0	-5.5
*	-31.5	5.5	20.0	41.0	*	-57.0	20.0	-48.5	5.5	-5.5
31.5	-20.0	*	-85.0	-57.0	20.0	5.5	-71.0	-37.0	48.5	-62.0
-48.5	-20.0	-62.0	-20.0	-78.0	-48.5	-88.0				

$\Sigma R- = 3525$ $n = 91$ non-zero ranks

$\Sigma R+ = 661$ check: $\Sigma R = n(n+1)/2 = 91(92)/2 = 4186$

$\Sigma R = 4186$

For $n=91$ ranks, use the z approximation with

$\mu_T = n(n+1)/4$

$= 91(92)/4 = 2093$

$\sigma_T = \sqrt{[n(n+1)(2n+1)]/24}$

$= \sqrt{[91(92)(183)]/24} = 252.66$

H_o: median = 98.6 °F
H_1: median $\neq$ 98.6 °F
$\alpha = 0.05$
C.V. $z = \pm z_{\alpha/2} = \pm 1.960$
calculations:

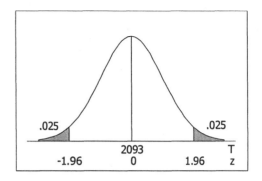

$\qquad T = 661$
$\qquad z = (T - \mu_T)/\sigma_T$
$\qquad\quad = (661 - 2093)/252.66$
$\qquad\quad = -1432/252.66 = -5.668$
$\quad$ P-value $= 2 \cdot P(z < -5.67) = 2(0.0001) = 0.0002$
conclusion:

$\qquad$ Reject H_o; there is sufficient evidence to reject the claim that the median is 98.6 and conclude that median $\neq$ 98.6 (in fact, that the median body temperature of healthy adults is less than 98.6°F).

13-4 Wilcoxon Rank-Sum Test for Two Independent Samples

1. The Wilcoxon signed-ranks test applies to data consisting of matched pairs – which means that there must be an equal number of observations from each population, and that the observations are paired. The Wilcoxon rank-sum test applies to independent samples – which means that there may or may not be the same number of observations from each population.

3. The tests in section 9-3 require that the original populations be approximately normal. Since the Wilcoxon rank-sum test places no restrictions on the original populations, it may be used in a wider variety of situations.

NOTE: For each exercise in this section, on the following page are given the ordered scores and the accompanying rank for each of the ordered scores. All the values necessary to work the exercises are found on that page. Make certain that you understand the format of that page, and how each of the ranks was obtained.

$\quad$ Always letting $R = \Sigma R_1$ (i.e., the sum of the ranks from the scores designated as group 1) guarantees that there will be agreement between H_1 and the C.R. (regardless of whether the test is two-tailed or one-tailed in either direction) and that the sign of the test statistic indicates in the natural manner which group has the larger median.

$\quad$ When working the problems by hand, it is strongly recommended that you calculate both ΣR_1 and ΣR_2 (even though only the former is necessary for the test) and that you verify that $\Sigma R_1 + \Sigma R_2 = n(n+1)/2$. This simple check can identify the presence of errors that must be corrected before proceeding with the calculations.

Below are the ordered scores and accompanying ranks for each exercise in this section.

| | exercise #5 | | | | exercise #6 | | | | exercise #7 | | | | exercise #10 | | | |
row	s1	R	s2	R	s1	R	s2	R	o-c	R	ctrl	R	m	R	f	R
1	2	1	3	2	8	3	3	1	0.210	1.0	0.334	7.5	56	1.0	60	3.0
2	7	4	4	3	15	7	5	2	0.287	2.0	0.349	11.0	60	3.0	64	6.0
3	10	5	11	6	27	9	9	4	0.288	3.0	0.402	12.0	60	3.0	68	9.5
4	16	8	14	7	39	11	11	5	0.304	4.0	0.413	14.0	64	6.0	68	9.5
5	20	9	28	14	45	13	14	6	0.305	5.0	0.429	15.0	64	6.0	68	9.5
6	22	10	35	17	62	15	21	8	0.308	6.0	0.445	16.0	68	9.5	72	14.0
7	23	11	40	18	68	16	33	10	0.334	7.5	0.460	18.0	72	14.0	72	14.0
8	26	12	46	19	72	18	44	12	0.340	9.0	0.476	20.0	72	14.0	76	18.0
9	27	13	47	20	77	19	61	14	0.344	10.0	0.483	21.0	72	14.0	76	18.0
10	30	15	52	21	80	20	70	17	0.307	13.0	0.501	22.0	76	18.0	80	20.5
11	33	16	53	22	87	22	85	21	0.455	17.0	0.519	23.0	88	23.0	80	20.5
12		104	60	23		153		100	0.463	19.0	0.594	24.0	88	23.0	88	23.0
13				172						96.5		203.5	96	25.0		165.5
														159.5		

| | exercise #8 | | | | exercise #9 | | | | exercise #11 | | | | exercise #12 | | | |
row	one	R	eight	R	c	R	t	R	m	R	f	R	wheat	R	pre-83	R
1	25	1.0	225	5.0	0	1.5	0	1.5	56	3.5	60	11.5	3.00500	3.0	2.95930	1.0
2	169	2.0	300	6.0	1	4.0	3	10.5	56	3.5	60	11.5	3.00540	4.0	2.98000	2.0
3	175	3.0	354	7.0	1	4.0	3	10.5	56	3.5	60	11.5	3.00640	5.0	3.01850	7.5
4	200	4.0	372	8.0	1	4.0	3	10.5	56	3.5	64	22.0	3.01310	6.0	3.03980	12.0
5	500	11.5	400	9.0	2	6.0	3	10.5	56	3.5	64	22.0	3.01850	7.5	3.04060	13.0
6	560	13.0	428	10.0	3	10.5	4	18.5	56	3.5	64	22.0	3.02900	9.0	3.05020	17.0
7	800	16.0	500	11.5	3	10.5	4	18.5	60	11.5	64	22.0	3.03570	10.0	3.05220	18.0
8	842	17.0	640	14.0	3	10.5	4	18.5	60	11.5	68	31.5	3.03770	11.0	3.05460	19.5
9	856	18.0	750	15.0	3	10.5	5	23.5	60	11.5	68	31.5	3.04080	14.0	3.05460	19.5
10	1110	19.0	1200	20.0	4	18.5	6	27.0	60	11.5	68	31.5	3.04760	15.0	3.05860	23.0
11	1252	21.0	1500	22.0	4	18.5	6	27.0	60	11.5	68	31.5	3.04800	16.0	3.06030	24.0
12	1560	23.0	1876	24.0	4	18.5	7	31.0	60	11.5	68	31.5	3.05610	21.0	3.06310	25.0
13	2040	26.0	2000	25.0	4	18.5	7	31.0	60	11.5	72	42.5	3.05700	22.0	3.07040	26.0
14	4000	29.5	2050	27.0	4	18.5	8	36.5	64	22.0	72	42.5	3.07550	30.0	3.07120	27.0
15	5000	31.0	3600	28.0	5	23.5	8	36.5	64	22.0	72	42.5	3.07650	33.0	3.07130	28.0
16	5635	32.0	4000	29.5	6	27.0	8	36.5	64	22.0	72	42.5	3.07860	35.0	3.07170	29.0
17	10000	33.0	23410	34.0	6	27.0	8	36.5	64	22.0	72	42.5	3.08160	37.0	3.07620	31.0
18	40320	35.0	42000	36.0	6	27.0	9	41.0	64	22.0	72	42.5	3.08620	40.0	3.07630	32.0
19	42200	37.0	49000	38.0	7	31.0	10	42.0	64	22.0	72	42.5	3.09360	45.0	3.07750	34.0
20	49654	39.0	52836	40.0	8	36.5	11	44.0	64	22.0	72	42.5	3.09650	46.0	3.07970	36.0
21		411	64582	41.0	8	36.5		511.5	68	31.5	76	53.5	3.09760	48.0	3.08170	38.0
22			100000	42.0	8	36.5			68	31.5	76	53.5	3.09940	49.0	3.08460	39.0
23				492	8	36.5			68	31.5	76	53.5	3.10290	51.0	3.08660	41.0
24					11	44.0			72	42.5	76	53.5	3.10310	52.0	3.08770	42.0
25					11	44.0			72	42.5	76	53.5	3.10380	53.5	3.09170	43.0
26					14	46.0			72	42.5	76	53.5	3.10830	56.0	3.09340	44.0
27					15	47.0			72	42.5	80	60.5	3.11140	58.0	3.09660	47.0
28						616.5			72	42.5	80	60.5	3.11410	59.0	3.10280	50.0
29									72	42.5	80	60.5	3.12670	60.0	3.10380	53.5
30									76	53.5	80	60.5	3.13660	64.0	3.10430	55.0
31									76	53.5	80	60.5	3.14610	65.0	3.10860	57.0
32									84	65.5	80	60.5	3.15240	66.0	3.12740	61.0
33									84	65.5	88	72.0	3.15350	67.0	3.12990	62.0
34									84	65.5	88	72.0	3.16920	69.0	3.13400	63.0
35									84	65.5	88	72.0	3.19340	70.0	3.15820	68.0
36									88	72.0	88	72.0		1297		1188
37									88	72.0	88	72.0				
38									88	72.0	96	77.5				
39									88	72.0	104	79.0				
40									96	77.5	124	80.0				
										1339.5		1900.5				

5. Let the sample 1 values be group 1.

$n_1 = 11$	$R = \Sigma R_1 = 104$	chk: $\Sigma R_i = n(n+1)/2$	$\mu_R = n_1(n+1)/2$	$\sigma^2_R = n_1 n_2(n+1)/12$	
$n_2 = 12$	$\Sigma R_2 = 172$	$= 23(24)/2$	$= 11(24)/2$	$= (11)(12)(24)/12$	
$n = \Sigma n = 23$	$\Sigma R_i = 276$	$= 276$	$= 132$	$= 264$	
				$\sigma_R = 16.248$	

H_0: median$_1$ = median$_2$
H_1: median$_1 \neq$ median$_2$
$\alpha = 0.05$
C.V. $z = \pm z_{\alpha/2} = \pm 1.960$
calculations:

$$z = (R - \mu_R)/\sigma_R$$
$$= (104 - 132)/\sqrt{264}$$
$$= -28/16.248$$
$$= -1.723$$

P-value $= 2 \cdot P(z < -1.72) = 2(0.0427) = 0.0854$
conclusion:

Do not reject H_0; there is not sufficient evidence to reject the claim that median$_1$ = median$_2$.

7. Let the obsessive-compulsive values be group 1.

$n_1 = 12$	$R = \Sigma R_1 = 96.5$	chk: $\Sigma R_i = n(n+1)/2$	$\mu_R = n_1(n+1)/2$	$\sigma^2_R = n_1 n_2(n+1)/12$	
$n_2 = 12$	$\Sigma R_2 = 203.5$	$= 24(25)/2$	$= 12(25)/2$	$= (12)(12)(25)/12$	
$n = \Sigma n = 24$	$\Sigma R_i = 300.0$	$= 300$	$= 150$	$= 300$	
				$\sigma_R = 17.321$	

H_0: median$_1$ = median$_2$
H_1: median$_1 \neq$ median$_2$
$\alpha = 0.01$
C.V. $z = \pm z_{\alpha/2} = \pm 2.575$
calculations:

$$z = (R - \mu_R)/\sigma_R$$
$$= (96.5 - 150)/\sqrt{300}$$
$$= -53.5/17.321$$
$$= -3.089$$

P-value $= 2 \cdot P(z < -3.09) = 2(0.0010) = 0.0020$
conclusion:

Reject H_0; there is sufficient evidence to reject the claim that median$_1$ = median$_2$ and conclude that median$_1 \neq$ median$_2$ (in fact, that obsessive compulsive patients have a smaller median brain volume).

Yes. Based on this result, it appears that obsessive-compulsive disorders have a biological basis – but not necessarily that the smaller brain volume *causes* the disorders.

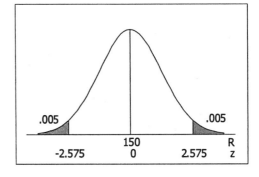

9. Let the cars be group 1.

$n_1 = 27$	$R = \Sigma R_1 = 616.5$	chk: $\Sigma R_i = n(n+1)/2$	$\mu_R = n_1(n+1)/2$	$\sigma^2_R = n_1 n_2(n+1)/12$	
$n_2 = 20$	$\Sigma R_2 = 511.5$	$= 47(48)/2$	$= 27(48)/2$	$= (27)(20)(48)/12$	
$n = \Sigma n = 47$	$\Sigma R_i = 1128.0$	$= 1128$	$= 648$	$= 2160$	
				$\sigma_R = 46.476$	

H_0: median$_1$ = median$_2$
H_1: median$_1 \neq$ median$_2$
$\alpha = 0.05$
C.V. $z = \pm z_{\alpha/2} = \pm 1.960$
calculations:

$z = (R - \mu_R)/\sigma_R$

$= (616.5 - 648)/\sqrt{2160}$

$= -31.5/26.476 = -0.678$

P-value $= 2 \cdot P(z < -0.68) = 2(0.2483) = 0.4966$

conclusion:

Do not reject H_0; there is not sufficient evidence to support the claim that
median$_1 \neq$ median$_2$.

The results do not suggest that taxis are newer – in fact in this sample data, the median age for
the cars was lower.

11. Let the men be group 1.

$n_1 = 40$	$R = \Sigma R_1 = 1339.5$	chk: $\Sigma R_i = n(n+1)/2$	$\mu_R = n_1(n+1)/2$	$\sigma^2{}_R = n_1 n_2(n+1)/12$
$n_2 = 40$	$\Sigma R_2 = 1900.5$	$= 80(81)/2$	$= 40(81)/2$	$= (40)(40)(81)/12$
$n = \Sigma n = 80$	$\Sigma R_i = 3240.0$	$= 3240$	$= 1620$	$= 10800$
				$\sigma_R = 103.9232$

H_0: median$_1$ = median$_2$
H_1: median$_1 \neq$ median$_2$
$\alpha = 0.05$ [assumed]
C.V. $z = \pm z_{\alpha/2} = \pm 1.960$
calculations:

$z = (R - \mu_R)/\sigma_R$

$= (1339.5 - 1620)/\sqrt{10800}$

$= -280.5/103.923 = -2.699$

P-value $= 2 \cdot P(z < -2.70) = 2(0.0035) = 0.0070$

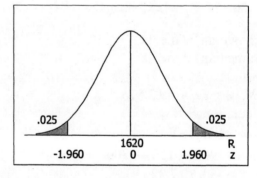

conclusion:

Reject H_0; there is sufficient evidence to reject the claim that median$_1$ = median$_2$ and to
conclude that median$_1 \neq$ median$_2$ (in fact, that men have a lower median pulse rate than
women).

13. The denominator of the Mann-Whitney and Wilcoxon statistics are the same.

Assuming $R = \Sigma R_1$, the numerator of the Wilcoxon statistic is

$R - \mu_R = R - n_1(n+1)/2$

Assuming $R = \Sigma R_1$, the numerator of the Mann-Whitney statistic is

$U - n_1 n_2/2 = [n_1 n_2 + n_1(n_1+1)/2 - R] - n_1 n_2/2$

$= n_1 n_2/2 + n_1(n_1+1)/2 - R$

$= (n_1/2)(n_2 + n_1 + 1) - R$

$= n_1(n+1)/2 - R = \mu_R - R = -[\text{numerator of the Wilcoxon statistic}]$

For the BMI measures listed in Table 13-5,

$U = n_1 n_2 + n_1(n_1+1)/2 - R = 13(12) + 13(14)/2 - 187 = 60$

$\mu_U = n_1 n_2/2 = 13(12)/2 = 78$

$\sigma^2{}_U = n_1 n_2(n_1+n_2+1)/12 = 13(12)(26)/12 = 338$

$z_U = (\mu_U - U)/\sigma^2{}_U = (60 - 78)/\sqrt{338} = -18/18.385 = -0.98$

As predicted, z_U has the same absolute value but the opposite sign of z_R.

13-5 Kruskal-Wallis Test

1. The one-way analysis of variance test requires that the original populations be approximately normal. Since the Kruskal-Wallis test places no restrictions on the original populations, it may be used in a wider variety of situations.

3. The efficiency of the Kruskal-Wallis test is 0.95. This means that for normal populations, the Kruskal-Wallis test requires a sample of size n=100 to identify departures from the null hypothesis that the one-way analysis of variance test could identify with a sample of size n=95.

NOTE: The critical value and accompanying graph for the exercises in this section are given in terms of χ^2, which approximates the distribution of the calculated statistic H. The exercises in this section also appeared previously in a parametric context in section 12-2. Because comparison of the parametric and non-parametric results are usually informative and provide statistical insights, we list those pairings.

this section	#5	#6	#7	#8	#9	#10	#11	#12
section 12-2	#11	#12	#10	#14	#13	#8	#15	#9

5. Below are the ordered scores and ranks for each group. The group listed first is group 1, etc.

sub	R	com	R	mid	R	full	R
420	4	442	6	259	1	360	2
428	5	514	9	454	7	384	3
681	16	525	10.5	469	8	602	12
898	19	643	13	525	10.5	656	15
917	20	655	14	727	18	687	17
	64		52.5		44.5		49

$n_1 = 5 \quad R_1 = 64$
$n_2 = 5 \quad R_2 = 52.5$
$n_3 = 5 \quad R_3 = 44.5$
$\underline{n_4 = 5} \quad \underline{R_4 = 49}$
$n = \Sigma n = 20 \quad \Sigma R = 210$

chk: $\Sigma R = n(n+1)/2 = 20(21)/2 = 210$

H_o: the populations have the same median
H_1: at least one median is different
$\alpha = 0.05$ and df =3
C.V. $H = \chi^2_\alpha = 7.815$

calculations:

$$H = \{12/[n(n+1)]\} \cdot [\Sigma (R_i^2/n_i)] - 3(n+1)$$
$$= \{12/[20(21)]\} \cdot [(64)^2/5 + (52.5)^2/5$$
$$+ (44.5)^2/5 + (49)^2/5] - 3(21)$$
$$= \{0.0286\} \cdot [2246.7] - 63$$
$$= 1.191$$

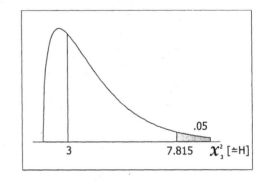

P-value $= \chi^2 cdf(1.191,99,3) = 0.7551$

conclusion:

Do not reject H_o; there is not sufficient evidence to support the claim that head injury measurements for the four car weight categories are not all the same.

No. By this measurement, the data do not show that heavier cars are safer in a crash.

7. Below are the ordered scores and ranks for each group. The group listed first is group 1, etc.

sunny	R	cloudy	R	rainy	R
13.0	12.0	12.5	7	11.6	1
13.2	14	12.6	8	11.9	2
13.5	15	12.7	9.5	12.1	3
13.8	16	12.7	9.5	12.2	4.5
13.9	17	13.0	12.0	12.2	4.5
14.0	18	13.0	12.0	12.3	6
	92		58		21

$n_1 = 6 \quad R_1 = 92$

$n_2 = 6 \quad R_2 = 58$

$n_3 = 6 \quad R_3 = 21$

$n = \Sigma n = 18 \quad \Sigma R = 171$

chk: $\Sigma R = n(n+1)/2 = 18(19)/2 = 171$

H_0: the populations have the same median

H_1: at least one median is different

$\alpha = 0.05$ and df = 2

C.V. $H = \chi_\alpha^2 = 5.991$

calculations:

$$H = \{12/[n(n+1)]\} \cdot [\Sigma (R_i^2/n_i)] - 3(n+1)$$
$$= \{12/[18(19)]\} \cdot [(92)^2/6 + (58)^2/6 + (21)^2/6] - 3(19)$$
$$= \{0.0351\} \cdot [2044.8] - 57$$
$$= 14.749$$

P-value = χ^2cdf $(14.749, 99, 2) = 0.0006$

conclusion:

Reject H_0; there is sufficient evidence to reject the claim that voltage readings have the same median for the three different kinds of days.

Sunny days appear to produce greater amounts of electrical energy, but subsequent analysis would be required to determine exactly which pairs of medians are significantly different.

9. Below are the ordered scores and ranks for each group. The group listed first is group 1, etc.

fema/b	R	male/b	R	fema/w	R	male/w	R
92.00	1	115.67	12	106.00	7	99.67	4
93.67	2	118.33	15	107.33	8	102.00	5
96.33	3	120.33	17	108.33	9	111.00	10
102.67	6	120.67	18	113.33	11	124.33	19
117.00	13.5	124.67	20	117.00	13.5	127.33	21
130.67	23	133.00	24	119.67	16	128.33	22
	48.5		106		64.5		81

$n_1 = 6 \quad R_1 = 48.5$

$n_2 = 6 \quad R_2 = 106$

$n_3 = 6 \quad R_3 = 64.5$

$n_4 = 6 \quad R_4 = 81$

$n = \Sigma n = 24 \quad \Sigma R = 300$

chk: $\Sigma R = n(n+1)/2 = 24(25)/2 = 300$

H_0: the populations have the same median

H_1: at least one median is different

$\alpha = 0.05$ and df = 3

C.V. $H = \chi_\alpha^2 = 7.815$

calculations:

$$H = \{12/[n(n+1)]\} \cdot [\Sigma (R_i^2/n_i)] - 3(n+1)$$
$$= \{12/[24(25)]\} \cdot [(48.5)^2/6 + (106)^2/6 + (64.5)^2/6 + (81)^2/6] - 3(25)$$
$$= \{0.02\} \cdot [4051.6] - 75 = 6.032$$

P-value = χ^2cdf $(6.032, 99, 3) = 0.1101$

conclusion:

Do not reject H_0; there is not sufficient evidence to reject the claim that the different groups of people have the same blood pressure.

Yes, the groups can be considered samples from the same blood pressure population.

11. Below are the ordered scores and ranks for each group. The group listed first is group 1, etc.

Indian	R	wheat	R	pre-83	R	post-83	R
2.91490	38	3.00500	42.0	2.95930	39.0	2.45290	1.0
2.97530	40	3.00540	43.0	2.98000	41.0	2.46580	2.0
3.04870	56	3.00640	44.0	3.01850	46.5	2.47620	3.0
3.06300	65	3.01310	45.0	3.03980	51.0	2.47970	4.0
3.13580	105	3.01850	46.5	3.04060	52.0	2.48060	5.0
	304	3.02900	48.0	3.05020	57.0	2.48230	6.0
		3.03570	49.0	3.05220	58.0	2.48480	7.0
		3.03770	50.0	3.05460	59.5	2.48760	8.0
		3.04080	53.0	3.05460	59.5	2.49000	9.0
		3.04760	54.0	3.05860	63.0	2.49070	10.5
		3.04800	55.0	3.06030	64.0	2.49070	10.5
		3.05610	61.0	3.06310	66.0	2.49250	12.0
		3.05700	62.0	3.07040	67.0	2.49330	13.0
		3.07550	71.0	3.07120	68.0	2.49500	14.5
		3.07650	74.0	3.07130	69.0	2.49500	14.5
		3.07860	76.0	3.07170	70.0	2.49730	16.0
		3.08160	78.0	3.07620	72.0	2.49780	17.0
		3.08620	81.0	3.07630	73.0	2.49980	18.0
		3.09360	86.0	3.07750	75.0	2.50040	19.5
		3.09650	87.0	3.07970	77.0	2.50040	19.5
		3.09760	89.0	3.08170	79.0	2.50160	21.0
		3.09940	90.0	3.08460	80.0	2.50170	22.0
		3.10290	92.0	3.08660	82.0	2.50240	23.0
		3.10310	93.0	3.08770	83.0	2.50580	24.0
		3.10380	94.5	3.09170	84.0	2.50670	25.0
		3.10830	97.0	3.09340	85.0	2.50680	26.0
		3.11140	99.0	3.09660	88.0	2.50730	27.0
		3.11410	100.0	3.10280	91.0	2.50850	28.0
		3.12670	101.0	3.10380	94.5	2.51130	29.0
		3.13660	106.0	3.10430	96.0	2.51270	30.0
		3.14610	107.0	3.10860	98.0	2.51390	31.0
		3.15240	108.0	3.12740	102.0	2.51630	32.0
		3.15350	109.0	3.12990	103.0	2.51700	33.0
		3.16920	111.0	3.13400	104.0	2.52220	34.0
		3.19340	112.0	3.15820	110.0	2.52480	35.0
			2714.0		2607.0	2.52520	36.0
						2.52980	37.0
							703.0

$n_1 = 5$ $R_1 = 304$
$n_2 = 35$ $R_2 = 2714$
$n_3 = 35$ $R_3 = 2607$
$n_4 = 37$ $R_4 = 703$
$\Sigma n = 112$ $\Sigma R = 6328$
   $= n$

chk: $\Sigma R = n(n+1)/2$
   $= 112(113)/2$
   $= 6328$

H_0: the populations have the same median
H_1: at least one median is different
$\alpha = 0.05$ [assumed] and df = 3
C.V. $H = \chi_\alpha^2 = 7.815$
calculations:

$H = \{12/[n(n+1)]\}\cdot[\Sigma(R_i^2/n_i)] - 3(n+1)$

$= \{12/[112(113)]\}\cdot[(304)^2/5 + (2714)^2/35$
$+ (2607)^2/35 + (703)^2/37] - 3(113)$

$= \{0.00094817\}\cdot[436475.8] - 339$

$= 74.852$

P-value $= \chi^2$cdf $(74.852,99,3) \approx 0$

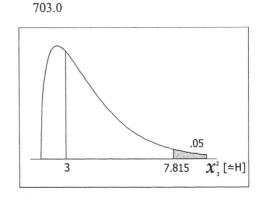

conclusion:

Reject H_0; there is sufficient evidence to reject the claim that the median weight is the same
for the four different penny categories.

No. Machines that consider a coin's weight cannot treat all pennies the same.

13. For the n=18 observations in exercise #7, the correction is determined as follows.

rank	t	$T = t^3 - t$
4.5	2	6
9.5	2	6
12.0	3	24
		36

correction factor:
$$1 - \Sigma T/(n^3 - n) = 1 - 36(18^3 - 18)$$
$$= 1 - 13/5418$$
$$= 0.99380805$$

The original calculated test statistic in exercise #7 is H = 14.74853801
The corrected calculated test statistic is H = 14.74853801/0.99380805 = 14.8404
No, the corrected value of H does not differ substantially from the original one.
NOTE: Be careful when identifying the tied ranks. In addition to the easily recognized ".5's"
that result from an even number of ties, there may be multiple whole number ranks resulting
from an odd number of ties. The notation used in exercise #7 helps to identify tied ranks and
to bring some order to the process of ranking across the different groups: let only the whole
number represent ranks that are not tied (e.g., 3 and 14), and let a decimal place represent
ranks that are tied (e.g., 9.5 and 12.0).

13-6 Rank Correlation

1. The linear correlation method of section 10-2 requires that the original populations be
approximately normal. Since the rank correlation places no restrictions on the original
populations, it may be used in a wider variety of situations.

3. The subscript "s" is necessary to distinguish rank correlation from the linear correlation of
section 10-2. The "s" stands for "Spearman" – and rank correlation is sometimes referred to as
Spearman's rank correlation. The subscript "s" used to designate Spearman's rank correlation
is not related to the lower case "s" used to designate standard deviation.

NOTE: This manual calculates $d = R_x - R_y$, thus preserving the sign of d. This convention means
that $\Sigma d = 0$ must always be true and provides a check for the assigning and differencing of the
ranks – and it must also be true that $\Sigma R_x = \Sigma R_y = n(n+1)/2$.

In addition, some exercises in this section appeared previously in a parametric context in
section 10-2. Because comparison of the parametric and non-parametric results are usually
informative and provide statistical insights, we list those pairings.

this section	#13	#15	#16	#17	#18
section 10-2	#25	#14	#15	#31	#32

5. To calculate r_s, the n=5 pairs of x and y values must be converted to ranks R_x and R_y

R_x	R_y	d	d^2
2	2	0	0
4	4	0	0
1	1	0	0
5	5	0	0
3	3	0	0
15	15	0	0

$r_s = 1 - [6(\Sigma d^2)]/[n(n^2-1)]$
$= 1 - [6(0)]/[5(24)]$
$= 1 - 0$
$= 1$

Yes. There appears to be a positive correlation between R_x and R_y – and, therefore, between x
and y.

7. a. Since n≤30, use Table A-9.
   CV: $r_s = \pm 0.587$

   b. Since n≤30, use Table A-9.
   CV: $r_s = \pm 0.570$

   c. Since n>30, use Formula 13-1.
   CV: $r_s = \pm 1.960/\sqrt{59} = \pm 0.255$

   d. Since n>30, use Formula 13-1.
   CV: $r_s = \pm 2.575/\sqrt{79} = \pm 0.290$

IMPORTANT NOTE: The rank correlation is correctly calculated using the ranks in the Pearson product moment correlation formula 10-1 to produce

$$r_s = [n\Sigma R_x R_y - (\Sigma R_x)(\Sigma R_y)]/[\sqrt{n(\Sigma R_x^2) - (\Sigma R_x)^2}\sqrt{n(\Sigma R_y^2) - (\Sigma R_y)^2}]$$

Since $\Sigma R_x = \Sigma R_y = 1 + 2 + \ldots + n = n(n+1)/2$ [always]
 and $\Sigma R_x^2 = \Sigma R_y^2 = 1^2 + 2^2 + \ldots + n^2 = n(n+1)(2n+1)/6$ [when there are no ties in the ranks]
it can be shown by algebra that the above formula can be shortened to
    $r_s = 1 - [6(\Sigma d^2)]/[n(n^2-1)]$ when there are no ties in the ranks.
As ties typically make no difference in the first 3 decimals of r_s, some texts define r_s using the shortened formula and recommend a "correction for ties" only when the number of ties is "excessive". This manual follows the text and uses the longer formula when there are ties – although the calculated value using the shortened formula is also given as a comparison.

9. The following table summarizes the calculations.

R_x	R_y	d	d^2
1	2	-1	1
2	3	-1	1
3	5	-2	4
4	10	-6	36
5	7	-2	4
6	6	0	0
7	4	3	9
8	1	7	49
10	8	2	4
9	9	0	0
55	55	0	108

$r_s = 1 - [6(\Sigma d^2)]/[n(n^2-1)]$
$= 1 - [6(108)]/[10(99)]$
$= 1 - 0.655$
$= 0.345$

H_0: $\rho_s = 0$
H_1: $\rho_s \neq 0$
$\alpha = 0.05$
C.V. $r_s = \pm 0.648$
calculations:
    $r_s = 0.345$

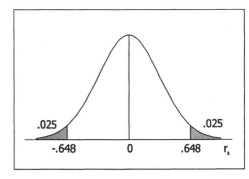

conclusion:
    Do not reject H_0; there is not sufficient evidence to reject the claim that $\rho_s = 0$.
    There does not appear to be a correlation between the DJIA and the numbers of cars sold.

11. The following table summarizes the calculations.

R_x	R_y	d	d^2
2	2	0	0
6	7	-1	1
3	6	-3	9
5	4	1	1
7	5	2	4
10	8	2	4
9	9	0	0
8	10	-2	4
4	3	1	1
1	1	0	0
55	55	0	24

$r_s = 1 - [6(\Sigma d^2)]/[n(n^2-1)]$
$= 1 - [6(24)]/[10(99)]$
$= 1 - 0.145$
$= 0.855$

$H_0: \rho_s = 0$
$H_1: \rho_s \neq 0$
$\alpha = 0.05$
C.V. $r_s = \pm 0.648$
calculations:
$r_s = 0.855$
conclusion:

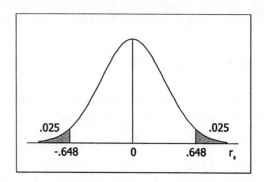

Reject H_0; there is sufficient evidence to reject the claim that $\rho_s = 0$ and conclude that $\rho_s \neq 0$ (in fact, that $\rho_s > 0$).

Yes, it does appear that salary increases as stress increases.

13. The following table summarizes the calculations.

x	R_x	y	R_y	d	d^2
882	2	69.7	1	1	1
1188	7	93.3	8	-1	1
1104	6	84.3	6	0	0
864	1	76.3	3	-2	4
1200	8	88.6	7	1	1
1032	5	82.6	5	0	0
960	4	71.6	2	2	4
900	3	79.6	4	-1	1
	36		36	0	12

$r_s = 1 - [6(\Sigma d^2)]/[n(n^2-1)]$
$= 1 - [6(12)]/[8(63)]$
$= 1 - 0.143$
$= 0.857$

$H_0: \rho_s = 0$
$H_1: \rho_s \neq 0$
$\alpha = 0.05$
C.V. $r_s = \pm 0.738$
calculations:
$r_s = 0.857$
conclusion:

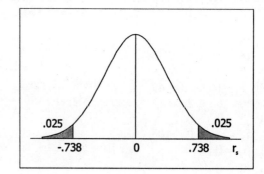

Reject H_0; there is sufficient evidence to reject the claim that $\rho_s = 0$ and conclude that $\rho_s \neq 0$ (in fact, that $\rho_s > 0$).

Yes, conclude that there is a relationship between the number of chirps in a minute and the temperature – the higher the number of chirps, the higher the temperature.

15. The following table summarizes the calculations.

x	R_x	y	R_y	d	d^2
28	9	19	7	2	4
13	1	7	4.5	-3.5	12.25
14	2.5	7	4.5	-2	4
24	7.5	20	8	-0.5	0.25
20	6	6	3	3	9
18	5	4	1	4	16
14	2.5	5	2	0.5	0.25
24	7.5	25	9	-1.5	2.25
17	4	12	6	-2	4
	45		45	0	52

$r_s = 1 - [6(\Sigma d^2)]/[n(n^2-1)]$
$= 1 - [6(52)]/[9(80)]$
$= 1 - 0.433$
$= 0.567$

[$r_s = 0.561$ using the longer formula because of the ties]

H_o: $\rho_s = 0$
H_1: $\rho_s \neq 0$
$\alpha = 0.05$
C.V. $r_s = \pm 0.700$

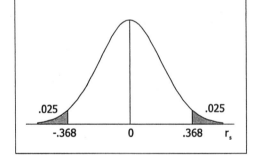

calculations:

$$r_s = [n\Sigma R_x R_y - (\Sigma R_x)(\Sigma R_y)]/ [\sqrt{n(\Sigma R_x^2) - (\Sigma R_x)^2} \sqrt{n(\Sigma R_y^2) - (\Sigma R_y)^2}]$$

$$= [9(258.25) - (45)(45)]/ [\sqrt{9(284.00) - (45)^2} \sqrt{9(284.50) - (45)^2}]$$

$$= [299.25]/[\sqrt{531}\sqrt{535.5}] = 0.561$$

conclusion:

   Do not reject H_o; there is not sufficient evidence to reject the claim that $\rho_s = 0$.

17. Refer to the preliminary calculations on page 268.

   a. H_o: $\rho_s = 0$
      H_1: $\rho_s \neq 0$
      $\alpha = 0.05$
      C.V. $r_s = \pm 0.368$
      calculations:
      $r_s = 53084.5/[\sqrt{57536}\sqrt{58145}]$
      $= 0.9178$

   conclusion:
      Reject H_o; there is sufficient evidence to support the claim that $\rho_s \neq 0$ (in fact, that $\rho_s > 0$).
   There a significant correlation between nicotine and tar – and researchers need not measure both of them, thus saving laboratory expenses.

   b. H_o: $\rho_s = 0$
      H_1: $\rho_s \neq 0$
      $\alpha = 0.05$
      C.V. $r_s = \pm 0.368$
      calculations:
      $r_s = 42608.25/[\sqrt{57536}\sqrt{57826}]$
      $= 0.7387$
   conclusion:
      Reject H_o; there is sufficient evidence to
      support the claim that $\rho_s \neq 0$ (in fact, that $\rho_s > 0$).
   There a significant correlation between nicotine and carbon monoxide – and researchers

need not measure both of them, thus saving laboratory expenses.

This chart gives the preliminary calculations necessary for exercises #17 and #18.

#	nic	R_n	tar	R_t	car	R_c	act	R_a	5dp	R_5	1dp	R_1
1	1.2	24.0	16	23.5	15	19.0	80	28.0	80	32.0	78	24.5
2	1.2	24.0	16	23.5	15	19.0	77	23.0	80	32.0	75	16.5
3	1.0	15.0	16	23.5	17	26.0	81	31.0	79	28.5	81	33.0
4	0.8	8.5	9	8.0	6	3.0	85	35.0	80	32.0	85	35.0
5	0.1	1.0	1	1.0	1	1.0	73	15.0	79	28.5	76	21.0
6	0.8	8.5	8	5.5	8	6.0	73	15.0	82	34.0	75	16.5
7	0.8	8.5	10	10.0	10	8.0	80	28.0	76	21.0	79	27.0
8	1.0	15.0	16	23.5	17	26.0	72	13.0	73	12.0	74	13.0
9	1.0	15.0	14	16.5	13	14.0	83	34.0	77	24.0	75	16.5
10	1.0	15.0	13	14.0	13	14.0	81	31.0	83	35.0	80	30.0
11	1.1	20.0	13	14.0	13	14.0	75	18.5	77	24.0	75	16.5
12	1.2	24.0	15	19.0	15	19.0	78	25.5	79	28.5	79	27.0
13	1.2	24.0	16	23.5	15	19.0	80	28.0	74	15.0	80	30.0
14	0.7	5.5	9	8.0	11	10.5	71	11.0	75	18.0	70	9.5
15	0.9	11.0	11	11.0	15	19.0	73	15.0	76	21.0	72	12.0
16	0.2	2.0	2	2.0	3	2.0	78	25.5	78	26.0	79	27.0
17	1.4	28.5	18	28.5	18	28.5	75	18.5	76	21.0	75	16.5
18	1.2	24.0	15	19.0	15	19.0	63	3.5	75	18.0	67	3.0
19	1.1	20.0	13	14.0	12	12.0	63	3.5	77	24.0	64	2.0
20	1.0	15.0	15	19.0	16	23.5	70	9.0	71	8.5	69	7.0
21	1.3	27.0	17	27.0	16	23.5	77	23.0	74	15.0	77	23.0
22	0.8	8.5	9	8.0	10	8.0	82	33.0	73	12.0	81	33.0
23	1.0	15.0	12	12.0	10	8.0	81	31.0	75	18.0	81	33.0
24	1.0	15.0	14	16.5	17	26.0	76	20.5	79	28.5	80	30.0
25	0.5	3.0	5	3.0	7	4.5	77	23.0	74	15.0	78	24.5
26	0.6	4.0	6	4.0	7	4.5	76	20.5	71	8.5	76	21.0
27	0.7	5.5	8	5.5	11	10.5	74	17.0	70	6.5	76	21.0
28	1.4	28.5	18	28.5	15	19.0	66	6.0	73	12.0	70	9.5
29	1.1	20.0	16	23.5	18	28.5	66	6.0	72	10.0	69	7.0
30		435		435		435	62	2.0	69	5.0	68	4.5
31							71	11.0	70	6.5	75	16.5
32							68	8.0	68	4.0	71	11.0
33							66	6.0	67	3.0	68	4.5
34							71	11.0	61	1.0	69	7.0
35							58	1.0	64	2.0	56	1.0
								630		630		630

exercise #17

$\Sigma R^2_n = 8509.00$

$\Sigma R^2_t = 8530.00$

$\Sigma R^2_c = 8519.00$

$\Sigma R_n R_t = 8355.50$

$\Sigma R_n R_c = 7994.25$

$n\Sigma R_n R_t - (\Sigma R_n)(\Sigma R_t) = 29(8355.5) - (435)(435)$
$= 53084.5$

$n\Sigma R_n R_c - (\Sigma R_n)(\Sigma R_c) = 29(7994.25) - (435)(435)$
$= 42608.25$

$n\Sigma R^2_n - (\Sigma R_n)^2 = 29(8509) - (435)^2 = 57536$

$n\Sigma R^2_t - (\Sigma R_t)^2 = 29(8530) - (435)^2 = 58145$

$n\Sigma R^2_c - (\Sigma R_c)^2 = 29(8519) - (435)^2 = 57826$

a. Use the nicotine (n) and tar (t) values.

b. Use the nicotine (n) and CO (c) values.

C.V. = ±0.368 from Table A-9

exercise #18

$\Sigma R^2_a = 14896.00$

$\Sigma R^2_5 = 14892.00$

$\Sigma R^2_1 = 14881.00$

$\Sigma R_a R_5 = 13386.25$

$\Sigma R_a R_1 = 14574.00$

$n\Sigma R_a R_5 - (\Sigma R_a)(\Sigma R_5) = 35(13386.25) - (630)(630)$
$= 71618.75$

$n\Sigma R_a R_1 - (\Sigma R_a)(\Sigma R_1) = 35(14574) - (630)(630)$
$= 113190$

$n\Sigma R^2_a - (\Sigma R_a)^2 = 35(14896) - (630)^2 = 124460$

$n\Sigma R^2_5 - (\Sigma R_5)^2 = 35(14892) - (630)^2 = 124320$

$n\Sigma R^2_1 - (\Sigma R_1)^2 = 35(14881) - (630)^2 = 123935$

a. Use the actual (a) and five-day (5) values.

b. Use the actual (a) and one-day (1) values.

C.V. = ±1.960/$\sqrt{34}$ = ±0.336 from Formula 13-1.

c. Tar is the better choice, since it has a higher correlation with nicotine.

19. This exercise requires three tasks: the scatterplot, the nonparametric test for Spearman's correlation (this section), the parametric test for Pearson's correlation (section 10-2).

a. The scatterplot at the right reveals the non-linear pattern typical for learning curves.

b. The following table summarizes the calculations necessary for both tests.

x	R_x	y	R_y	d	d^2
1	1	1	1	0	0
25	2	2	2	0	0
31	3	20	3	0	0
33	4	36	4	0	0
36	5	51	5	0	0
38	6	72	6	0	0
48	7	74	7	0	0
55	8	75	8	0	0
95	9	77	9	0	0
165	10	78	10	0	0
300	11	79	11	0	0
827	66	565	66	0	0

$r_s = 1 - [6(\Sigma d^2)]/[n(n^2-1)]$
$= 1 - [6(0)]/[11(120)]$
$= 1 - 0$
$= 1$

$\Sigma xy = 57993$
$\Sigma x^2 = 136995$
$\Sigma y^2 = 38841$
$n(\Sigma xy) - (\Sigma x)(\Sigma y) = 11(57993) - (827)(565) = 170668$
$n(\Sigma x^2) - (\Sigma x)^2 = 11(136995) - (827)^2 = 823016$
$n(\Sigma y^2) - (\Sigma y)^2 = 11(38841) - (565)^2 = 108026$
$r = [n(\Sigma xy) - (\Sigma x)(\Sigma y)]/[\sqrt{n(\Sigma x^2) - (\Sigma x)^2} \sqrt{n(\Sigma y^2) - (\Sigma y)^2}]$
$= 170668/[\sqrt{823016} \cdot \sqrt{108026}] = 0.5724$

H_0: $\rho_s = 0$
H_1: $\rho_s \neq 0$
$\alpha = 0.05$ [assumed]
C.V. $r_s = \pm 0.618$
calculations:
  $r_s = 1.000$

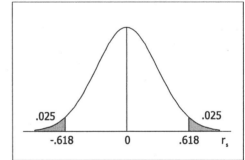

conclusion:
  Reject H_0; there is sufficient evidence to reject the claim that $\rho_s = 0$ and conclude that $\rho_s \neq 0$ (in fact, that $\rho_s > 0$).

c. H_0: $\rho = 0$
  H_1: $\rho \neq 0$
  $\alpha = 0.05$ [assumed]
  C.V. $t = \pm t_{9,\alpha/2} = \pm 2.262$ [or $r = \pm 0.602$]
  calculations:
    $t_r = (r - \mu_r)/s_r$
    $= (0.572 - 0)/\sqrt{[1 - (0.572)^2]/9}$
    $= 0.572/0.2733$
    $= 2.094$

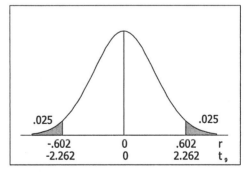

  P-value $= 2 \cdot tcdf(2.094,99,9) = 2(0.0329) = 0.0657$
  conclusion:
    Do not reject H_0; there is not sufficient evidence to reject the claim that $\rho \neq 0$.
The nonparametric test is able to detect the nonlinear relationship.

13-7 Runs Test for Randomness

1. No. The runs test for randomness requires the order in which the subjects were selected, and shuffling the papers does not preserve that order.

3. No. The runs test determines only whether the order of the subjects in the sample is random according to a particular criterion – not whether the sample is a convenience sample, a voluntary response sample, etc.

NOTE: In each exercise, the item that appears first in the sequence is considered to be of the first type and its count is designated by n_1.

5. $n_1 = 4$ (# of Y's)   G = 8 (# of runs)
   $n_2 = 8$ (# of N's)   CV: 3,10 (from Table A-10)
   Yes. According to the runs test, the sequence is random in that it does not have significantly fewer or significantly more runs than expected by chance – but there appears to be a pattern of one Y followed by three N's.

7. $n_1 = 6$ (# of M's)   G = 7 (# of runs)
   $n_2 = 6$ (# of F's)   CV: 3,11 (from Table A-10)
   Yes. According to the runs test, the sequence is random in that it does not have significantly fewer or significantly more runs than expected by chance.

9. Since $n_1 = 13$ and $n_2 = 7$, use Table A-10.
   H_o: the sequence is random
   H_1: the sequence is not random
   $\alpha = 0.05$
   C.V. G = 5, 15
   calculations:
   $\quad$ G = 11

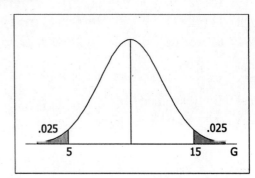

   conclusion:
   $\quad$ Do not reject Ho; there is not sufficient evidence to reject the claim that the values occur in a random sequence.

11. Since $n_1 = 14$ and $n_2 = 19$, use Table A-10.

   H_o: the sequence is random
   H_1: the sequence is not random
   $\alpha = 0.05$
   C.V. G = 11,23
   calculations:
   $\quad$ G = 20

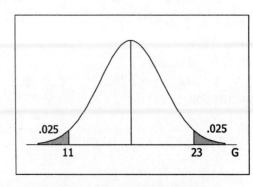

   conclusion:
   $\quad$ Do not reject Ho; there is not sufficient evidence to reject the claim that the values occur in a random sequence.
   The test for randomness does not address the relative abilities of the two leagues, but $n_1 = 14 \approx 19 = n_2$ suggests that neither league is dominant over the other.

13. The median of the 17 values arranged in numerical order from $x_1 = 20{,}595$ to $x_{17} = 36{,}448$ is $x_9 = 26{,}995$. Passing through the values chronologically and assigning A's and B's as directed (and ignoring values equal to the median), yields the sequence

    B B B B B B B B A A A A A A A A

Since $n_1 = 8$ and $n_2 = 8$, use Table A-10.

H_0: the sequence is random

H_1: the sequence is not random

$\alpha = 0.05$

C.V. $G = 4, 14$

  calculations:

    $G = 2$

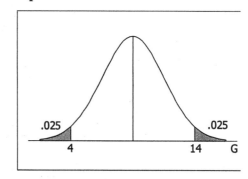

conclusion:

    Reject Ho; there is sufficient evidence to reject the claim that the values occur in a random sequence.

It appears that since 1987 there has been an upward trend in the number of indoor movie theaters.

15. The genders of the 54 bears (1=male, 2=female) are as follows.

    1 1 1 1 2 2 1 1 2 2 1 1 2 1 2 1 1 2 1 1 2 2 1 1 1 1 1 2 2 1 1 2 2 1 1 1 1 1 2 1 1 1 1 2 2 1 1 1 1 1 2 2 2 1 1

Since $n_1 = 35$ and $n_2 = 19$, use the normal approximation.

  $\mu_G = 2n_1n_2/(n_1 + n_2) + 1$

    $= 2(35)(19)/54 + 1 = 25.630$

  $\sigma^2_G = [2n_1n_2(2n_1n_2 - n_1 - n_2)]/[(n_1 + n_2)^2(n_1 + n_2 - 1)]$

    $= [2(35)(19)(1276)]/[(54)^2(53)] = 10.981$

H_0: the sequence is random

H_1: the sequence is not random

$\alpha = 0.05$

C.V. $z = \pm z_{\alpha/2} = \pm 1.96$

  calculations:

    $G = 23$

    $z_G = (G - \mu_G)/\sigma^2_G$

      $= (23 - 25.630)/\sqrt{10.981}$

      $= -2.630/3.314 = -0.794$

    P-value $= 2 \cdot P(z < -0.79) = 2(0.2148) = 0.4296$

conclusion:

    Do not reject Ho; there is not sufficient evidence to reject the claim that the values occur in a random sequence.

17. The sequence of the 100 digits in O's and E's is follows.

    O E O O O E E O O O E O O O O E O E E E E E E O O

    E O E O O O E E E E E O O O O E O O O O O O O O E

    O E E E O O E O E E O O E O E O E O E E E E E E E

    E E E O O E E E E E O E E E O O E E O O O E E O O

Since $n_1 = 49$ and $n_2 = 51$, use the normal approximation.

  $\mu_G = 2n_1n_2/(n_1 + n_2) + 1$

    $= 2(49)(51)/100 + 1 = 50.980$

  $\sigma^2_G = [2n_1n_2(2n_1n_2 - n_1 - n_2)]/[(n_1 + n_2)^2(n_1 + n_2 - 1)]$

    $= [2(49)(51)(4898)]/[(100)^2(99)] = 24.727$

H_o: the sequence is random

H_1: the sequence is not random

$\alpha = 0.05$

C.V. $z = \pm z_{\alpha/2} = \pm 1.96$

calculations:

$G = 43$

$z_G = (G - \mu_G)/\sigma^2_G$

$= (43 - 50.980)/\sqrt{24.727}$

$= -7.980/4.973 = -1.605$

P-value $= 2 \cdot P(z<-1.60) = 2(0.0548) = 0.1096$

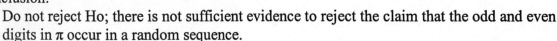

conclusion:

Do not reject Ho; there is not sufficient evidence to reject the claim that the odd and even digits in π occur in a random sequence.

19. The minimum possible number of runs is $G = 2$, and occurs when all the A's are together and all the B's are together (e.g., AABB).

The maximum possible number of runs is $G = 4$, and occurs when the A's and B's alternate (e.g., ABAB).

Because the critical values for $n_1 = n_2 = 2$ are

C.V. $G = 1,6$

the null hypothesis of the sequence being random can never be rejected at the 0.05 level. This means that it is not possible for such a small sample to provide 95% certainty that a non-random phenomenon is occurring. Table A-10 could have given that information more clearly by using an asterisk, as in Table A-7 and Table A-8, to indicate those sample sizes for which it is not possible to achieve statistical significance no matter what the data.

Statistical Literacy and Critical Thinking

1. A nonparametric test is a test not requiring that the sample data come from a population that has a particular distribution (usually the normal distribution) defined by certain parameters.

2. There is no difference. "Nonparametric" and "distribution-free" are different names for the same category of tests.

3. A rank is a number assigned to a value according to its order in a list of values sorted by magnitude. The smallest value is typically assigned rank 1, the next smallest is assigned rank 2, etc.

4. The efficiency of a nonparametric test refers to the sample size the corresponding parametric test requires to obtain the same precision that a nonparametric reaches with a sample of size 100 – whenever all the requirements of the parametric test are met. The efficiency of a non-parametric test is a number less than 1.00 and can be interpreted as "percent of effectiveness" in the following sense: *The efficiency of the sign test for detecting the difference between matched pairs is 0.63. This means that when the parametric assumptions are met (i.e., when the original populations have normal distributions), the sign test requires a sample of size 100 to detect the same differences that the parametric test can detect with a sample of size 63. And so there is a sense in which the nonparametric test is using only 63% of the information available in the data.*

Review Exercises

1. Use the sign test of section 13-2. Let the first trial be group 1.

child	A	B	C	D	E	F	G	H	I	J	K	L	M	N	O
1st-2nd	0	+	+	+	+	+	+	−	−	+	+	+	+	+	+

n = 14: 12 +'s and 2 −'s

H_o: median difference = 0

H_1: median difference $\neq$ 0

α = 0.05

C.V. $x = x_{L,\alpha/2} = 2$

$x = x_{U,\alpha/2} = 14 - 2 = 12$

calculations:

x = 2 (using less frequent count)

x = 12 (using + count)

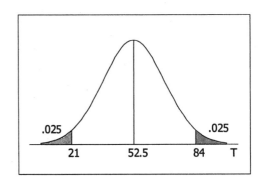

conclusion:

Reject H_o; there is sufficient evidence to reject the claim that the median difference is 0 and to conclude that the median times are different (in fact, that the times for the first trial are greater).

2. Use the Wilcoxon signed-rank test of section 13-3. Let the first trial be group 1.

1st-2nd	0	13	5	15	15	126	28	-2	-5	31	3	51	3	14
R	-	6	4.5	8.5	8.5	14	10	-1	-4.5	11	2.5	13	2.5	7

ΣR– =   5.5   n = 14 non-zero ranks

ΣR+ = 99.5   check: $\Sigma R = n(n+1)/2 = 14(15)/2 = 105$

ΣR = 105

H_o: median difference = 0

H_1: median difference $\neq$ 0

α = 0.05

C.V. $T = T_{L,\alpha/2} = 21$

$T = T_{U,\alpha/2} = 105 - 21 = 84$

calculations:

T = 5.5 (using smaller ranks)

T = 99.5 (using positive ranks)

conclusion:

Reject H_o; there is sufficient evidence to reject the claim that the median difference is 0 and to conclude that the median times are different (in fact, that the times for the first trial are greater).

3. Use the rank correlation test of section 13-6. The following table summarizes the calculations.

R_x	R_y	d	d^2
1	2	-1	1
6	1	5	25
2	3	-1	1
5	4	1	1
4	5	-1	1
3	6	-3	9
10	9	1	1
8	10	-2	4
9	7	2	4
7	8	-1	1
55	55	0	48

$r_s = 1 - [6(\Sigma d^2)]/[n(n^2-1)]$
$= 1 - [6(48)]/[10(99)]$
$= 1 - 0.291$
$= 0.709$

H_0: $\rho_s = 0$
H_1: $\rho_s \neq 0$
$\alpha = 0.05$
C.V. $r_s = \pm 0.648$
calculations:
   $r_s = 0.709$
conclusion:

Reject H_0; there is sufficient evidence to reject the claim that $\rho_s = 0$ and conclude
that $\rho_s \neq 0$ (in fact, that $\rho_s > 0$).
Yes. There appears to be a correlation between the business school rankings and the law
school rankings.

4. NOTE: It is arbitrary whether one assigns the ranks 1,2,…,n from low to high or from high to
low. Usually the ranks are assigned from low to high so that low values have low ranks and
high values have high ranks. To preserve the direction of the relationship, however, the same
system should be used for both x and y. Since the schools were ranked with lower numbers
indicating better ranks (i.e., rank #1 is better than rank #8), rank the MCAT scores from high
to low so that the lower numbers also indicate better ranks for the y variable 0.

Use the rank correlation test of section 13-6. The following table summarizes the calculations.

R_x	y	R_y	d	d^2
1	11.3	2	-1	1
4	11.0	3	1	1
2	11.4	1	1	1
3	10.3	6	-3	9
5	10.8	4	1	1
6	10.5	5	1	1
7	9.5	8	-1	1
8	9.6	7	1	1
36		36	0	16

$r_s = 1 - [6(\Sigma d^2)]/[n(n^2-1)]$
$= 1 - [6(16)]/[8(63)]$
$= 1 - 0.190$
$= 0.810$

H_0: $\rho_s = 0$
H_1: $\rho_s \neq 0$
$\alpha = 0.05$
C.V. $r_s = \pm 0.738$
calculations:
   $r_s = 0.810$
conclusion:

Reject H_0; there is sufficient evidence to reject the claim that $\rho_s = 0$ and conclude that
   $\rho_s \neq 0$ (in fact, that $\rho_s > 0$).
Yes, the schools with the better rankings are the schools with the better MCAT scores.

5. Use the sign test of section 13-2. Let x = the number of females hired.
   claim: $p < 0.5$ [this is the charge that there is bias against women (i.e., in favor of men)]
   n = 40: 15 +'s and 25 –'s [the data are in the proper direction to proceed]
   Since n>25, use z with

   $\mu_x = n/2 = 40/2 = 20$

   $\sigma_x = \sqrt{n}/2 = \sqrt{40}/2 = 3.162$

   H_0: $p = 0.5$
   H_1: $p < 0.5$
   $\alpha = 0.05$
   C.V. $z = -z_\alpha = -1.645$
   calculations:

   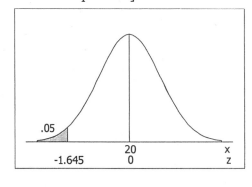

   x = 15

   $z = [(x+.5) - n/2]/[\sqrt{n}/2]$
   $= [15.5 - 20]/3.162 = -4.5/3.162 = -1.423$
   P-value = $P(z<-1.42) = 0.0778$

   conclusion:
   Do not reject H_0; there is not sufficient evidence to support the charge of bias against women (i.e., in favor of men).

6. Use the Wilcoxon rank-sum test of section 13-4. Let the beer drinkers be group 1.
   Below are the ordered scores for each group.

Beer	R	liquor	R
0.129	1	0.182	9
0.146	2	0.185	10
0.148	3	0.190	12.5
0.152	4	0.205	15
0.154	5	0.220	17
0.155	6	0.224	18
0.164	7	0.225	19
0.165	8	0.226	20
0.187	11	0.117	21
0.190	12.5	0.234	22
0.203	14	0.241	23
0.212	16	0.247	24
	89.5	0.253	25
		0.257	26
			261.5

   $n_1 = 12$ $R = \Sigma R_1 = 89.5$ chk: $\Sigma R_i = n(n+1)/2$
   $n_2 = 14$ $\Sigma R_2 = 261.5$ $= 26(27)/2$
   $n = \Sigma n = 26$ $\Sigma R_i = 351.0$ $= 351$

   $\mu_R = n_1(n+1)/2$ $\sigma^2_R = n_1 n_2(n+1)/12$
   $= 12(27)/2$ $= (12)(14)(27)/12$
   $= 162$ $= 378$

   H_0: median$_1$ = median$_2$
   H_1: median$_1 \neq$ median$_2$
   $\alpha = 0.05$
   C.V. $z = \pm z_{\alpha/2} = \pm 1.960$
   calculations:

   $z = (R - \mu_R)/\sigma_R$
   $= (89.5 - 162)/\sqrt{378}$
   $= -72.5/19.442 = -3.729$

   P-value $2 \cdot P(z<-3.73) = 2(0.0001) = 0.0002$
   conclusion:
   Reject H_0; there is sufficient evidence to reject the claim that median$_1$ = median$_2$ and to conclude that median$_1 \neq$ median$_2$ (in fact, that beer drinkers have a lower BAC than liquor drinkers.

   By this criterion, it appears that liquor drinkers are more dangerous.

7. Use the runs test of section 13-7. The sequence of the 40 digits in O's and E's is follows.

O O E O O O O O E E E O E E O E O E E O

O O O E E E E O O E O O O O O E O E E E

Since $n_1 = 22$ and $n_2 = 18$, use the normal approximation.

$\mu_G = 2n_1n_2/(n_1 + n_2) + 1$

$= 2(22)(18)/40 + 1 = 20.800$

$\sigma^2_G = [2n_1n_2(2n_1n_2 - n_1 - n_2)]/[(n_1 + n_2)^2(n_1 + n_2 - 1)]$

$= [2(22)(18)(752)]/[(40)^2(39)] = 9.545$

H_o: the sequence is random

H_1: the sequence is not random

$\alpha = 0.05$

C.V. $z = \pm z_{\alpha/2} = \pm 1.96$

calculations:

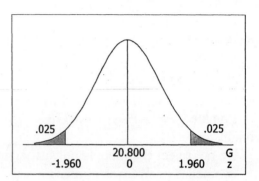

G = 18

$z_G = (G - \mu_G)/\sigma^2_G$

$= (18 - 20.800)/\sqrt{9.545}$

$= -2.800/3.089 = -0.906$

P-value $= 2 \cdot P(z < -0.91) = 2(0.1814) = 0.3628$

conclusion:

Do not reject Ho; there is not sufficient evidence to reject the claim that the odd and even digits in the lottery occur in a random sequence.

8. Use the Kruskal-Wallis test of section 13-5. The group listed first is group 1, etc.

Below are the ordered scores and ranks for each group.

sub	R	com	R	mid	R	full	R
422	2	584	4	181	1	804	8
519	3	946	11	629	6	971	12
595	5	984	13	645	7	996	14
885	10	1051	15	880	9	1085	17
1063	16	1193	18	1686	20	1376	19
	36		61		43		70

$n_1 = 5$ $R_1 = 36$
$n_2 = 5$ $R_2 = 61$
$n_3 = 5$ $R_3 = 43$
$n_4 = 5$ $R_4 = 70$
$n = \Sigma n = 20$ $\Sigma R = 210$

chk: $\Sigma R = n(n+1)/2 = 20(21)/2 = 210$

H_o: the populations have the same median

H_1: at least one median is different

$\alpha = 0.05$ and df = 3

C.V. $H = \chi^2_\alpha = 7.815$

calculations:

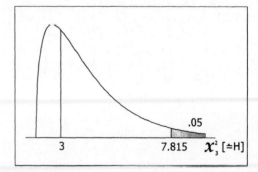

$H = \{12/[n(n+1)]\} \cdot [\Sigma(R_i^2/n_i)] - 3(n+1)$

$= \{12/[20(21)]\} \cdot [(36)^2/5 + (61)^2/5$

$+ (43)^2/5 + (70)^2/5] - 3(21)$

$= \{0.0286\} \cdot [2353.2] - 63$

$= 4.234$

P-value $= \chi^2 cdf(4.234,99,3) = 0.2373$

conclusion:

Do not reject H_o; there is not sufficient evidence to support the claim that leg injury measurements for the four car weight categories are not all the same.

No. By this measurement, the data do not show that heavier cars are safer in a crash.

Cumulative Review Exercises

1. scores in order: 460 470 490 490 510 510 600 610 620

$n = 9$ $\bar{x} = 4760/9 = 528.9$ $R = x_9 - x_1 = 620 - 460 = 160$

$\Sigma x = 4760$ $\tilde{x} = x_5 = 510$ $s^2 = [9(2549400) - (4760)^2]/[9(8)] = 3986.1$

$\Sigma x^2 = 2549400$ $s = 63.1$

For the remaining exercises, let x = the before scores and let y = the after scores.
The following table contains all the preliminary calculations for n=9 pairs of scores.

x	R_x	y	R_y	d=x-y	R_d	sign	other needed calculations
460	1.0	480	1.0	-20	-2.0	−	$\Sigma xy = 2834700$
470	2.0	510	3.0	-40	-4.5	−	$\Sigma R_x R_y = 278.75$
490	3.5	500	2.0	-10	-1.0	−	$n\Sigma xy - (\Sigma x)(\Sigma y) = 284300$
490	3.5	610	5.0	-120	-8.5	−	$n\Sigma x^2 - (\Sigma x)^2 = 28700$
510	5.5	590	4.0	-80	-6.5	−	$n\Sigma y^2 - (\Sigma y)^2 = 405800$
510	5.5	630	6.5	-120	-8.5	−	$n\Sigma R_x R_y - (\Sigma R_x)(\Sigma R_y) = 483.75$
600	7.0	630	6.5	-30	-3.0	−	$n\Sigma (R_x)^2 - (\Sigma R_x)^2 = 531.0$
620	9.0	660	8.0	-40	-4.5	−	$n\Sigma (R_y)^2 - (\Sigma R_y)^2 = 535.5$
610	8.0	690	9.0	-80	-6.5	−	$n\Sigma d^2 - (\Sigma d)^2 = 124200$
Σv 4760	45.0	5300	45.0	-540			
Σv² 2549400	284.0	3166200	284.5	46200			

2. The scatterplot is given at the right.

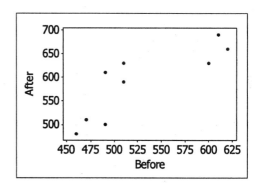

Yes, there appears to be a correlation between the
before and after scores. The correlation appears
to be a positive one – meaning that those who got the lower scores before the program tended
to have the lower scores after the program, and those who got the higher scores before the
program tended to have the higher scores after the program.

3. $r = [n(\Sigma xy) - (\Sigma x)(\Sigma y)]/[\sqrt{n(\Sigma x^2) - (\Sigma x)^2} \sqrt{n(\Sigma y^2) - (\Sigma y)^2}]$

     $= 284300/[\sqrt{287000} \sqrt{405800}] = 0.833$

$H_0: \rho = 0$

$H_1: \rho \neq 0$

$\alpha = 0.05$ [assumed] and df = 7

C.V. $t = \pm t_{\alpha/2} = -2.365$ [or r = ±0.666]

calculations:

     $t_r = (r - \mu_r)/s_r$

        $= (0.833 - 0)/\sqrt{[1 - (0.833)^2]/7}$

        $= 0.833/0.209$

        $= 3.984$

     P-value = 2·tcdf(3.984,99,7) = 2(0.0026) = 0.0053

conclusion:

     Reject H_0; there is sufficient evidence to conclude that $\rho \neq 0$ (in fact, that $\rho > 0$).

No, a significant linear correlation does not mean the preparation course is effective – it means only that you can predict the after score from the before score (and vice-versa). If every student gets exactly 10 points <u>lower</u> on the after test, for example, there would be a perfect positive correlation of r = 1.00.

4. $r_s = [n\Sigma R_x R_y - (\Sigma R_x)(\Sigma R_y)]/[\sqrt{n(\Sigma R_x^2) - (\Sigma R_x)^2}\sqrt{n(\Sigma R_y^2) - (\Sigma R_y)^2}]$

$= 483.75/[\sqrt{531.0}\sqrt{535.5}] = 0.907$

H$_o$: $\rho_s = 0$
H$_1$: $\rho_s \neq 0$
$\alpha = 0.05$ [assumed]
C.V. $r_s = \pm 0.700$
calculations:
   $r_s = 0.907$

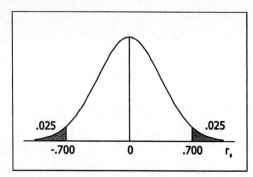

conclusion:
   Reject H$_o$; there is sufficient evidence to conclude that $\rho_s \neq 0$ (in fact, that $\rho_s > 0$). Yes, there is a positive correlation between the ranking and the MCAT score.

5. Refer to the table to calculate the following preliminary statistics.
   $\bar{d} = \Sigma d/n = -540/9 = -60$
   $s_d^2 = [n(\Sigma d^2) - (\Sigma d)^2]/[n(n-1)] = 124200/72 = 1725$
   $s_d = 41.533$
original claim: $\mu_d = 0$
H$_o$: $\mu_d = 0$
H$_1$: $\mu_d \neq 0$
$\alpha = 0.05$ [assumed] and df = 8
C.V. $t = \pm t_{\alpha/2} = \pm 2.306$
calculations

$t_{\bar{d}} = (\bar{d} - \mu_{\bar{d}})/s_{\bar{d}}$

$= (-60 - 0)/(41.533/\sqrt{9})$

$= -60/13.844 = -4.334$

P-value = $2 \cdot$tcdf(-99,-4.334,8) = 2(0.0012) = 0.0025

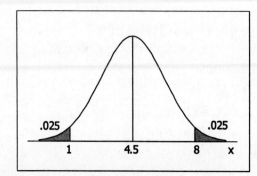

conclusion:
   Reject H$_o$; there is sufficient evidence to reject the claim that $\mu_d = 0$ and conclude that $\mu_d \neq 0$ (in fact, that $\mu_d < 0$ – i.e., that the before scores are lower).

6. The "sign" column indicates the following.
   n = 9 pairs: 0 +'s and 9 –'s
H$_o$: median difference = 0
H$_1$: median difference $\neq$ 0
$\alpha = 0.05$ [assumed]
C.V. x = $x_{L,\alpha/2} = 1$
         x = $x_{U,\alpha/2} = 9 - 1 = 8$
calculations:
   x = 0

conclusion:
   Reject H$_o$; there is sufficient evidence to reject the claim that there is no difference between the before and after scores and to conclude median difference $\neq$ 0 (in fact, that median difference < 0 – i.e., that the before scores are lower).

7. The R_d column of the table gives the signed ranks of the differences. Refer to the table to calculate the following preliminary statistics.

    $\Sigma R- = 45$ $n = 9$ non-zero ranks

    $\Sigma R+ = 0$ check: $\Sigma R = n(n+1)/2 = 9(10)/2 = 45$

    $\Sigma R\ = 45$

H_o: median difference $= 0$

H_1: median difference $\neq 0$

$\alpha = 0.05$

C.V. $T = T_{L,\alpha/2} = 6$

     $T = T_{U,\alpha/2} = 45 - 6 = 39$

calculations:

    $T = 0$

conclusion:

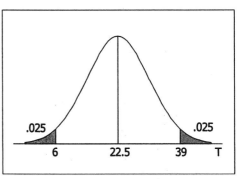

    Reject H_o; there is sufficient evidence to reject the claim that there is no difference between the before and after scores and to conclude median difference $\neq 0$ (in fact, that median difference < 0 – i.e., that the before scores are lower).

8. The main question of interest is whether the coaching program is effective in raising math SAT test scores. Four of the preceding exercises (viz., #1-#4) do not address that question, and the three exercises (viz., #5-#7) that do are not equally effective. The following paragraphs summarize the relevance of each of the preceding exercises to this question.

  a. exercise #1 – descriptive statistics for the before scores. If these results were compared to similar results for the after scores, one could make a crude comparisons between the two populations. Such comparisons would not take advantage of the individual pairings, and any differences between the before and after scores would probably be hidden by the greater differences between individual students.

  b. exercise #2 – scatterplot. Visualize the best straight line through the points. The tightness of the points around that line indicate the consistency of the effect of the program (the program is consistent from student to student if the points are tightly clustered around the line), and the slope of the line indicates the effect (the program is effective if the slope is significantly greater than 1.00). But such detailed examination requires formal regression analysis, not merely a scatterplot.

  c. exercise #3 – linear correlation using the raw scores. This measures only how well the after scores can be predicted from the before score (or vice-versa), and does not address the issue of which score is higher.

  d. exercise #4 – linear correlation using the ranks. This measures only how well the after scores can be predicted from the before score (or vice-versa), and does not address the issue of which score is higher. It is more effective than exercise #3 in detecting a relationship that is not linear in the raw scores but that is linear in ranks.

  e. exercise #5 – matched pairs using the difference in the raw scores. This is the most effective measure. It addresses the question directly and uses all the available information – both the actual raw scores and the individual pairings.

  f. exercise #6 – matched pairs using only the sign of the difference in the raw scores. This addresses the question, but in using only the sign of the change and not the magnitude of the change it does not use all the available information. In general, it is the least efficient of the three appropriate tests in exercises #5, #6 and #7.

  g. exercise #7 – matched pairs using the signed rank of the difference in the raw scores. This addresses the question, but in using only the (signed) rank of the change and not the actual magnitude of the change it does not use all the available information.

Chapter 14

Statistical Process Control

14-2 Control Charts for Variation and Mean

1. Process data are values in chronological sequence that measure a characteristic of a controllable activity.

3. A control chart is a plot of values over time, with a center line representing the standard value and upper and lower lines indicating limits beyond which values are considered unacceptable. An R chart is a control chart that plots the ranges of consecutive samples of size n in order to monitor the variation of a process. An $\bar{x}$ chart is a control chart that plots the means of consecutive samples of size n in order to monitor the central tendency of a process.

5. a. and b. Within statistical control.
   c. No. The variation is too large. The runs chart indicates a variation of about ±1.2 ounces, which is about ±10% of the target value, and this is not acceptable for 12 ounce cans of soda.

7. a. and b. Not within statistical control, because there is one unusually low value.
   c. No. The dramatic departure of the extremely low value from the acceptable values suggests that there was a problem with the machine or the operator – and that steps to identify the cause of the departure are in order.

NOTE: In this section, k = number of sample subgroups
                     n = number of observations per sample subgroup

Although the $\bar{\bar{x}}$ line is not specified in the text as a necessary part of a run chart, such a line is helpful. It is included in figures generated by many statistical software packages, and it is included in this manual.

9. There are k=25 samples of size n=7 each. $UCL = D_4\bar{R} = 1.924(54.96) = 105.74$
   $\bar{R} = \Sigma R/k = 1374/25 = 54.96$ $LCL = D_3\bar{R} = 0.076(54.96) = 4.18$

The process variation appears to be within statistical control.

11. There are k·n = 20·5 = 100 total observations on the run chart.
For convenience, the problem is worked in milligrams (i.e., the original weights x 1000).
$\bar{\bar{x}} = \Sigma \bar{x}/k = \Sigma x/(k \cdot n) = 570980/100 = 5709.8$

Yes, there appears to be a pattern that suggests the process is not within statistical control –
viz., the variability is increasing. In practical terms, this suggests that the equipment is not
holding its tolerance and that the process needs to be shut down for an adjustment.

13. For convenience, the problem is worked in milligrams (i.e., the original weights x 1000).
There are k=20 samples of size n=5 each. $\bar{\bar{x}} = \Sigma \bar{x}/k = 114196.0/20 = 5709.8$

$\bar{R} = \Sigma R/k = 7292/20 = 364.6$

$UCL = \bar{\bar{x}} + A_2\bar{R} = 5709.8 + (0.577)(364.6) = 5709.8 + 210.4 = 5920.2$
$LCL = \bar{\bar{x}} - A_2\bar{R} = 5709.8 - (0.577)(364.6) = 5709.8 - 210.4 = 5499.4$

The process variation is not within statistical control. To be out of control, the process must
meet at least one of three criteria – and this process actually meets all three!
    (1) There is a pattern to the $\bar{x}$ values – in this case, an upward drift.
    (2) There is a point beyond the upper or lower control limit – in this case, there are 2 points
        above the UCL.
    (3) There are at least 8 consecutive points on the same side of the center line – in this case,
        the first 8 points are below the center line.
Yes, corrective action is needed.

NOTE: Application of the techniques of this section in exercises #14, #15 and #16 is interesting – but probably not appropriate. It is questionable whether the Boston rainfall data meet the definition of process data. In general, process data are values in chronological sequence that measure a characteristic of a controllable activity. If the activity (rainfall) is not controllable, it is really irrelevant to discuss whether or not it is within statistical control.

The following chart summarizes the information necessary for exercises #15 and #16.

week	$\bar{x}$	R	week	$\bar{x}$	R	week	$\bar{x}$	R	week	$\bar{x}$	R
1	0.019	0.05	14	0.000	0.00	27	0.036	0.14	40	0.067	0.24
2	0.027	0.10	15	0.097	0.40	28	0.027	0.11	41	0.003	0.02
3	0.101	0.71	16	0.237	0.87	29	0.010	0.05	42	0.003	0.02
4	0.186	0.64	17	0.067	0.47	30	0.017	0.12	43	0.181	0.68
5	0.016	0.05	18	0.054	0.24	31	0.083	0.44	44	0.243	1.48
6	0.091	0.64	19	0.036	0.14	32	0.044	0.18	45	0.290	1.28
7	0.051	0.30	20	0.237	0.92	33	0.164	0.64	46	0.139	0.96
8	0.001	0.01	21	0.049	0.27	34	0.134	0.85	47	0.116	0.79
9	0.039	0.16	22	0.003	0.01	35	0.009	0.03	48	0.109	0.41
10	0.121	0.39	23	0.000	0.00	36	0.017	0.12	49	0.017	0.08
11	0.181	0.78	24	0.101	0.71	37	0.050	0.26	50	0.000	0.00
12	0.029	0.17	25	0.097	0.33	38	0.057	0.40	51	0.106	0.74
13	0.416	1.41	26	0.000	0.00	39	0.017	0.12	52	0.146	0.43

NOTE: The year begins and ends on Wednesday. Removing the last Wednesday gives 52 weeks of data that begin on Wednesday and end on Tuesday – and each row in Data Set 10 is that type of week. To construct a run chart for those $k \cdot n = 52 \cdot 7 = 364$ days, one would have to move the Monday and Tuesday columns to the end (i.e., following Sunday) before reading through the data by rows – but this is irrelevant to calculating $\bar{x}$ and R for each Wednesday to Tuesday week.

15. There are k=52 samples of size n=7 each. $UCL = D_4 \bar{R} = 1.924(0.3915) = 0.753$
$\bar{R} = \Sigma R/k = 20.36/52 = 0.3915$ $LCL = D_3 \bar{R} = 0.076(0.3915) = 0.030$

The process variation is not within statistical control – there are points above the upper control limit and below the lower control limit. While only one point above or below the appropriate control limit is sufficient to declare the process out of control, these data have multiple points beyond each limit. A point above the upper control limit represents a large range for that sample – which indicates a potential problem within the time frame of that particular sample, as well as the fact that the variance from sample to sample exhibits more than acceptable fluctuation. A point below the lower control limit represents a small range for that sample – which does not indicate a problem with that particular sample, but does indicate that the variance from sample to sample exhibits more than acceptable fluctuation. See also the NOTE at the top of the page.

17. There are k=20 samples of size n=4 each.

$\bar{s} = \Sigma s/k = 193.85/20 = 9.6925$

$UCL = B_4 \bar{s} = 2.266(9.6925) = 21.963$

$LCL = B_3 \bar{s} = 0.000(9.6925) = 0.000$

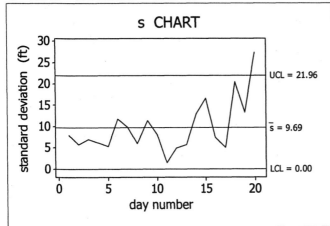

This is very similar to the R chart given in the text, and both charts indicate that the process variation is not within statistical control. For small n's in the sub-samples (n=4 in this example), there will be a very high correlation between the values of R and s – and so the two charts will be almost identical, but with different labels on the vertical axis.

14-3 Control Charts for Attributes

1. A p chart is a control chart to monitor the proportion of items having a particular attribute over time. It is used to determine whether the corresponding process is within statistical control.

3. When the formula determines the lower control limit in a p chart to be less than zero, use the LCL = 0.00. In any problem with upper and lower physical boundaries, the control limits should be truncated to match those boundaries – as having a limit outside the realm of possibility is not informative, and adjusting the control limits to match the physical boundaries may help to identify errors resulting in otherwise impossible values. In the given scenario, the UCL and LCL should be 0.250 and 0.000 respectively.

5. This process is within statistical control. Since the first third of the values are generally less than the overall mean, the middle third are generally more than the overall mean, and the final third are generally less than the overall mean, however, one may wish to check future analyses to see whether such a pattern of clustering tends to repeat itself. The final seven values being less than the overall mean is just one away from meeting the criterion for being declared out of statistical control whenever 8 consecutive values fall above or below the center line.

7. This process is out of statistical control. There is an upward tend, and there is a point above the upper control limit.

9. $\bar{p} = (\Sigma x)/(\Sigma n) = (155+152+\ldots+139)/(10)(10,000) = 1462/100,000 = 0.01462$

$\sqrt{\bar{p}\,\bar{q}/n} = \sqrt{(0.01462)(0.98538)/10000} = 0.001200$

$UCL = \bar{p} + 3\sqrt{\bar{p}\,\bar{q}/n} = 0.01462 + 3(0.001200) = 0.01462 + 0.00360 = 0.01822$

$LCL = \bar{p} - 3\sqrt{\bar{p}\,\bar{q}/n} = 0.01462 - 3(0.001200) = 0.01462 - 0.00360 = 0.01102$

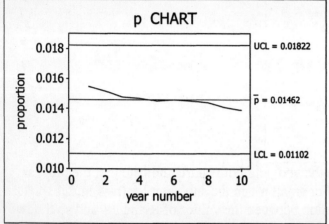

The process is out of statistical control because there is a downward trend. One possible explanation for a decreasing U.S. birth rate might be an increasing U.S. life expectancy. The 10,000 persons were selected each year from the entire population – as life expectancy increases, the proportion of persons in the child-bearing years (and hence the relative number of births to the entire population) would be expected to decrease.

11. $\bar{p} = (\Sigma x)/(\Sigma n) = (8+6+\ldots+8)/(12)(1,000) = 86/12000 = 0.00717$

$\sqrt{\bar{p}\,\bar{q}/n} = \sqrt{(0.00717)(0.99283)/1000} = 0.002667$

$UCL = \bar{p} + 3\sqrt{\bar{p}\,\bar{q}/n} = 0.00717 + 3(0.002667) = 0.00717 + 0.00800 = 0.01517$

$LCL = \bar{p} - 3\sqrt{\bar{p}\,\bar{q}/n} = 0.00717 - 3(0.002667) = 0.00717 - 0.00800 = -0.00084,$

truncated at 0

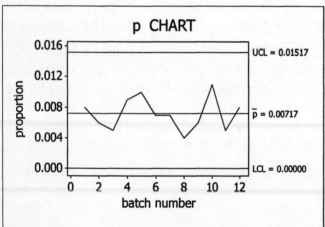

The process is within statistical control. Assuming that the level of $0.00717 = 0.717\%$ defects is acceptable, no corrective action is required.

13. $\bar{p} = (\Sigma x)/(\Sigma n) = (3+3+\ldots+5)/(52)(7) = 127/364 = 0.3489$

$\sqrt{\bar{p}\,\bar{q}/n} = \sqrt{(0.3489)(0.6511)/7} = 0.18015$

$\text{UCL} = \bar{p} + 3\sqrt{\bar{p}\,\bar{q}/n} = 0.3489 + 3(0.18015) = 0.3489 + 0.5404 = 0.8893$

$\text{LCL} = \bar{p} - 3\sqrt{\bar{p}\,\bar{q}/n} = 0.3489 - 3(0.18015) = 0.3489 - 0.5404 = -0.1915$, truncated at 0

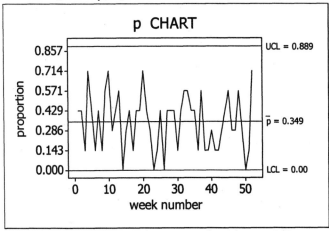

The 52 sample proportions are 3/7, 3/7,…, 5/7.
The vertical labels correspond to 0/7 = 0.000, 1/7 = 0.143, 2/7 = 0.286, etc.
The process is within statistical control.

Statistical Literacy and Critical Thinking

1. A process is designed to result in a product that meet certain specifications. Because conditions change over time, a process meeting those specifications at one given point in time may not always be meeting those specifications. For that reason it is important to monitor the process over time. Oftentimes detection of a changing pattern can lead to implementation of the necessary adjustments before the product strays so far from the specifications that the process output is unsatisfactory.

2. Statistical process control is the monitoring of data over time in order to detect trends or problems that indicate the production process is out of statistical control and/or that lead to the production of a product that does not meet prescribed specifications.

3. If a bottling plant runs a process that is not monitored, product not meeting the proper specifications could be leaving the plant and entering the market. The resulting financial and/or public relations loss in dealing with the defective products could be devastating.

4. In general, a process can go out of control because of a change in mean or a change in variation. The fact that the mean is within statistical control does not indicate that the variation is within statistical control, and vice-versa. For this reason, one should monitor a process with both an $\bar{x}$ chart and an R chart.

Review Exercises

Refer to the following tables for exercises #1–#3.

	kwh1	kwh2	kwh3	$\overline{x}$	R
year1a	3375	2661	2073	2703.00	1302
year1b	2579	2858	2296	2577.67	562
year2a	2812	2433	2266	2503.67	546
year2b	3128	3286	2749	3054.33	537
year3a	3427	578	3792	2599.00	3214

For exercise #1, use the 15 kwh values by row.
For exercise #2, use the range values in the R column.
For exercise #3, use the mean values in the $\overline{x}$ column.

1. The run chart, using the 15 kwh values by row, is given below.

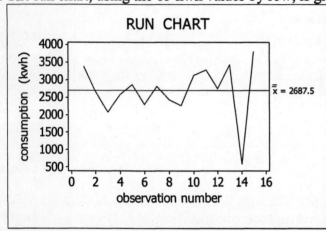

The existence of the extremely low value suggests that the process is not within statistical control – i.e., there appears to be a value that cannot be explained by normal variation, and that was caused by some identifiable circumstances.

2. There are k=5 samples of size n=3 each. $UCL = D_4\overline{R} = 2.574(1232.2) = 3171.7$
   $\overline{R} = \Sigma R/k = 6161/5 = 1232.2$ $LCL = D_3\overline{R} = 0.000(1232.2) = 0$

The process variation is out of statistical control – because there is a point above the upper control limit.

3. There are k=5 samples of size n=3 each. $\bar{\bar{x}} = \Sigma\bar{x}/k = 13427.67/5 = 2687.5$

$\bar{R} = \Sigma R/k = 6161/5 = 1232.2$

UCL $= \bar{\bar{x}} + A_2\bar{R} = 2687.5 + (1.023)(1232.2) = 2687.5 + 1260.5 = 3948.1$

LCL $= \bar{\bar{x}} - A_2\bar{R} = 2687.5 - (1.023)(1232.2) = 2687.5 - 1260.5 = 1427.0$

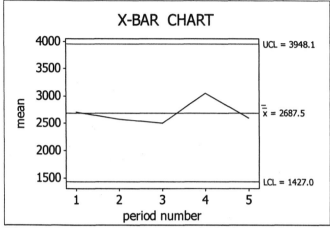

The process mean is within statistical control.

4. $\bar{p} = (\Sigma x)/(\Sigma n) = (270+264+...+343)/(13)(100,000) = 4183/1,300,000 = 0.00322$

$\sqrt{\bar{p}\,\bar{q}/n} = \sqrt{(0.00322)(0.99678)/100000} = 0.000179$

UCL $= \bar{p} + 3\sqrt{\bar{p}\,\bar{q}/n} = 0.00322 + 3(0.000179) = 0.00322 + 0.00053 = 0.00375$

LCL $= \bar{p} - 3\sqrt{\bar{p}\,\bar{q}/n} = 0.00322 - 3(0.000179) = 0.00322 - 0.00053 = 0.00268$

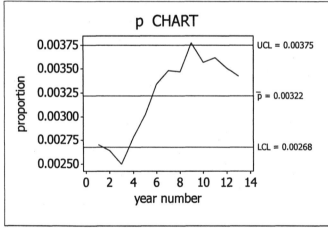

There are three out-of-control criteria, and meeting any one of them means that the process is not within statistical control. This process meets all three criteria.

(1) There is a pattern – in this case, an upward trend.

(2) There is a point outside the upper or lower control limit – in this case, there are 2 points below the LCL and 1 point above the UCL.

(3) The run of 8 rule applies – in this case, the last 8 points are all above the center line.

5. $\bar{p} = (\Sigma x)/(\Sigma n) = (4+2+...+14)/(14)(100) = 82/1400 = 0.0586$

   $\sqrt{\bar{p}\,\bar{q}/n} = \sqrt{(0.0586)(0.9414)/100} = 0.02348$

   $UCL = \bar{p} + 3\sqrt{\bar{p}\,\bar{q}/n} = 0.0586 + 3(0.02348) = 0.0586 + 0.0704 = 0.1290$

   $LCL = \bar{p} - 3\sqrt{\bar{p}\,\bar{q}/n} = 0.0586 - 3(0.02348) = 0.0596 - 0.0704 = -0.0119$, truncated at 0

The process is not within statistical control for two reasons – there is a pattern of increasing variation, and there is a point beyond the upper control limit.

Cumulative Review Exercises

1. $\bar{p} = (\Sigma x)/(\Sigma n) = (6+4+...+2)/(10)(200) = 48/2000 = 0.0240$

   $\sqrt{\bar{p}\,\bar{q}/n} = \sqrt{(0.0240)(0.9760)/200} = 0.01082$

   $UCL = \bar{p} + 3\sqrt{\bar{p}\,\bar{q}/n} = 0.0240 + 3(0.01082) = 0.0240 + 0.0325 = 0.0565$

   $LCL = \bar{p} - 3\sqrt{\bar{p}\,\bar{q}/n} = 0.0240 - 3(0.01082) = 0.0240 - 0.0325 = -0.0085$, truncated at 0

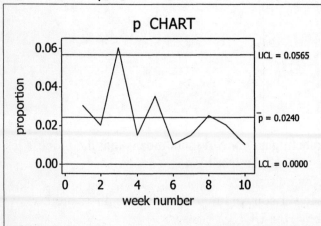

The process is not within statistical control because there is a point beyond the upper control limit.

2. $\alpha = 0.05$ and $\hat{p} = x/n = 48/2000 = 0.0240$

$$\hat{p} \pm z_{\alpha/2}\sqrt{\hat{p}\hat{q}/n}$$
$$0.0240 \pm 1.96\sqrt{(0.0240)(0.9760)/2000}$$
$$0.0240 \pm .0067$$
$$0.0173 < p < 0.0307$$

3. original claim: $p > 0.01$

$\hat{p} = x/n = 48/2000 = 0.024$

H_o: $p = 0.01$
H_1: $p > 0.01$
$\alpha = 0.05$
C.V. $z = z_\alpha = 1.645$
calculations:

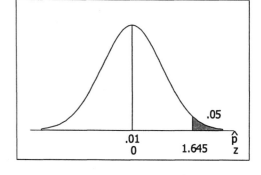

$$z\hat{p} = (\hat{p} - \mu\hat{p})/\sigma\hat{p}$$
$$= (0.024 - 0.01)/\sqrt{(0.01)(0.99)/2000}$$
$$= 0.014/0.00222 = 6.295$$

P-value = $P(z>6.29) = 1 - P(z<6.29) = 1 - 0.9999 = 0.0001$

conclusion:

Reject H_o; there is sufficient evidence to support the claim that $p > 0.01$.

4. a. $P(A) = P(A_1 \text{ and } A_2 \text{ and}\ldots\text{and } A_8)$
$$= P(A_1) \cdot P(A_2) \cdot \ldots \cdot P(A_8) = (\tfrac{1}{2})^8 = 1/256 = 0.0039$$
   b. $P(B) = P(B_1 \text{ and } B_2 \text{ and}\ldots\text{and } B_8)$
$$= P(B_1) \cdot P(B_2) \cdot \ldots \cdot P(B_8) = (\tfrac{1}{2})^8 = 1/256 = 0.0039$$
   c. $P(A \text{ or } B) = P(A) + P(B) - P(A \text{ and } B)$
$$= 0.0039 + 0.0039 - 0$$
$$= 0.0078$$

FINAL NOTE: Congratulations! You have completed statistics – the course that everybody likes to hate. I trust that this manual has helped to make the course a little more understandable – and that you leave the course with an appreciation of broad principles, and mot merely memories of manipulating formulas. I wish you well in your continued studies, and that you achieve your full potential wherever your journey of life may lead.